34th European Peptide Symposium and the 8th International Peptide Symposium 2016

Leipzig, Germany
4-9 September 2016

ISBN: 978-1-5108-5248-8

Printed from e-media with permission by:

Curran Associates, Inc.
57 Morehouse Lane
Red Hook, NY 12571

Printed by Curran Associates, Inc. (2017)

European Peptide Society
c/o Professor Paula Gomes
Rua do Campo Alegre, s/n
PT-4169-007 Porto
Portugal

Phone: +351 220 402 563

pgomes@fc.up.pt

Additional copies of this publication are available from:

Curran Associates, Inc.
57 Morehouse Lane
Red Hook, NY 12571 USA
Phone: 845-758-0400
Fax: 845-758-2633
Email: curran@proceedings.com
Web: www.proceedings.com

TABLE OF CONTENT

TABLE OF CONTENT – ORAL PRESENTATION

TABLE OF CONTENT – POSTER PRESENTATION I

TABLE OF CONTENT – POSTER PRESENTATION I

TABLE OF CONTENT – POSTER PRESENTATION II

TABLE OF CONTENT – POSTER PRESENTATION VI

TABLE OF CONTENT – POSTER PRESENTATION VII

TABLE OF CONTENT – POSTER PRESENTATION VIII

TABLE OF CONTENT - POSTER PRESENTATION VIII

TABLE OF CONTENT – POSTER PRESENTATION IX

TABLE OF CONTENT – POSTER PRESENTATION IX

OP
PP I
PP II
PP VI
PP VII
PP VIII
PP IX
PP X
PP XI
PP XII
PP XIII
PP XIV

PP IX 292 – PP XI 316 ►

TABLE OF CONTENT – POSTER PRESENTATION IX

TABLE OF CONTENT – POSTER PRESENTATION X

TABLE OF CONTENT – POSTER PRESENTATION XI

TABLE OF CONTENT – POSTER PRESENTATION XII

TABLE OF CONTENT – POSTER PRESENTATION XIV

OP
PP I
PP II
PP VI
PP VII
PP VIII
PP IX
PP X
PP XI
PP XII
PP XIII
PP XIV

34TH European Peptide Symposium

8TH International Peptide Symposium

Germany - Leipzig

04 September 2016 to

09 September 2016

EPS

Proceedings

www.34eps-2016.org

34TH European Peptide Symposium

8TH International Peptide Symposium

Germany – Leipzig
04 September 2016 to
09 September 2016

Supported by

DFG

UNIVERSITÄT LEIPZIG

EPS

THE EUROPEAN PEPTIDE SOCIETY

www.eurpepsoc.com/

The European Peptide Society was founded in 1989, primarily in order to ensure that the valuable but informal European Peptide Symposia should be continued on a sound basis. Its most important activity is the organisation in Europe of the biennial international symposium which regularly attracts about 1000 participants from all over the world. The Society also supports financially smaller local meetings and workshops. The Society circulates a newsletter which contains brief reports of meetings and other news, book reviews, and lists, and a calendar of relevant symposia. Society members may subscribe to the Journal of Peptide Science at a greatly reduced rate. The Society has a membership of about 1600 (from some 30 countries) who pay no subscription at present. The principle was established at the outset that there would be no subscription in order to ensure that all peptide scientists in Europe would be able to enroll. The Society administers the Josef Rudinger Memorial Lecture Award and the Leonidas Zervas Award, and a fund to assist younger members to attend symposia.

European Peptide Society Executive Committee

Prof. David Andreu
Pompeu Fabra University, Spain
President

Prof. em. Dirk Tourwé
Free University of Brussels, Belgium
Secretary

Prof. Anna Maria Papini
University of Florence, Italy
Treasurer

Prof. Norbert Sewald
University of Bielefeld, Germany
Scientific Affairs Officer

Prof. Krzysztof Rolka
University of Gdansk, Poland
Communications Officer

PROCEEDINGS OF THE 34TH EUROPEAN PEPTIDE SYMPOSIUM AND THE 8TH INTERNATIONAL PEPTIDE SYMPOSIUM SEPTEMBER 04 – 09, 2016 | LEIPZIG, GERMANY

Edited by

Prof. Dr. Annette Beck-Sickinger
Dr. Karin Mörl
Dr. Kathrin Bellmann-Sickert
Dr. Sylvia Els-Heindl

Leipzig University
Institute of Biochemistry
Brüderstr. 34
04103 Leipzig
Germany

Prof. Dr. Ulf Diederichsen

Georg-August-Universität Göttingen
Dept. of Organic and Biomolecular Chemistry
Tammannstrasse 2
37077 Göttingen
Germany

European Peptide Society

WELCOME PROCEEDINGS

Dear Friends, Partners and Colleagues,

The 34th European Peptide Symposium and the 8th International Peptide Symposium that took place in Leipzig, Germany from Sept.4th-Sept 9th, 2016 were a great success!
Participants from 40 countries and all continents have attended this truly very international conference. Close to 700 participants gathered from all over the world and discussed all aspects of this prosperous field of peptide science –

– from chemical synthesis to biology!
– from Academia to Industry!

Please, find here the Proceedings of this event. The topics addressed all relevant aspects, including

- New synthetic approaches to address current challenges
- Foldamers - from structure to biomaterials
- Peptides to Therapy
- Imaging and modulation of enzymes and receptors
- Peptides of the Future: Targeting intracellular systems and protein-protein interaction

In addition to 80 lectures, more than 300 posters have been presented and some of the presentations are now summarized in the Proceedings.

The European and International Peptide Symposium 2016 provided an excellent platform to exchange ideas on latest approaches and discoveries in all aspects of peptide science.
In addition to the excellent program the conference provided a superb opportunity to meet peers, build-up relationships and exchange views on recent scientific developments.
Thus, this meeting was interesting, challenging and informative.

Thank you for joining us! And remember Leipzig, this beautiful city in the heart of Germany, with a long tradition in science, culture and trade.

Annette G. Beck-Sickinger & Ulf Diederichsen

34TH EUROPEAN PEPTIDE SYMPOSIUM
AND THE 8TH INTERNATIONAL PEPTIDE SYMPOSIUM
SEPTEMBER 04 – 09, 2016 | LEIPZIG, GERMANY

CHAIRPERSONS

Prof. Dr. Annette G. Beck-Sickinger
Leipzig University | Leipzig, Germany

Prof. Dr. Ulf Diederichsen
University of Göttingen | Göttingen, Germany

SCIENTIFIC COMMITTEE

Kenichi Akaji
Department of Medicinal Chemistry, Kyoto Pharmaceutical University | Kyoto, Japan

Samuel Gellman
Department of Chemistry, University of Wisconsin | Madison, Wisconsin, USA

Ernest Giralt
Institut de Recerca Biomèdica Parc Cientific de Barcelona | Barcelona, Spain

Christian Hackenberger
Leibniz-Institute for Molecular Pharmacology (FMP) | Berlin, Germany

Knud Jensen
Center for Synthetic Biology, Nanobioscience Group, Nano Science Center, Department of Chemistry, Faculty of Science, University of Copenhagen | Frederiksberg, Denmark

Yoshiaki Kiso
Professor, Nagahama Institute of Bio-Science & Technology | Nagahama, Japan

Lei Liu
Tsinghua-Peking Center for Life Sciences, Key Laboratory of Bioorganic Phosphorus, Chemistry & Chemical Biology (Ministry of Education), Department of Chemistry, Tsinghua University | Beijing, China

William Lubell
Départment de chimie, Université de Montréal | Montréal, Québec, Canada

Anna Maria Papini
Laboratorio Interdipartimentale di Chimica e Biologia di Peptidi e Proteine Dipartimento di Chimica Ugo Schiff Polo Scientifico e Tecnologico, Università di Firenze | Sesto Fiorentino (Firenze), Italy

Richard Payne
Faculty of Science, School of Chemistry, University of Sydney | Sydney, Australia

Bernd Riedl
Bayer Pharma AG, Forschungszentrum | Wuppertal, Germany

Norbert Sewald
Department of Chemistry, Bielefeld University | Bielefeld, Germany

Helma Wennemers
ETH Zürich, Laboratory for Organic Chemistry | Zürich, Switzerland

WE THANK OUR SPONSORS AND EXHIBITORS!

SILVER SPONSORS

BACHEM

CEM

BRONZE SPONSORS

Biotage

GLS

GYROS PROTEIN Technologies

PolyPeptide GROUP

RAPP POLYMERE

SUNRESIN Driving the innovation

EXHIBITORS

aapptec

Activotec

Advion

Kromasil

Axonlab connecting ideas

biametrics biomolecular interaction analysis

3CN PEPTIDES

CAT

CBL

CORDENPHARMA

CS PEOPLE WHO LOVE PEPTIDES · PEOPLE WHO KNOW PEPTIDES

HORIBA Scientific

EXHIBITORS

FURTHER SUPPORTERS

EUROPEAN PEPTIDE SYMPOSIA

1st	1958	Prague, Czechoslovakia
2nd	1959	Munich, Germany
3rd	1960	Basel, Switzerland
4th	1961	Moscow, Russia
5th	1962	Oxford, UK
6th	1963	Athens, Greece
7th	1964	Budapest, Hungary
8th	1966	Noordwijk, The Netherlands
9th	1968	Orsay, France
10th	1969	Abano Terme, Italy
11th	1971	Vienna, Austria
12th	1972	Reinhardsbrunn, Germany
13th	1974	Kiriyat Anavim, Isreal
14th	1976	Wepion, Belgium
15th	1978	Gdansk, Poland
16th	1980	Helsingor, Denmark
17th	1982	Prague, Czechoslovakia

INTERNATIONAL PEPTIDE SYMPOSIUM

8th	2016	Leipzig, Germany
7th	2015	Singapore, Singapore
6th	2013	Hawaii, USA
5th	2010	Kyoto, Japan
4th	2007	Cairns, Australia
3rd	2004	Prague, Czech Republic
2nd	2001	San Diego, USA
1st	1997	Kyoto, Japan

18th	1984	Djuronaset, Sweden
19th	1986	Porto Carras, Greece
20th	1988	Tubingen, GFR
21st	1990	Barcelona, Spain
22nd	1992	Interlaken, Switzerland
23rd	1994	Braga, Portugal
24th	1996	Edinburgh, UK
25th	1998	Budapest, Hungary
26th	2000	Montpellier, France
27th	2002	Sorrento, Italy
28th	2004	Prague, Czech Republic
29th	2006	Gdansk, Poland
30th	2008	Helsinki, Finnland
31st	2010	Copenhagen, Denmark
32nd	2012	Athens, Greece
33rd	2014	Sofia, Bulgaria
34th	2016	Leipzig, Germany

THE JOSEF RUDINGER AWARD

This award was established by Ferring Pharmaceuticals in 1984, as a commemoration of Josef Rudinger's role in the foundation of European Peptide Symposia and diverse contributions he made to peptide chemistry. There is no restriction as to nationality, age or position of those nominated, but they must be distinguished.
The winners are listed below.

2016 Jean Martinez
University Montpellier, France

2014 Ernest Giralt
University of Barcelona, Barcelona, Spain

2012 David J. Craik
The University of Queensland, Australia

2010 Stephen B. H. Kent
University of Chicago, USA

2008 Horst Kessler and Manfred Mutter
Technical University of Munich, Germany and
University of Lausanne, Switzerland

2006 Ettore Benedetti and Claudio Toniolo
University of Napoli "Frederico II", Italy and
University of Padova, Italy

2004 Luis Moroder
Max-Planck-Institute fur Biochemie, Martinsried, Germany

2002 Sandor Bajusz and Kalman Medzihradszky
IVAX-Institute of Drug Research, Budapest, Hungary and Department of Organic Chemistry, Eotvos L. University, Budapest, Hungary

2000 Bernard P. Roques
INSERM, CNRS, Paris, France

1998 Shumpei Sakakibara
Peptide Institute, Osaka, Japan

1996 Ralph Hirschmann
University of Pennsylvania, Philadelphia, USA

1994 Robert C. Sheppard
MRC, Cambridge, United Kingdom

1992 Viktor Mutt
Karolinska Institute, Stockholm, Sweden

1990 R. Bruce Merrifield
The Rockefeller University, New York, USA

1988 Erich Wunsch
Max-Planck-Institut fur Biochemie, Munich, Germany

1986 Robert Schwyzer
ETH Zurich, Switzerland

JEAN MARTINEZ

Jean Martinez received both his PhD in 1972 from the University of Montpellier 2, at the National School of Chemistry and his Dr Sciences Degree in 1976 from the same University, under the supervision of Professor F. Winternitz. The same year, he joined Dr E. Bricas group at Orsay, University of Paris Sud, as a post-doctorate fellow and in 1977 the laboratory of Professor M. Bodanszky at Case Western Reserve University in Cleveland, Ohio, USA, where he stayed till mid-1979.

Jean Martinez is a Full Professor at the University of Montpellier. In 2007, he created the « Institut des Biomolécules Max Mousseron » (IBMM), which he has been the Director until December 2014. He is actually Head of the department of Amino Acids, Peptides and Proteins at IBMM.

He served the University of Montpellier 1 as a Member of the Scientific Council for 8 years, and as Vice-President for 6 years (2009-2014). Prof. Martinez is recognized for his important contributions, at the interface of chemistry and biology, to the development of methodology in organic and peptide synthesis, design and synthesis of various potent and selective neuropeptide analogues, and biomaterials containing biomolecules.

He has published over 800 original papers, 50 patents, numerous review articles, and he has been editor of several books. In 2012, he was accepted into the « Académie Nationale de Pharmacie », France, and in 2014 into the « Real Academia Nacional de Farmacia », Spain. In 2015, he has been nominated Docteur Honoris causa of the University of Kracow, Poland. He has received various prizes including the Leonidas Zervas Award (1990), the Silver Medal CNRS (1996), the Max Bergmann Medal (2004), the Cathay Award (2006), the Akabori Award (2006), the Ehrlich Award (2011), the Roche « Johannes Meinhofer Award » (2011), the Léon Velluz Award (2015). He is « Chevalier dans l'Ordre des Palmes Académiques » (2010), and Chevalier dans l'Ordre de la Légion d'Honneur (2011).

Jean Martinez served the European Peptide Society as French Representative, Member of the Executive Committee (1991-1998), Scientific Officer (1998-2001), and President (2001-2010).

THE LEONIDAS ZERVAS AWARDS

This award was established by Bachem Inc. USA in 1984, in commemoration of Leonidas Zervas and the outstanding contributions he made to peptide chemistry. The winners are listed below.

2016 Christian Becker
University of Vienna, Austria

2014 Miguel Castanho and Phillip Dawson
University of Lisbon , Portugal and
The Scripps Research Institute, La Jolla,USA

2012 Knud J. Jensen
University of Copenhagen, Denmark

2010 Helma Wennemers
University of Basel, Switzerland

2008 Anna Maria Papini
University of Florence, Italy

2006 Carlos GarcHa-EcheverrHa
Novartis Institutes for BioMedical Research, Basel, Switzerland

2004 Helene Gras-Masse
Institut Pasteur de Lille, France

2002 Thomas W. Muir
Rockefeller University, New York, USA

2000 Antonello Pessi
Instituto di Ricerche di Biolog ica Moleculare P. Angeletti, Rome, Italy

1998 Annette G. Beck-Sickinger
ETH Zurich, Switzerland

1996 Morten Meldal
Carlsberg Laboratory, Valby, Denmark

1994 Ernest Giralt and Fernando Albericio
University of Barcelona, Barcelona, Spain

1992 Gunther Jung
University of Tubingen, Tubingen, Germany

1990 Michal Lebl and Jean Martinez
Czechoslovak Academy of Sciences, Prague and
CNRS, Montpellier, France

1988 Alex Eberle
University of Basel, Basel, Switzerland

CHRISTIAN BECKER

Christian F.W. Becker obtained a Ph.D. in Chemistry in 2001 from the University of Dortmund (Germany), and was a postdoctoral fellow with Gryphon Therapeutics in South San Francisco (USA) from 2002 to 2003. He became a group leader at the Max-Planck Institute of Molecular Physiology in Dortmund, Germany in 2004 and was appointed as associate professor for protein chemistry at the Technische Universität München and the Center of Integrated Protein Sciences Munich (CIPSM) in November 2007. In 2011 he accepted a call to the University of Vienna (Austria) as Full Professor and Head of the Institute of Biological Chemistry. His research interests include the chemical synthesis/semisynthesis and site-specific modification of proteins.

THE MIKLÓS BODANSZKY AWARD

This award was established by BCN Peptides in 2014, in commemoration of Miklós Bodanszky's outstanding contributions to peptide science. The award is given to the scientist who made significant contributions to peptide-based drug research in the period of ten years after having obtained the PhD degree.
The winners are listed below.

2016
Maja Köhn
European Molecular Biology Laboratory, Germany

Markus Muttenthaler
Institute for Molecular Bioscience, The University of Queensland

MAJA KÖHN

Maja Köhn studied chemistry at the University of Kiel (Germany). She finished her PhD under the supervision of Prof. H. Waldmann at the Max-Planck-Institute of Molecular Physiology in Dortmund (Germany) in 2005. Afterwards, she joined Professor G. Verdine's group at Harvard University (USA) supported by a postdoctoral fellowship of the German Academic Exchange Service.
In November 2007 she started her current position as a group leader at the European Molecular Biology Laboratory in Heidelberg (Germany). In 2010, she was awarded an Emmy Noether Research group funded by the German Research Foundation, and in 2014 she became a European Research Council Investigator. She was awarded with the Friedmund Neumann prize of the Schering Foundation and the Tetrahedron Young Investigator Award in Bioorganic and Medicinal Chemistry. Her research interests are to control and investigate protein phosphatases using interdisciplinary approaches comprising synthetic chemistry, biochemistry, and molecular cell biology.

MARKUS MUTTENTHALER

Markus Muttenthaler is a Biological and Medicinal Chemist working at the interface of Chemistry and Biology with a passion for translational research. His interests lie in the exploration of nature's diversity (particularly venom peptides) to develop molecular tools, diagnostics and therapeutics. His background in drug discovery, design and development, and interdisciplinary training in the fields of chemistry, molecular biology, and pharmacology assist him in the characterization of these highly potent and selective compounds and allow him to study their interactions with human physiology with applications in neuropathic pain, cancer, autism and neurodegenerative diseases.

Markus Muttenthaler trained as an applied synthetic chemist in Vienna/Austria and completed his PhD in Biological and Medicinal Chemistry at the University of Queensland (UQ), where he focused on the chemical engineering and controlled folding of cysteine-rich peptides. This was followed by 1.5 years of postdoctoral work at the Institute for Molecular Bioscience (IMB) at UQ, in the group led by Prof. Alewood, renowned for his contributions to the development of solid-phase peptide synthesis and his pioneering efforts in unravelling the complexity of venom peptides. He was then awarded the prestigious and competitive Marie Curie International Outgoing Fellowship, which allowed him to pursue independent research at The Scripps Research Institute, La Jolla, USA, under the guidance of Prof. Dawson, distinguished for the invention of native chemical ligation, a strategy that allows the chemical synthesis of proteins. This period was followed by further independent research at the Institute for Research in Biomedicine Barcelona, Spain, in the group of Prof. Albericio, an international leader in combinatorial chemistry and peptide chemistry. During this time he was able to secure another Marie Curie Fellowship, a Spanish project grant and an Austrian project grant.

Recently, he was awarded the prestigious Australian Research Council Discovery Early Career Researcher Award, which allowed him to move back to UQ to set up his team and to continue his work on oxytocin/vasopressin probe development, venom drug discovery and gastrointestinal wound healing.

YOUNG INVESTIGATORS' AWARDS

Dr. Bert L. Schram Oral presentation Awards:

The young investigator symposium featured 8 participants, who were preselected from 76 requests for oral presentations. The quality of the oral presentations was first rate and ranged on a variety of subjects featuring the chemical, biological and medicinal aspects of peptide science. The young investigators presented their research effectively and responded to questions in a clear and informative manner, which inspired further discussion. Judging the presentations was a difficult task in light of the excellent performance from all of the participants. The judges were relieved to be able to compliment the Dr. Bert L. Schram Young Investigator Awards, which were generously given by ESCOM.

Two young participants were awarded with Dr. Bert Schram certificate for the best oral presentation. **Sunbum Kwon** from Institut Européen de Chimie et Biologie, France was awarded for the lecture entitled "ribosomal translation of foldamer-peptide hybrids by flexible in vitro translation system" and **Emma Watson** from the University of Sydney, Australia with the lecture entitled "sansanmycin analogues as potent and selective anti-tubercular agents with a novel mode of action".

Special thanks goes to Professor Ernest Giralt and Professor Christian Birr, who helped to preside over the session and assist in the judging.

Dr. Bert L. Schram Poster Awards:

The poster award winners were selected by the chairs of the Poster Appetizer Sessions, which acted as poster referees.

Brunello Nardonne from University of St. Andrews, UK was awarded for best poster presentation in the Session Synthesis with entitled "synthesis of patellamide analogues by a one-pot processing of solid-phase substrates", **Fabien Thoreau** from Université Grenoble Alpes, France was awarded for best poster in the Session Chemical Modification and Protein Interaction entitled "Design and synthesis of peptide vectors for tumoral periphery targeting and imaging: a tool for surgery.", **Masato Kaneda** from Kyoto University, Japan was awarded for best poster presentation in the Session Peptide Therapeutics I entitled "synthetic study of odoamide and its derivatives", **Bastian Franke** from School of Biomedical Sciences, Australia was awarded for best poster presentation in the Session Peptide Structure and Folding entitled "Structural features of an unusual sunflower proalbumin protein that gives rise to a small cyclic peptide and a seed storage protein", **Cristina Díaz-Perlas** from IRB Barcelona, Spain was awarded for best poster presentation in the Session Peptide Carrier and Analytics entitled "phage display as a tool to discover bbb-shuttle peptides: panning against a human blood-brain barrier cellular model" and **Lenka Maletínská** from Institute of Organic Chemistry and Biochemistry, Czech Republic was awarded for best poster presentation in the Session Peptide Therapeutics II entitled "Lipidized prolactin-releasing peptide analogs: a new tool for obesity treatment".

We would like to express our special thanks to Ms. Elisabeth Schram and Dr. Johan Elgersma for their participation and support of young participants at the 34th EPS.

TRAVEL GRANTS

The travel grant is obtained from the European Peptide Society to support the possibility of young scientists to attend the EPS conference and results in a free registration to the conference. The European Peptide Society and the Universities of Leipzig and Göttingen selected 43 young scientists from the following countries for receipt of a travel grant.

Country	Grants
Australia	4
Austria	2
Belgium	2
Brasil	2
Bulgaria	1
Canada	1
France	5
Germany	6
Hungary	1
Greece	1
India	3
Israel	1
Italy	1
Japan	1
Netherlands	3
Poland	3
Portugal	1
Russia	1
Spain	1
Sweden	1
Swotzerland	1
USA	1

NUMBER OF ACTIVE PARTICIPANTS

Australia	26
Austria	10
Belgium	16
Brasil	2
Bulgaria	3
Chile	2
China	10
Czech Republic	7
Denmark	8
Germany	160
Finland	1
France	42
Greece	11
Hungary	14
India	5
Ireland	1
Israel	8
Italy	19
Japan	58
Jordan	1
Canada	8
Latvia	4
New Zealand	3
Netherlands	14
Norway	2
Austria	10
Poland	34
Portugal	4
Russia	13
Sweden	9
Switzerland	34
Singapore	1
Slowenia	1
Spain	12
South Africa	3
Taiwan	1
United Kingdom	21
United States	40

THE LATENT SELENOESTER SEEA FACILITATES THE ASSEMBLY OF CHALLENGING PROTEINS. APPLICATION TO THE TOTAL SYNTHESIS OF HEPATOCYTE GROWTH FACTOR NK1 DOMAIN

Marine Cargoët, Laurent Raibaut, Nathalie Ollivier, Yun Min Chang, Hervé Drobecq, Emmanuelle Boll, Rémi Desmet, Oleg Melnyk
UMR CNRS 8161, Université de Lille, Institut Pasteur de Lille, 59021 Lille, France.

OP
PP I
PP II
PP VI
PP VII
PP VIII
PP IX
PP X
PP XI
PP XII
PP XIII
PP XIV

HGF and MET play a crucial role during embryonic development and tissue regeneration in the adult. HGF-MET signaling when deregulated contributes to tumorigenesis and cancer progression. We have explored since several years the chemical properties of the bis(2-sulfanylethyl)amido group[1,2] (Fig. 1A) and its interest for the total synthesis of functional proteins such as the N[3] or K1[4] domains of hepatocyte growth factor (HGF). These proteins enabled exploring the mechanism of HGF binding to MET tyrosine kinase receptor. In particular, dimerization or multimerization of K1 domain was found to be sufficient for triggering most of the phenotypes displayed by HGF.[5] This work clarified also the role of N domain, which does not bind MET but helps presenting K1 domain as a dimer to its receptor (Fig. 1B). To further explore the mechanism of binding of HGF to MET, we embarked upon the total synthesis of the biotinylated NK1 protein (NK1-B), a natural variant of HGF and a good agonist of MET receptor.[6]

A)

Bis(2-sulfanylethyl)amido
SEA

Bis(2-selenylethyl)amido
SeEA

B)

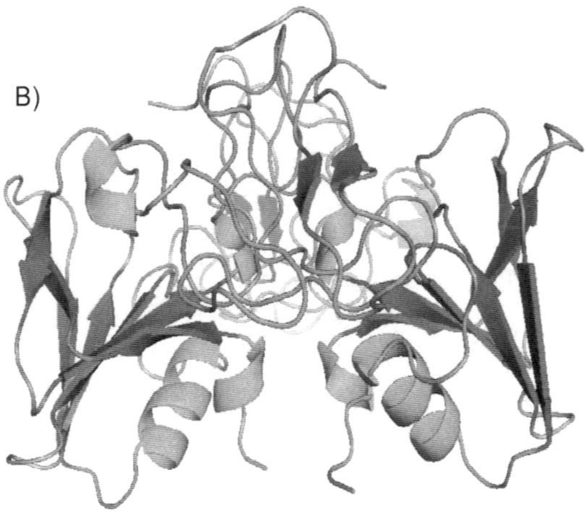

NK1 non-covalent homodimer

PDB : 1BHT, Ultsch, M. *Structure* **1998**, *6*, 1383-1393

1
QRKRRNTIHEFKKSAKTTLIKIDPALKIKTKKVNTADQ-SEA
2
CANRCTRNKGLPFTCKAFVFDKARKQ-SEA
3
CLWFPFNSMSSGVKKEFGHEFDLYENKDYIRA-SEA
4
CIIGKGRSYKGTVSITKSGIK-SeEA
5
CQPWSSMIPHEHSFLPSSYRGKDLQENY-SEA
6
CRNPRGEEGGPWCFTSNPEVRYEVCDIPQCSEVLK(Biotin)-NH_2

Fig. 1. Structure of the SEA and SeEA functionalities (A). X-ray crystal structure of NK1 non-covalent homodimer and peptide segments used for the assembly of NK1-B. An additional Lys residue was placed at the C-terminus of NK1 to attach the biotin group through its side-chain (B).

NK1 protein is 176 AA in length and can potentially be assembled from 6 peptide segments due to the presence of several cysteine (Cys) residues in its sequence (Fig. 1B, 4 cysteines in N domain, 6 cysteines in K1 domain). To facilitate the assembly of NK1 we explored the chemical properties of the bis(2-selenylethyl)amido group (Fig.1A), i.e. the selenium analog of the SEA group.[6,7] Indeed, we showed that the combination of SEA and SeEA chemistries allowed reducing the number of chemical steps and intermediate purifications needed to produce NK1 in comparison with classical assembly schemes based on NCL and standard protecting groups for Cys. In particular, we could design three one-pot assembly schemes that were combined for accessing NK1-B with a global yield of 3%.

Segments 1-6 were produced by Fmoc SPPS starting from a SEA solid support.[1,8,9] SEA segments 1,2,3,5 and 6 were used as such for the assembly of NK1-B. The SEA group of segment 4 was further exchanged in solution by the bis(2-selenylethyl)amine to access the corresponding SeEA segment, a reaction which is reminiscent of the capacity of the SEA group to be exchanged by alkylthiols at mildly acidic pH.[10] The production of segment 4-SeEA required the temporary protection of the N-terminal Cys residue using acetoacetyl (AcA) group,[11] which was removed after the exchange reaction by the addition of hydroxylamine. The SEA→SeEA exchange reaction and the removal of the AcA group were performed in one-pot (not shown).

The assembly of NK1-B was performed as described in Scheme 1. The first one-pot process produced segment 1-2 equipped at the C-terminus with an alkylthioester functionality. It was then used to produce intermediate 1-2-3-4-SeEA by a sequential NCL and SEA ligation process (one-pot process II). The later reaction was performed in the presence of DTT which has the capacity to reduce and thus to activate the SEA group, while keeping the SeEA analog in off state.[7] The last one-pot process III leading to the full length NK1-B protein was performed in kinetically controlled conditions by exploiting the higher reactivity of the SeEA group in comparison with the SEA one.[6] Overall, the assembly of NK1-B required only three resolutive purifications (after each one-pot process).

In conclusion, the selenium SeEA analog of the SEA group is a useful addition to the SEA chemical tool box. SEA ligation can be performed in the presence of the SeEA cyclic diselenide by using DTT as a selective reducing agent of the SEA cyclic disulfide. Alternatively, the simultaneous reduction of SEA and SeEA groups by TCEP permitted performing a kinetically controlled ligation due to the higher reactivity of the SeEA selenoester surrogate in comparison with SEA thioester surrogate. The combination of SEA and SeEA chemistries simplifies considerably the assembly of challenging proteins as shown by the total synthesis of NK1-B.

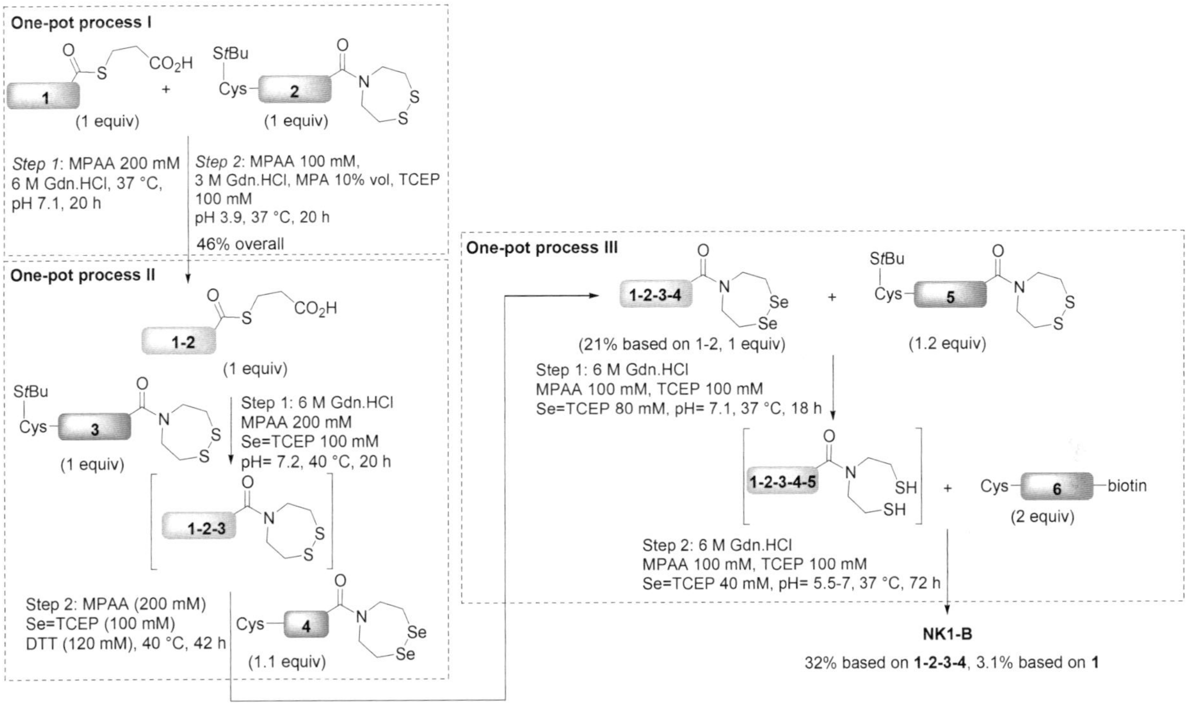

Scheme 1. Total synthesis of NK1-B.

References

01. Ollivier, N.; Dheur, J.; Mhidia, R.; Blanpain, A.; Melnyk, O. Org. Lett. 2010, 12, 5238-5241.
02. Raibaut, L.; Ollivier, N.; Melnyk, O. Chem. Soc. Rev. 2012, 41, 7001-7015.
03. Raibaut, L.; Vicogne, J.; Leclercq, B.; Drobecq, H.; Desmet, R.; Melnyk, O. Bioorg. Med. Chem. 2013, 21, 3486-3494.
04. Ollivier, N.; Vicogne, J.; Vallin, A.; Drobecq, H.; Desmet, R.; El-Mahdi, O.; Leclercq, B.; Goormachtigh, G.; Fafeur, V.; Melnyk, O. Angew. Chem. Int. Ed. 2012, 51, 209-213.
05. Simonneau, C.; Berenice, L.; Mougel, A.; Adriaenssens, E.; Paquet, C.; Raibaut, L.; Ollivier, N.; Drobecq, H.; Marcoux, J.; Cianferani, S.; Tulasne, D.; de Jonge, H.; Melnyk, O.; Vicogne, J. Chem. Sci. 2015, 6, 2110-2121.
06. Raibaut, L.; Cargoet, M.; Ollivier, N.; Chang, Y. M.; Drobecq, H.; Boll, E.; Desmet, R.; Monbaliu, J.-C. M.; Melnyk, O. Chem. Sci. 20 16, 7, 2657-2665
07. Raibaut, L.; Drobecq, H.; Melnyk, O. Org. Lett. 2015, 17, 3636-3639.
(8) Boll, E.; Drobecq, H.; Ollivier, N.; Raibaut, L.; Desmet, R.; Vicogne, J.; Melnyk, O. Chem. Sci. 2014, 5, 2017-2022.
09. Boll, E.; Drobecq, H.; Ollivier, N.; Blanpain, A.; Raibaut, L.; Desmet, R.; Vicogne, J.; Melnyk, O. Nat. Protoc. 2015, 10, 269-292.
10. Dheur, J.; Ollivier, N.; Vallin, A.; Melnyk, O. J. Org. Chem. 2011, 76, 3194-3202.
11. Boll, E.; Ebran, J. P.; Drobecq, H.; El-Mahdi, O.; Raibaut, L.; Ollivier, N.; Melnyk, O. Org. Lett. 2015, 17, 130-133.

Anaïs Terrien[1], Keyvan Rahgoshay[2], Nathalie Lensen[2], Thierry Brigaud[2], Grégory Chaume[2*], Emeric Miclet[1*]
[1] Laboratoire des Biomolécules, UMR 7203, Université Pierre et Marie Curie, 4 place Jussieu, Paris Cedex 05, France
[2] Laboratoire LCB, EA 4505, Université de Cergy-Pontoise, 5 mail Gay Lussac, Cergy-Pontoise, France, gregory.chaume@u-cergy.fr

OP – 07

Introduction

Collagen is the most abundant protein in the animal kingdom. Its tertiary structure consists of three individual left-handed polyproline II (PPII) helices folded into a right-handed triple helix which is stabilized by inter-strand hydrogen bonds. Each strand comprises the repeat of the primary triplet sequence (Xaa-Yaa-Gly) where the proline (Pro) and the (4R)-hydroxyproline (Hyp) are the most prevalent at Xaa and Yaa positions, respectively.[1] C4-endo pucker at Xaa-site and C4-exo at Yaa-site have been proposed to be a prerequisite for formation of the triple helix since they preorganize the main-chain dihedral angles.[2] Because both Pro and Hyp residues exhibit the expected puckering, their prevalence imparts an excellent thermal stability to the collagen triple helix. Over the past decades, numerous studies have been conducted for understanding its self-assembling properties as well as for the development of new collagen-related biomaterials. One common strategy consists in using Collagen Model Peptides (CMP) based on the repetition of the triplet Pro-Hyp-Gly. Side chain modifications have been used to modulate or enhance the stability of the triple helix. (4R)- or (4S)-substituents (F, Cl, Me, N_3, NH_2, NHCOR, OMe, SH, CN, triazole) have been used to replace Pro or Hyp residue, respectively.[3] Our group has developed convenient methodologies for the synthesis of various trifluoromethylated pseudoprolines bearing the CF_3 group at the C5 position and their incorporation into peptides.[4] We demonstrated that the trifluoromethylated pseudoprolines, compared to their non-fluorinated analogues, proved to be completely stable when incorporated into peptides and therefore can be used as proline surrogates. NMR studies as well as theoretical calculations showed that the (5R)-CF_3-ψPro (FYPro) i) substantially decreases the cis-trans rotational barriers with only moderate effects on the cis/trans population ratio; ii) stabilizes the ψ-polyproline backbone conformation in water and iii) efficiently constrains the C4-exo puckering. Here, we propose to assess the structural features induced by the incorporation of FYPro into 21-mer CMPs. Replacement of one or several Hyp residues have been performed at different position of the peptide sequence.

Results and discussion

The synthesis of the fluorinated CMPs via solid-phase peptide synthesis (SPPS) by the direct incorporation of the FYPro residue 1 has been excluded because of the very low nucleophilicity of its amino group and the steric bulkiness of the vicinal CF_3-group. We demonstrated that an acyl chloride activation is required to promote the peptide coupling of FYPro residue 1 in good yield.[4d] This condition is however incompatible with SPPS. Therefore, we have first considered the preparation of the ready-to-use Fmoc-Pro-FYPro-Gly-OH building block 3 which has been then incorporated into the CMP sequence using SPPS. We decided to perform the peptide chain elongation from the N- to the C-terminus since we recently observed that this strategy was more effective for the preparation of tripeptides bearing FYPro residue at the central position.[5] Thus, the coupling reaction of a diastereomeric mixture of (5S)- and (5R)-CF_3-pseudoprolines 1 with a stoichiometric amount of Fmoc-proline chloride in base free conditions gave the dipeptide 2 in 77% yield as a single diastereomer (Scheme 1). We demonstrated that the reaction involved a dynamic kinetic resolution process explaining the observed diastereoselectivity. Selective hydrogenolysis of the benzyl ester group of dipeptide 5 under hydrogen atmosphere (1 bar) in the presence of Pd/C catalyst gave the corresponding acid without any trace of early Fmoc deprotection. Finally, coupling reaction with the glycine tert-butylester using standard conditions (EDCI/HOBt) followed by the deprotection of the C-terminal acid function gave the building block 3 in 37% yield over three steps. The efficiency of the synthesis allowed its preparation on several grams.

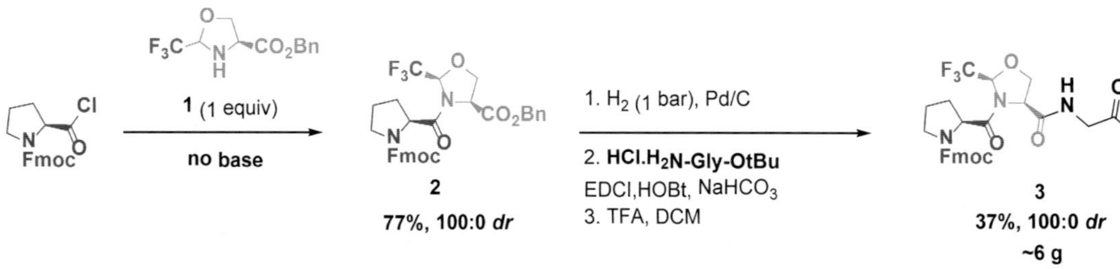

Scheme 1. Synthesis of the building block 3

We then focused on the synthesis of three fluorinated 21-mer CMPs (CMPFs) (Figure 1). As mentioned before, we were interested to assess the impact of the FΨPro residue onto the structural features of the corresponding CMPFs. For this purpose, we have considered the synthesis of peptide 4 bearing a single FΨPro residue located at the central position of the CMPF sequence. In addition, peptides 5 and 6 bearing three consecutive or skip-spaced FΨPro respectively have been synthesized to investigate the effect of both the number and the position of these residues. The peptides 4-6 were synthesized in good yield (ca 15-30%) by SPPS starting from a pre-loaded Fmoc-Gly-Wang resin using HATU/DIEA coupling reagents.

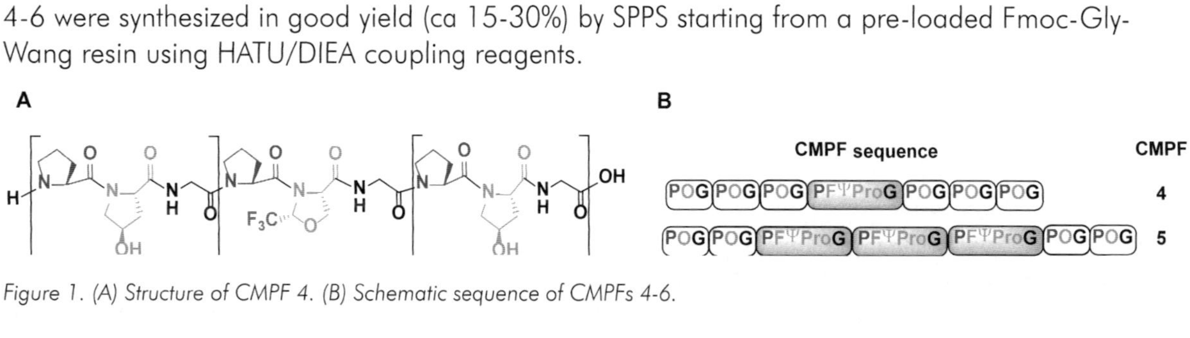

Figure 1. (A) Structure of CMPF 4. (B) Schematic sequence of CMPFs 4-6.

We next performed the structural analysis of our three fluorinated peptides by Circular dichroism (CD). Spectra of 10 mM aqueous solutions of CMPFs 4-6 have been recorded after a 24 hours incubation at 4°C. In these studies, we used $(POG)_7$ as a reference peptide (Figure 2A). Only the CMPF-4 peptide which incorporates a single FYPro residue exhibits a PPII signature as proved by the characteristic positive band at 225 nm. Compared to the $(POG)_7$, this signal was attenuated, indicating a weaker propensity to adopt the PPII extended conformation. Thermal denaturation experiments of both the CMPF-4 and $(POG)_7$ peptides showed cooperative unfolding transitions (Figure 2B). However, the replacement of the central Hyp residue by a FYPro residue leads to a significant decrease of the melting temperature (DTm = 26°C). This result is in apparent contradiction with previous studies that have established a direct correlation between the stabilization of the C4-exo puckering at the Yaa position and the increase of the triple helix Tm.[3a] However, our recent NMR structural studies showed that the C4-exo pucker was observed for both the Pro and the FYPro residues in the triplet model Ac-Pro-FYPro-Gly-NH_2, while a fast exchanging C4-endo/exo pucker was found the Pro residue in the Ac-Pro-Hyp-Gly-NH_2 sequence.[4e] Since the C4-endo pucker at Xaa-site and C4-exo at Yaa-site have been ascribed for the stabilization of the triple helix, the FYPro may have detrimental effects on the triple helix formation by perturbing the conformation of the preceding Pro residue. The low triple helix content in CMPFs 5 and 6 could be a direct consequence of additive effects of this local destabilization. In summary, our CD analysis has shown that collagen-like triple helix can accommodate bulky CF_3 groups in close vicinity to the peptide backbone. MD calculations have confirmed this result and will be reported elsewhere. The effect of the FYPro residue on the preceding residue is however harmful for the triple helix formation but this could be circumvented by also replacing the Pro by the (5S)-CF_3-ψPro (fYPro) which displayed a strong propensity for the C4-endo pucker.[4e]

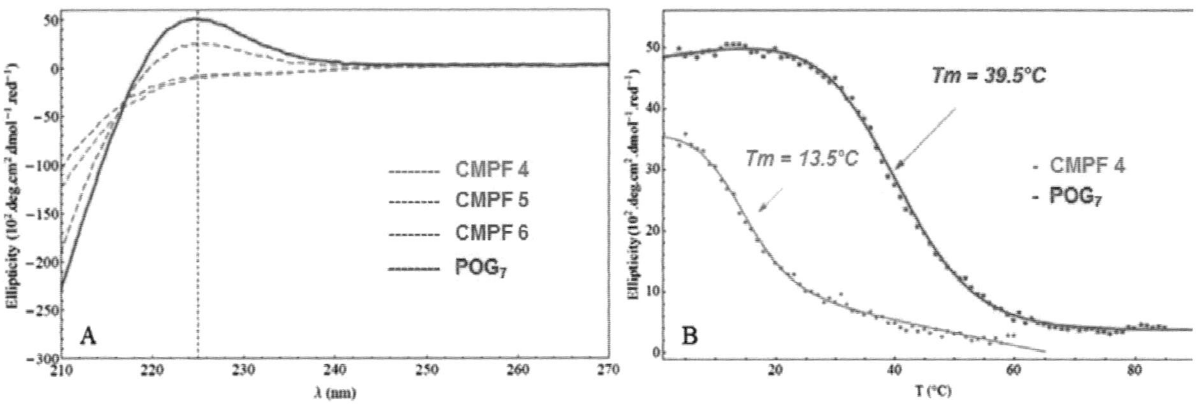

Figure 2. (A) CD spectra of POG_7 and CMPFs (4°C, 50 mM phosphate buffer, pH = 7); (B) Ellipticity at 225 nm depending of temperature of POG_7 and CMPF 4 (pH = 7)

References

01. Ramshaw, J. A. M.; Shah, N. K.; Brodsky, B. J. Struct. Biol. 1998, 122, 86-91.

02. (a) Berisio, R.; Vitagliano, L.; Mazzarella, L.; Zagari, A. Protein Sci. 2002, 11, 262-270; (b) Hongo, C.; Noguchi, K.; Okuyama, K.; Tanaka, Y.; Norikazu, Nishino, N. J. Biochem. 2005, 138, 135-144; (c) Improta, R.; Benzi, C.; Barone, V. J. Am. Chem. Soc. 2001, 123, 12568-12577; (d) Improta, R.; Mele, F.; Crecenzi, O.; Benzi, C.; Barone, V. J. Am. Chem. Soc. 2002, 124, 7857-7865.

03. Shoulders, M. D.; Raines, R. T. Annu. Rev. Biochem. 2009, 78, 929-958 and references therein; (b) Erdmann, R. S.; Wennemers, H. Angew. Chem., Int. Ed. 2011, 50, 6835-6838; (c) Siebler, C.; Erdmann, R. S.; Wennemers, H. Angew. Chem., Int. Ed. 2014, 53, 10340-10344; (d) Umashankara, M.; Sonar, M. V.; Bansode, N. D.; Ganesh, K. N. J. Org. Chem. 2015, 80, 8552-8560; (e) Erdmann, R. S.; Wennemers, H. J. Am. Chem. Soc. 2010, 132, 13957-13959; (f) Erdmann, R. S. Wennemers, H. Bioorg. Med. Chem. 2013, 21, 3565-3568; (g) Erdmann, R. S.; Wennemers, H. Org. Biomol. Chem. 2012, 10, 1982-1986.

04. (a) Chaume, G.; Barbeau, O.; Lesot, P.; Brigaud, T. J. Org. Chem. 2010, 75, 4135-4145; (b) Feytens, D.; Chaume, G.; Chassaing, G.; Lavielle, S.; Brigaud, T.; Byun, B. J.; Kang, Y. K.; Miclet, E. J. Phys. Chem. B 2012, 116, 4069-4079; (c) Chaume, G.; Feytens, D.; Chassaing, G.; Lavielle, S.; Brigaud, T.; Miclet, E. New J. Chem. 2013, 37, 1336-13; (d) Chaume, G.; Simon, J.; Caupène, C.; Lensen, N.; Miclet, E.; Brigaud, T. J. Org. Chem. 2013, 78, 10144; (e) Chaume, G; Terrien, A; Renaglia, E; Marquant, R; Jacquot, Y; Brigaud, T; Miclet, E J. Pept. Sci. 2014, 20(S1), 187-188.

05. Simon, J.; Pytkowicz, J.; Lensen, N.; Chaume, G.; Brigaud, T. J. Org. Chem. 2016, 81, 5381-5392.

MAX BERGMANN AND BRUCE MERRIFIELD AS PIONEERS IN PEPTIDE AND PROTEIN SYNTHESIS AT THE ROCKEFELLER INSTITUTE FOR MEDICAL RESEARCH

Alexander R. Mitchell[1]
[1] Lawrence Livermore National Laboratory (retired), Livermore, California, United States

OP – 11

Introduction

Max Bergmann and Bruce Merrifield significantly contributed to the peptide and protein chemistry we know and use today. Both overcame significant challenges, not always scientific, in their respective careers.

Max Bergmann in Germany (1886-1933)

The curriculum vitae (Lebenslauf) from Max Bergmann's doctoral dissertation, "Über Acylpolysulfide" (1911), informs us that Bergmann was born on February 12, 1886 in Fürth, Bavaria as the son of a merchant (1). Also, consistent with practice at that time, he is self-described as a member of the Jewish faith. Bergmann obtained his doctorate in chemistry under Ignaz Bloch working in the Chemical Institute of the University of Berlin headed by the brilliant and eminently renowned Emil Fischer. Bergmann subsequently served as Fischer's scientific assistant until Fischer's death in 1919. Bergmann later became the first director of the newly formed Kaiser Wilhelm Institute for Leather Research in Dresden and professor of chemistry at the Dresden Technical University (1922-1933). In Dresden Bergmann established a productive laboratory balancing applied research (leather chemistry) with basic research (carbohydrate, peptide and protein chemistry) that resulted in the publication of 227 papers and 29 patents (2). The discovery in 1932 and use of the carbobenzoxy protecting group (benzyloxycarbonyl or Z group in honor of Zervas) revolutionized peptide synthesis (3).

Changing Status of Jews in the German States and Germany

Briefly stated, in contrast to many other parts of Europe, the ascent of German Jews from the mid-18th century until 1933 represents one of the most spectacular social leaps in European history. The Jews in the German states evolved from a persecuted clan of cattle dealers and wandering peddlers into a stunningly successful community of writers, philosophers, scientists, tycoons and activists (4). Almost one-third of all German Nobel laureates during the first four decades of the twentieth century were Jews. In addition, nearly one-third of the directors of the Kaiser Wilhelm Institutes were Jews when the National Socialists came to power in early 1933 (5). The appointment of Adolf Hitler as chancellor of Germany on January 30, 1933 signaled the beginning of a concerted campaign to return German Jews to the status of pariahs that prevailed two hundred years earlier (6).

Max Bergmann in the United States (1933-1944)

Although Bergmann's position as director of a Kaiser Wilhelm Institute seemed secure in 1933, there was no assurance that future dismissal based on National Socialist racial policies could be discounted. In late 1933 Bergmann departed for the United States to give a lecture tour that included a talk, "Synthesis and Degradation of Proteins in the Laboratory," at The Rockefeller Institute for

Medical Research (RIMR) on November 17, 1933 (7). In April 1934 Bergmann was granted a US immigration visa and appointed Associate Member of the RIMR.

Remarkably, Bergmann's output of scientific work was little affected by the circumstances that forced him in middle age to leave Germany. Leonidas Zervas, Bergmann's productive colleague from Dresden (35 papers), joined him at the RIMR for two years (1934-35) to continue work initiated in Dresden. Two major goals defined Bergmann's research at the RIMR: (1) use of synthetic peptides to investigate the specificities of proteolytic enzymes, and (2) elucidation of protein constituents and structure. Joseph Fruton, initially with the aid of Zervas, admirably achieved the first goal. The second goal was met with limited success. This, however, provided the impetus for Stanford Moore and William Stein to later develop the first successful quantitative analyses of amino acids in protein hydrolysates. Bergmann's laboratory at the RIMR featured many noteworthy postdoctoral associates that included eight future members of the US National Academy of Sciences (J.S Fruton, C.G. Niemann, H. Fraenkel-Conrat, W.H. Stein, K.H. Hofmann, S. Moore, E.L. Smith and P.C. Zamecnik) and two Nobel Laureates (S. Moore and W.H. Stein, 1972).

Bruce Merrifield in New York (1949-2006)

How did such a mild-mannered and relatively unknown biochemist in the mid twentieth century evolve to a global chemical icon by the end of the twentieth century? The answer to this question is found in Bruce Merrifield's scientific autobiography that provides a detailed and incisive history of solid-phase peptide synthesis from 1959 to 1993 (8). A former member of the Merrifield laboratory has provided additional insights (9).

Bruce Merrifield received his doctorate in biochemistry from UCLA (1949) with a strong background in microbiology. He arrived at the RIMR in 1949, five years after the death of Bergmann, to work in the laboratory of D.W. Woolley. Woolley is best known for his work on vitamins, growth factors and antimetabolites. One of Merrifield's projects was to investigate the strepogenin class of peptides that served as bacterial growth factors. Partial acid hydrolysis of crystalline beef insulin and subsequent fractionation procedures yielded a pentapeptide with strepogenin activity. The structure was determined by amino acid analysis and sequencing to be Ser-His-Leu-Val-Gly.

Proof of structure and unequivocal analysis of biological activity would require peptide synthesis, a new endeavor for Merrifield. Although lacking in synthetic experience, Merrifield was in the very laboratories that Bergmann, Zervas, Fruton and others had done their pioneering studies on the synthesis of peptide substrates for proteolytic enzymes. This was an inspiration and incentive to synthesize the strepogenin pentapeptide. Suffice it to say that the project proved more challenging than anticipated. The synthesis took 11 months to provide the pentapeptide in an overall yield of 7%. Merrifield noted that an experienced peptide chemist would have done better, but not without considerable effort. There must be a better way!

Solid-Phase Peptide Synthesis (SPPS)

The use of an insoluble polymer covalently linked to a growing peptide chain was without chemical precedent when Merrifield formulated his concept of SPPS in 1959. The search for an acceptable polymer support and appropriate chemistry was especially challenging (p. 90, ref. 8): "At the end of the first two years the results were so poor, I wonder what made me think that this approach would ever succeed. But from the outset I had a strong conviction that this was a good idea, and I am glad that I stayed with it long enough." Finally, after 3 years of examining numerous polymer supports, reaction conditions and protecting groups, including the Z group introduced 30 years earlier by Bergmann and Zervas, a model tetrapeptide was prepared. Merrifield described the solid-phase synthesis of Leu-Ala-Gly-Val at the meeting of the Federation of American Societies for Experimental Biology in 1962 and a full paper appeared in 1963 (p.87, ref. 8). The response of non-specialists (scientists employing peptides in biological investigations) was quite enthusiastic while many specialists (synthetic peptide chemists) were highly critical. The idea of conducting a multistep synthesis, without isolating, purifying, and characterizing intermediates was clearly beyond the pale for those trained in synthetic organic chemistry. This summarizes the early days of SPPS. Bruce Merrifield's original intent was simply to make the task of peptide synthesis less onerous. He could not have imagined, especially in the early years, that his work would result in a paradigm shift in how synthetic chemistry is now used in molecular biology, biotechnology, chemistry and materials science. Also, the impact of solid-phase synthesis on combinatorial chemistry, a field not yet conceived in 1959, could not have been predicted. Merrifield was recognized as the sole recipient of the Nobel Prize in chemistry (1984) for his invention of SPPS.

Conclusions

Max Bergmann was an organic chemist uprooted from an outstanding career in his native country. He quickly reestablished himself as an exceptional scientist at the RIMR where he set up a pioneering laboratory in peptide and protein chemistry (1934-1944). Bruce Merrifield, initially a relatively unknown biochemist, spent his entire scientific career at the RIMR (1949-2006). His invention of SPPS transformed the use of synthetic chemistry in many areas. Together, Bergmann and Merrifield, separated by years and training, profoundly influenced the peptide and protein chemistry we know and use today. Their impact on the scientific endeavors and careers of so many people goes far beyond their initial inventions.

References

01. Bergmann, M. Über Acylpolysulfide. Ph.D. Dissertation, Berlin University, 1911.
02. Helferich, B. Chem. Ber. 1969, 102, I-XXVI.
03. Bergmann, M.; Zervas, L. Ber. 1932, 65, 1192-1197.
04. Elon, A. The Pity of It All. A Portrait of the German-Jewish Epoch, 1743-1933; Picador: New York, 2003.
05. Rürup, R; Schüring, M. Schicksale und Karrieren. Gedenkbuch für die von den Nationalsozialisten aus der Kaiser-Wilhelm Gesellschaft vertriebenen Forscherinnen und Forscher; Wallstein, Göttingen, 2008.
06. Friedländer, S. Nazi Germany and the Jews: The Years of Persecution, 1933-1939; HarperCollins: New York, 1997.
07. Bergmann, M. Science. 1934, 79, 439-445.
08. Merrifield, B. Life During a Golden Age of Peptide Chemistry: The Concept and Development of Solid-Phase Peptide Synthesis; American Chemical Society, Washington, DC, 1993.
09. Mitchell, A.R. Biopolymers (Peptide Science), 2008, 90, 175-184.

MITOCRYPTIDES INDUCE NEUTROPHIL MIGRATION IN VIVO AS MITOCHONDRIAL DAMPS PROMOTE

Tatsuya Hattori, Hiroki Morikawa, Koki Tsutsumi, Takayuki Marutani, Yoshiaki Kiso, Hidehito Mukai
Nagahama Institute of Bio-Science and Technology, Graduate school of Bio-Science, Nagahama, Japan

OP - 39

Introduction

Innate immune responses protect our bodies from bacterial infection and non-infective tissue injury. Neutrophil is a type of leukocytes that plays pivotal roles in innate immunity. They immediately migrate and infiltrate into tissue injury sites from bloodstream. Subsequently, they produce superoxide to sterilize infected bacteria and phagocytose toxic cell debris. These roles of neutrophils are important for initial innate defense mechanisms, but their excessive accumulation and activation often cause irreparable tissue damage in ischemia-reparfusion injury and fulminant hepatic failure. Recently, mitochondrial damaged-associated molecular patterns (mtDAMPs) are focused as proinflammatory factors [1]. The mtDAMPs that comprised mitochondrial DNA and formyl peptides were suggested to induce innate immune responses including neutrophil migration and activation. However, it is demonstrated that highly purified mitochondrial DNA does not promote neutrophil activation [5]. Moreover, formyl peptides in mtDAMPs are not specified yet.

As endogenous neutrophil-activating substances, we have isolated and identified mitocryptide-1 (MCT-1) and mitocryptide-2 (MCT-2), from healthy porcine hearts [2, 3]. MCT-1 and MCT-2 are found to be fragmented peptides derived from mitochondrial proteins, and it is also suggested that there are many unidentified neutrophil-activating peptides derived from various mitochondrial proteins. Since MCT-2 is an only identified N-formyl peptide from mammalian sources in present, MCT-2 is expected for an activating factor in mtDAMPs. In addition, not only MCT-2 but also MCT-1 that is a non-formylated peptide may involve innate immunity as an mtDAMPs factor. In this way, a family of mitocryptides (MCTs) has similar characteristic to mtDAMPs factors, but their physiological functions are still obscure. Here, in order to elucidate physiological and pathophysiological roles of mitocryptides as mtDAMPs factors in vivo, we examined the influences of mitocryptides to immune cells in mice.

DRIAL D

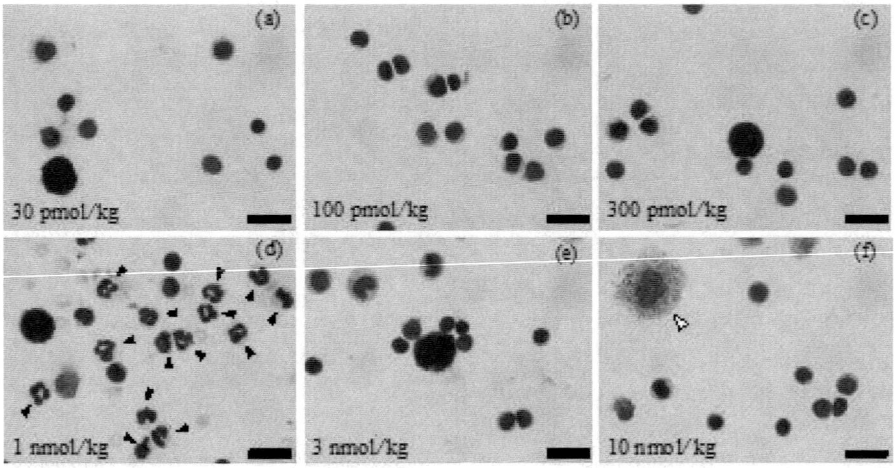

e cells in peritoneal cavity under stimulation by mouse MCT-2. Black arrows and a ell, respectively. Scale bars, 20 mm.

ɟ human MCT-2 into male C57BL/6 mice (8-10 weeks) at several mulation by MCT-2, cells in peritoneal exudate were harvested by ited cells were prepared for smear, and stained with Diff-Quik to ally. As a result, mouse MCT-2 (mMCT-2) induced neutrophil mi- ol/kg but not other concentrations (Fig. 1 a-f). On the other hand, rved by human MCT-2 in same condition (data not shown). These difference of MCT-2 is strictly recognized by formyl-peptide recep- tion, we also investigated the effects of mMCT-2 to mast cells, and kg mMCT-2 induced degranulation of mast cells without influence itoneal cavity (Fig. 1 f). We previously reported that porcine MCT- ıd degranulation of mast cells in vivo [6]. These facts demonstrate o mtDAMPs in vivo, proposing that MCTs that are released from nnate immune responses (Fig. 2).

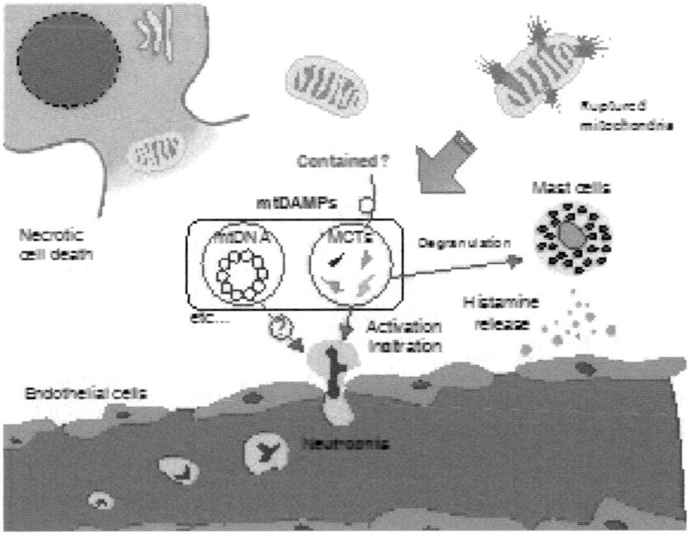

Fig. 2. The model of innate immune responses induced by MCTs.

References

01. Zhang, Q.; Raoof, M.; Chen, Y.; Sumi, Y.; Sursal, T.; Junger, W.; Brohi, K.; Itagaki, K.; Hauser, C. J. (2010) Nature, 464, 104-107.
02. Mukai, H.; Seki, T.; Nakano, H.; Hokari, Y.; Takao, T.; Kawanami, M.; Tsukagoshi, H.; Kimura, H.; Kiso, Y.; Shimonishi, Y.; Nishi, Y.; Munekata, E. (2009) J. Immunol., 182, 5072-5080.
03. Mukai, H.; Hokari, Y.; Seki, T.; Takao, T.; Kubota, M.; Matsuo, Y.; Tsukagoshi, H.; Kato, M.; Kimura, H.; Shimonishi, Y.; Kiso, Y.; Nishi, Y.; Wakamatsu, K.; Munekata, E. (2008) J. Biol. Chem., 283, 30596-30605.
04. Marutani, T.; Hattori, T.; Tsutsumi, K.; Koike, Y.; Harada, A.; Noguchi, K.; Kiso, Y.; Muaki, H. (2016) Biopolymers (Peptide Science), 106, 580-587.
05. Prikhodko, A. S.; Shabanov, A. K.; Zinovkina, L. A.; Popova, E. N.; Aznauryan, M. A.; Lanina, N. O.; Vitushkina, M. V.; Zinovkin, R. A. (2015) Biochemistry (Moscow), 80, 629-635.
06. Hattori, T.; Morikawa, H.; Marutani, T.; Tsutsumi, K.; Kiso, Y.; Mukai, H. (2016) Peptide Science, 2015, 253-254.

OP – 44

Introduction

The recently the studies of the principles for designing ideal protein structure are energetically investing, however the shorter peptide possessing proteolytic activity including both natural sources and synthetic one has not found yet (1, 2). Although it is generally accepted that enzyme should be a large molecular protein consisted of more than thousands of amino acids, we found the intrinsic hydrolase-like peptide named JAL-TA9 (YKGSGFRMI) which is consisted of 9 amino acid (3, 4). The Cys residue in BoxA domain of Tob1 protein was substitute to Met residue in JAL-TA9. Tob/BTG family proteins are involved in cell cycles and regulation in a variety of cells such as T lymphocytes, fibroblasts, epithelial cell, and germ cells (5, 6). Although there are many reports concerning with Tob/BTG family proteins including function and structure analysis, any research concerning with the catalytic activity of these proteins has not been reported yet. We show herein the evidence of proving the hydrolase-like activity of JAL-TA9 and its digesting activity to Aβ42.

Results and Discussion

The peptides using in this experiment including JAL-TA9 were synthesized by a solid phase automatic peptide synthesizer with F-moc method and purified by HPLC. The molecular weight of objective peptide was confirmed by MS analysis. JAL-TA9 (final conc.: 0.2 mM) was individually incubated with Aβ-derived fragment peptides, insoluble solid form Aβ42 or authentic soluble Aβ42 (final conc.: 0.05 mM) in the presence of human serum albumin (HSA) (final conc.: 0.025% w/v) in PBS at 37 °C. A portion of reaction mixture was analyzed with time dependent manner by an analytical HPLC system. The appearing peaks monitoring with 220 nm were collected into a micro-tube. After lyophilization, a proper quantity of 36 % CH_3CN containing 0.1% HCOOH was added judging from the peak height on chromatography, and stirred with automatic mixer. The determination of cleavage site was performed by ESI-MS by flow injection method using a Qstar Hybrid LC/MS/MS system (ABI). We first analyzed auto-degradation activity of JAL-TA9. The reaction mixture was monitored at every 1hr up to 8hrs, and then successively continued to 22 hrs by an analytical HPLC monitoring with photo-diode array detector. Although only JAL-TA9 was identified as single peak on day 0, eight peaks were appeared after 6 hr incubation. To determine the cleavage site by auto-digestion of JAL-TA9, we collected these peaks, and then applied to MS analysis. Seven kinds of newly appearing peaks were identified as fragment peptides from JAL-TA9, but any fragment from HSA was not. This auto-degradation of JAL-TA9 was inhibited with 4-(2-Aminoethyl) benzenesulfonyl fluoride hydrochloride (AEBSF) of serine protease inhibitor. The chromatogram changing was appeared in 1 hr incubation. The JAL-TA9 was decreased in time dependent manner, in contrast four newly appearing fragment peptides, especially JAL12-17 (YKGSGF) which is identified at earlier retention time than JAL-TA9, were increasing in inverse proportion with JAL-TA9. This result indicates that JAL-TA-9 might possess proteolytic activity.

To confirm the proteolytic activity of JAL-TA9, we next examined whether JAL-TA9 cleaved Aβ-derived fragment peptides. Base on the 3D structure (7), three kinds of Aβ-derived fragment peptides, Aβ1-20, Aβ11-29 and Aβ28-42 were synthesized and incubated with JAL-TA9 up to 5 days according to the same manner described in. In the case of Aβ1-29, this fragment peptide forms β-sheet structure and contains essential region to form the oligomer/aggregate of Aβ42 (7), nine and ten peptides were identified as the fragment peptides derived from JAL-TA9 and Aβ11-29, respectively. This cleavage reaction was also inhibited by AEBSF as expected. These data suggest that JAL-TA9 may possess serine protease-like activity. In the case of Aβ1-20 of soluble N-terminus region and Aβ28-42 of insoluble region, three and two fragment peptides were identified as Aβ-derived peptides, respectively. These results suggest that JAL-TA9 has higher affinity to Aβ11-29 than both Aβ1-20 and Aβ28-42. Judging from the chromatogram of the time dependent analyses, Aβ11-29 is the most potency substrate of JAL-TA9. These data prove that JAL-TA9 possess the serine protease-like activity, and cleaves Aβ fragment peptides, especially Aβ11-29. The cleavage sites of Aβ42 fragment peptides by JAL-TA9 were shown in Fig. c.
Although any peptide hydrolase has still not identified, our results indicate that the shorter peptide such as JAL-TA9 can show the protease activity. This is the first finding which nobody have never thought. Thus, we next planned to examine whether JAL-TA9 can degrade Aβ42. To test the cleavage activity of JAL-TA9 to Aβ42, we tried to synthesized and purified Aβ42. As well known, the preparation of synthetic Aβ42 including deprotection and purification processes is very difficult, since Aβ42 is insoluble in any solvent except trifluoroacetic acid (TFA) of strong acid and formed polymerized or aggregated form which interfere the study of Alzehimer's disease (8). Therefore, we used only TFA for deprotection and cleavage reaction which method is different from the manufacture recommended method. The reaction mixture filtrated and ten diluted water. After lyophilization, the slightly yellow solid material, which was insoluble in CH_3CN, CH_3OH or CH_3COOH, was obtained. Thus, we tried to use this solid material without purification procedure as substrate. Since this reaction conditions are unusual manner, we were afraid that the hydrolytic reaction would not occur. After washing well with CH_3CN and CH_3OH to remove the protect groups of amino acids, the solid material was incubated with the JAL-TA9 by the same manner described above. A portion of reaction mixtures was analyzed by using an analytical HPLC every day up to 7days. Although any major peak except JAL-TA9 appearing at 10.5 min was not identified on day 0, some new peaks were appeared with decreasing of JAL-TA9 on day 1. Almost of all newly appearing peaks on day 3 were getting higher, in contrast, JAL-TA9 was getting lower. The peak appearing at 9.5 min was distinguishably increased. JAL-TA9 was disappeared at day 4, however, the chromatogram patterns were still changing up to day 7. The newly appearing peaks on day 7 were collected (Fig. a) and applied to MS analysis. The collecting peaks were identified as fragment peptides derived from JAL-TA9 (J1 to J9) and Aβ42 (A1 to A10). Such changing was not appeared in the case of Aβ42 alone.

OP – 44

Only three fragment peptides containing Ala residue of the C-terminal end in Aβ42, those peptides are thought to be a side-product of Aβ42 synthesis, were identified. These data suggest that the solid material contains the Aβ42, and JAL-TA9 cleaves the solid type of Aβ42 (s-Aβ42).

However, it is very hard to have fully confidence for us, because we never seen the any report about the peptide protease or peptidase. Moreover, it is incredible story that JAL-TA9 hydrolyzes the s-Aβ42. If our finding is true, we might open the novel door to new aspect of peptide and enzyme chemistry loading to Alzheimer's disease therapy. Therefore, we next examine the degradation effects of JAL-TA9 to authentic Aβ42 (a-Aβ42) purchased from Peptide Institute (Osaka, Japan) with great expectation. A portion of the reaction mixtures of JAL-TA9 and the a-Aβ42 were analyzed every day up to day 5. On day 0, three peaks, JAL-TA9, a-Aβ42 and HSA, were identified. The peak height of JAL-TA9 was decreasing on time dependent manner, in contrast the new peaks were appearing. On day 5, JAL-TA9 was not identified any more, in contrast, the newly appearing peaks were identified between 6 and 10 min. To identify the fragment peptides of a-Aβ42, we collected all of appearing peaks on day 7, and then analyzed by MS. Six peaks (A1 to A5) were identified as the fragment peptide derived from a-Aβ42, in addition to the JAL-TA9 fragment peptides (J1 to J9) (Fig. b). On the other hand, in presence of AEBSF, only two peaks were identified as JAL-TA9 derived fragment peptides (Fig. c). Moreover, no fragment peptide was identified as a-Aβ42 alone. Thus, we concluded that

JAL-TA9 is the serine protease-like peptide, and can digest Aβ42. The cleavage sites of both s- and a-Aβ42 were shown in Fig. d.

Furthermore, a 27-MHz quartz crystal microbalance (QCM) was used to quantify the affinity between immobilized JAL-TA9 and injected Aβ42 or its fragment peptides. In a result, authentic Aβ42 did not show frequency changes against JAL-TA9 in a half-hour measurement, but a frequency decrease after injection of Aβ42 was observed in 12 hour later. The conformational study using 2D NMR also provided that JAL-TA9 formed the compact structure.

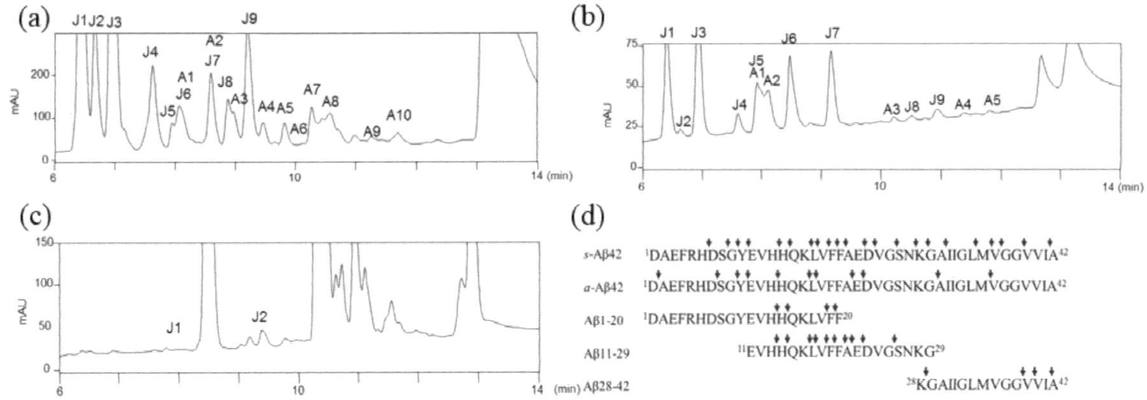

Fig. Determination of cleavage site on Aβ42 by JAL-TA9

Column: Capcell Pak C18 MGII (4.6 mm i.d. × 150 mm), Flow rate: 1mL/min,

Column temp.: 40°C, Elution: 0 – 70% CH3CN containing 0.1% TFA for 15 min, Detection: UV 220 nm

(a) Solid type Aβ42, (b) Authentic Aβ42, (c) Authentic Aβ42 in the presence of AEBSF, (d) Cleavage site of Aβ42 and its fragment peptide

Up to date, no peptide enzyme is found, however, we conclude that JAL-TA9 is well qualified as fundamental properties of protein enzyme, even if it can defy conventional wisdom. We propose that the peptide possessing the hydrolase-like activity such as JAL-TA9 termed Catalytide (catalytic peptide). The concept of Catalytide has opened the unknown door loading to the new aspect of peptide science, and is a new seed for the development of peptide drug for not only Alzheimer's disease but also another neurotoxic disease such as amyotrophic lateral sclerosis (ALS), Parkinson's disease, or Prion disease, etc.

References

01. Koga, N., Tatsumi-Koga, R., Liu, et al., Nature, 2012, 491(7423): 222.
02. Winkler, G.S., J. Cell. Physiol. 2010, 222, 66.
03. Yamamoto, T., Akizawa, T., 2016 NOVEL PEPTIDE Patent Pending No US62/275,599
04. Yamamoto, T., Akizawa, T., 2016 NOVEL HYDROLASE-LIKE PEPTIDE AND ITS USE, No JP2016-0668496
05. Yoshida, Y., Matsuda, S., Ikematsu, N., Kawamura-Tsuzuku, J., Inazawa, J., Umemori, H., and Yamamoto, T., Oncogene, 1998, 16, 2687-2693 [PubMed: 9632145]
06. Matsuda, S., Kawamura-Tsuzuki, J., Ohsugi, M., Yoshida, M., Nakamura, Y., Onda, M., Yoshida, Y., Nishiyama, A., and Yamamoto, T., Oncogene, 1996, 12, 705-713 [PubMed: 8632892]
07. Sun, X., Chen, W-D., and Wang, Y-D., Frontiers in Pharmacology, 2015, 6, 1-9, doil: 10.3389/fphar.2015.00221
08. Ahmed, M., Davis, J., Aucoin, D., Sato, T., Ahuja, S., Amimoto, S., Elliot, I. J., Van Nostrand, E. W., and Smith, O. S., Nat Struct Mol Biol. 2010, 17(5), 561-567, doil: 10.1038/nsmb.1799w

DOUBLE STRAND DNA RECOGNITION BY PEPTIDES CONSISTING OF PYRROLE AND IMIDAZOLE MOIETY, DESIGNATED PIPA, FOR MOLECULAR PROBE AND DRUG CANDIDATES

Kiyoshi Nokihara[1], Akiyoshi Hirata[1], Atsushi Kitagawa[1] and Yuki Tominaga[1]
[1] HiPep Laboratories, Kyoto, 602-8158, Japan
noki@hipep.jp

OP - 46

Biomolecular recognition can be applied to research tools, diagnostics and drug-development. Peptides containing N-methylpyrrole (Py) and N-methylimidazole (Im) as building blocks, designated PIPA, bind to specific nucleotide sequences in the minor groove of double-helical DNA through hydrogen bonding with high affinity and specificity [1]; PIPA blocks binding of transcription factors inhibiting gene expression, thus PIPA can be used for gene-control. PIPA consists of Py, Im, bAla for distance optimization and gAbu for the hairpin motif, which are all non-proteinogenic amino acids. Binding site specificity is dependent on the side-by-side pairing of Py and Im corresponding to the combination of Py and Im and binding (dsDNA-PIPA) can be optimized by the combinatorial library construction. Hence 3-(dimethylamino)propylamine (Dp) at the C-terminus was found to increase affinity to DNA. Advantages of PIPA over siRNA or PNA technologies are follows:
(1) PIPAs are alternative gene silencers other than siRNA or PNA; (2) PIUPAs are stable in cells or bodies because of nuclease resistance; (3) PIPAs bind specifically to the target double strand DNA, designable against any gene and prepared by chemical syntheses; (4) PIPAs enter into cell nucleus without any DDS, that is, PIPA doesn't require CPP such as PNA-delivery; (5) No-toxicities have been found in the preliminary cell and animal experiments; (6) PIPA excreted into the urine for 3~4 weeks. Sustainable and reproducible industrial production is indispensable for clinical trial and FDA-approval, although practical production is not yet envisaged in world wide. In fact PIPA-syntheses have been reported only by academic groups. Recently a system for larger scale production of PIPAs with standard operation protocols of all materials, assembly and quality control methods have been successfully established in consideration with less manpower (automated solvent delivery, removal and resin washing), contamination free process, batch wise ca. 15~150 gram/reactor of crude PIPA using 300 ~3000 mL reactors and reproducible with high quality.
Thus 7~8 cycles a day by one operator could be performed. In fact PIPA-production is more difficult and complicated in synthesis, purification and characterization comparing to the conventional peptides with natural amino acids because of the following reasons. Building unites Py and Im are rather expensive in large scale production, PIPA possess structures similar to the peptides having difficult sequences. The chemical syntheses of target molecules are generally known, thus de novo sequencing is not necessary, although in the stepwise solid-phase syntheses theoretically numerous deletion compounds can be generated and by-products in the syntheses of PIPA have similar properties, as Im and Py residues differ only one mass unit, therefore deletion (by-products) have similar properties and gave often overlapping peak-patterns in HPLC.
Additionally by-products often disturb ionization of the target compounds in MS analyses, and monitoring during assembly of PIPA is favorable for production economics. Automated assignment software cannot be applied therefore manual de novo sequencing are indispensable (MS^n Mode).

In the present report two compounds have been chosen as targets, based on demands and market-size or compete products for similar purposes. PIPA targeting on human TGF-b1, which causes fibrosis such as renal failure [2, 3], in fact life style diseases (40% patients in Japan) causes chronic renal failure and 37,000 patients are received dialyses which arise 125 Million USD/year in Japan. The first target molecule designated HPTB-01, Ac-Im-Py-Py-βAla-Im-Py-Py-γAbu-Py-Py-Py-βAla-Py-Py-βAla–Dp (mol. wt. 1787.9), which was structurally optimized for binding to human TGF-β1 promoter AP-1 region. HPTB-01 is expected as drugs for progressive renal diseases, pulmonary fibrosis and/or liver cirrhosis and they can be used without any DDS. Preclinical tests of HPTB-01 are underway. The second target molecule is a molecular-probe, designated HPTH59 (Fig. 1), a tandem hairpin PIPA targeting telomeres. Telomere has the repeat of simple nucleotide sequence (TTAGGG) at the terminal region of chromosomes.

The length of telomere is believed to be associated with aging or cancer diseases. Telomere visualization by PIPA had been reported initially in 2001 [4], although no details of compounds had been described and a tandem polyamide recognizing repetitive sequences of vertebrate telomere has not been confirmed until 2013 [5]. Recently we have successfully optimized the structure of the compounds [6] as HPTH59 and commercialized [http://hipep.jp/?cat=95]. During quality assurance micro-heterogeneity of PIPA has been explored. Even tiny amounts oxidized compounds have been identified. While mass differences was 15.995, oxygen was adducted at different residues and this was envisaged at Py but not at Im residues [6]. HPTH59 allows simple with lower background v.s. conventional methods: telomerase based method, Southern blotting and/or PNA-FISH. Applicable for both intact Cells and Fixed samples. Telomere length can also be quantitatively determined. Protocols for practical applications of HPTH59 is published recently [7]. Quantitative measurements of telomeres length is expected for the novel diagnostics. In conclusion PIPAs have functions for gene regulation through DNA binding and applied as novel chemical probes (diagnostics) or therapeutic agents.

The present study was partially supported by Japan Science and Technology Agency 2013, Innovative Medical Technologies, Kyoto City, 2013 and Innovation Commercialization Venture Support Project, New Energy Development Organization 2014. An author (KN) appreciates the Alexander von Humboldt-Stiftung (Germany) for the support to attend this symposium.

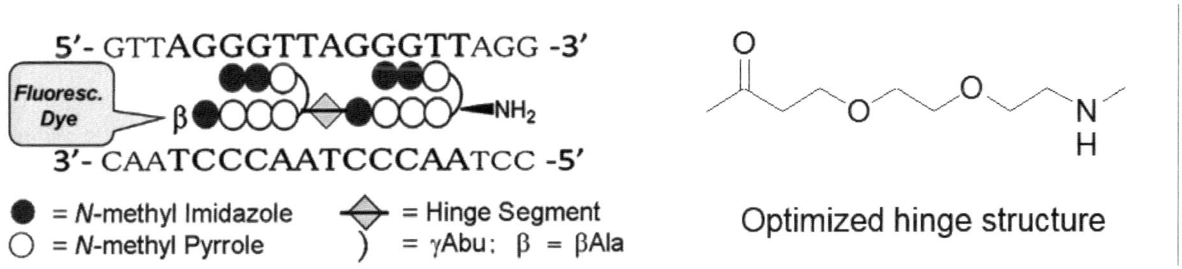

Fig. 1. Recognition Model for HPTH59 with telomere.

References

01. Dervan, P.B., et al. (1998) Science, 282, 111-115 ; (1998) Nature, 391, 468-471.
02. Matsuda, H., et al. (2011) Kidney Int., 79, 46-56.
03. Washio, H., et al. (2011) J. Invest. Dermat., 131, 1987-1995.
04. Maeshima K, et al. (2001) EMBO J., 20, 3218-3228.
05. Kawamoto, K., et al. (2013) J. Am. Chem. Soc., 135, 16468-16477.
06. Hirata, A., et al. (2014) J. Am. Chem. Soc., 136, 11546–11554.
07. Sasaki, A., et al. (2016) Sci. Rep., 6, 29261.

THE RENAISSANCE OF OXYTOCIN: A NEW FRONTIER IN TRANSLATIONAL RESEARCH

B. Chini, M. Busnelli[1], O. Burton[2], S. Stoev[2] and M. Manning[2]
[1]CNR, Institute of Neuroscience, Milan 20129, Italy
[2]Department of Biochemistry & Cancer Biology, University of Toledo College of Medicine & Life Sciences, 3000 Arlington Ave., Toledo, OH 43614, U.S.A.

OP - 62

Oxytocin (OT) is well known as the first biologically active peptide to be synthesized in the laboratory (1), as a therapeutic in obstetrics for its use in the induction of labor and for the treatment of postpartum hemorrhage (2). OT has also been the focus of intensive structure/activity studies aimed at the design and solid phase synthesis (3) of selective agonists and antagonists of the OT and vasopressin (VP), OTR, V_{1a}, V_{1b} and V_2 receptors (4,5). OT has recently entered an exciting new era of research and therapeutic interest focused on its central actions (6-11). This research has been sparked by the tantalizing promise of its potential as a therapeutic for the treatment of neuropsychiatric diseases such as autism spectrum disorder (ASD) (6), schizophrenia (9) and anxiety disorders (6-8, 10, 11) and has led to a burgeoning number of clinical trials on OT.

In parallel to the recent human studies, a myriad of translational investigations have been carried out in animals to clarify which receptor subtype(s) (OTR, V_{1a}, V_{1b} or V_2) and in which brain region(s), mediate the central effects of OT in a wide variety of behaviors, such as pair bonding, maternal care, social recognition, pain, addictions, fear and aggression (7). These translational studies have been greatly facilitated by the availability of OT and VP selective agonists and antagonists (4,5) (see refs 90, 117, 195-245 in ref (5) for studies prior to 2012). Here, we will comment on the current status of the human studies and clinical trials on OT. We will illustrate that, sixty years after the first laboratory synthesis of OT by du Vigneaud and colleagues (1), functionally selective OT agonists and bivalent OT agonists being developed in our laboratories (12-14), are promising new leads to the development of new therapies for neuropsychiatric disorders. We will also briefly mention the key findings of more recent translational studies carried out with peptides from this lab (14-16).

Clinical Trials on OT

The number of clinical trials on OT for psychiatric conditions has increased 8-fold over the past four years (source: NIH Clinical Trials.GOV). However, to date, none of these trials have unambiguously demonstrated that OT is an effective treatment for ASD, schizophrenia, stress and post-traumatic stress disorders, depression, pain or drug-dependence (11). Among the possible causes of such shortcomings are the very short OT half-life and its poor capability to cross the blood brain barrier. Furthermore, when OT is administered intranasally, it is not clear where it acts and what doses are effective (11,17). As a consequence, no standard treatment protocols have been optimized so far. Potential solutions include the development of new more potent and selective OT ligands and their validation in animal models. In this respect, it is noteworthy to mention that, in animals, OT analogs can be injected intracerebroventricularly (i.c.v.) and/or in selected brain regions, allowing to test for region specific behavioral effects (refs 14-16).

Progress in the Development of New OT Analogs: Bivalent Ligands for the OTR

The challenges of designing OTR selective ligands are reviewed in (4,5). We have uncovered a highly selective OT agonist (Thr^4Gly^7OT) and antagonists for the mouse OTR (13). We have also identified functionally selective OT ligands able to activate specific signaling pathways (12, 18-19). In 2006, we reported the first bivalent antagonists for the human OTR (see pp 491-492 in 4). These were designed by linking two molecules of an OT antagonist at their Orn^8 and Lys^8 residues with suberic acid (C8) (see 14 for ref). At that time, we also synthesized a bivalent C8 agonist of dLVT, recently referred as $dOTK_2$-C8 (14), whose totally unexpected enhancement in affinity for the OTR, investigated with the help of multiple collaborators, were recently published (14) and reported as an ACS Editors' Choice. The $dOTK_2$-C8 bivalent ligand, in addition to being 1,000-fold more potent than OT in vitro, is 100- and 40-fold more potent than OT in vivo in social behavior tests in mice and zebrafish (14). Thus $dOTK_2$-C8 represents a powerful tool to identify and investigate dimeric OTR in the brain and other tissues.

Conclusions

More work needs to be done to clarify the molecular and functional actions of OTR in the brain and other tissues. Studies in animals using new selective agonists and antagonists, biased analogs and bivalent compounds, offer promising insights (12-16, 18). Of note, the recently reported bivalent OTR agonists (14) represent very promising new research tools and a potential new therapy for neuropsychiatric illnesses.

Acknowledgments

We thank all of our co-authors on ref. 14 for their contributions. We are also grateful to all our collaborators and co-workers (noted in ref. 5) for their longstanding contributions. We also thank Ms. Jenny Zak for expert help in the preparation of this manuscript. M.B. is the recipient of an Umberto Veronesi Postdoctoral fellow. This work was supported by NIH Grant GM-25280 (to M.M.). Research support (to MM) from Mr. and Mrs. Robert Tyner, Dr. Rao Makineni, Mr. Frederik Paulsen (Ferring International), the Dept. Biochemistry & Cancer Biology, UTCOMLS.

References

01. du Vigneaud, V., Ressier, C., Swan, J.M., Roberts, C.W., Katsoyannis, P.G. J. Am. Chem. Soc. 1954; 76:3115-3121.
02. Prankerd, R.J., Nguyen, T.H., Ibrahim, J.P., Bischof, R.J., Nassta, G.C., Olerile, L.D., Russell, A.S., Meiser, F., Parkington, H.C., Coleman, H.A., Morton, D.A., McIntosh, M.P. PLoS One 2013; 8(12):e82965
03. Merrifield, R.B., J. Am. Chem. Soc. 1963; 85:2149-2154.
04. Manning, M., Stoev, S., Chini, B., Mouillac, B., Guillon, G. In: Landgraf, R., Neumann, I., eds. Prog Brain Res Amsterdam: Elsevier, 2008: 473-512.
05. Manning, M., Misicka, A., Olma, A, Bankowski, K., Stoev, S., Chini, B., Durroux, T., Mouillac, B., Corbani, M., Guillon, G. J. Neuroendocrinol. 2012; 24:609-628
06. Young, L.J., Barrett, C.E. Science 2015; 347(6224):825-826.
07. Freeman, S.M., Young, L.J. J. Neuroendocrinol. 2016; 284(4):1-22.
08.Neumann, I.G., Slattery, D.A. Biolog. Psych. 2016; 79:213-221.
09. Shilling, P.O., Feifel, D. CNS Drugs 2016; 30:193-208.
10. Insel, T.R., O'Brien, D.J., Leckman, J.F. Biol. Psychiatry 1999; 45(2):145-157.
11. Insel, T.R. Biol. Psychiatry 2016; 79(3):153-154.
12. Busnelli, M., Saulière, A., Manning, M., Bouvier, M., Galés, C., Chini, B. J. Biol. Chem. 2012; 287(6): 3617-3629.
13. Busnelli, M., Bulgheroni, E., Manning, M., Kleinau, G., Chini, B. J. Pharmacol. Exp. Ther. 2013; 346(2): 318-327.
14.Busnelli, M., Kleinau, G., Muttenthaler, M., Stoev, S., Manning, M., Bibic, L., Howell, L.A., McCormick, P.J., Di Lascio, S., Braida, D., Sala, M., Rovati, G.E., Bellini, T., Chini, B. J. Med. Chem. 2016; 59(15):7152-7166.
15. Anacker, A.J.J., Christensen, J.D., LaFlamme, E.M., Grunbert, D.M., Beery, A.K. Psychoneuroendocrinol. 2016; 68:156-162.
16. Song, Z., Larking, T.E., O'Malley, M.E., Albers, H.E. Hormones and Behavior 2016; 81:20-27.
17. Leng, G., Ludwig, M. Biol. Psychiatry 2015; 79(8)e55-e56.
18. Passoni I, Leonzino M, Gigliucci V, Chini B, Busnelli M. J. Neuroendocrinol. 2016; 28(4).
19. Reversi A, Rimoldi V, Marrocco T, Cassoni P, Bussolati G, Parenti M, Chini B. J. Biol. Chem. 2005; 280

Ferenc Hudecz[1,2]
[1]MTA-ELTE Research Group of Peptide Chemistry, Eötvös Loránd University, Hungarian Academy of Sciences, P.O.Box 32, Budapest 112, H-1518 Hungary,
[2]Department of Organic Chemistry, Eötvös L. University, Pázmány P. stny. 1/A, H-1117 Budapest, Hungary
email: fhudecz@elte.hu

OP – 71

This review illustrates the design, structural and functional characterization of two- or three-party bioconjugates of proteins (monoclonal antibodies), polymeric polypeptides, oligopep-tides and of PLGA nanoparticles containing targeting oligopeptide with complement activating peptide from HIV1 gp120 protein, B-cell eitope from fibrin, "reporter molecule" (biotin), or with antitumour drugs [daunomycin, vindoline derivatives]). As potential biolo-gicals, the biomedicinal applications of these conjugates for diagnosis/therapyof rheumatoid artritis or for studying the cellular mecha-nism of action in tumour treatment are reported.

Keywords: bioconjugates, B-cell epitope of fibrin, rheumatoid arthritis, drug delivery, diagnosis of autoimmune disease

Introduction

Bioconjugation chemistry could represent a special and highly multidisciplinary field of organic chemistry. The reaction strategy must consider that the partners, covalently attached in the bio-conjugate, must preserve their functional properties after the completion of the synthesis and pu-rification. In peptide/polypeptide/protein based conjugates the functional properties could mean various bioactivities (e.g. recognition, drug properties, cellular uptake, intracellular trafficking) or "reporter" properties (e.g. fluorescence, radionuclide, biotin)[1].
Here with selected cases we also illustrate the potential of bioconjugates as tools in immundiagno-sis, immune- and chemotherapy and in studying the relevant mechanism: two- or three-party pep-tide bioconjugates were designed, prepared and characterized for a) targeting autoreactive B-cells producing ACPA in rheumatoid arthritis using B-cell epitope recognition based elimination with a three-party nanoconstruct, b) the comparative analysis of the protein expression profiles of tumour cells after the treatment with a two-party conjugate in which an oligopeptide ligand of ErbB2 recep-tor is conjugated with daunomycin (Dau).

Results and Discussion

Conjugates of antitumour compound (e.g. Dau, folate antagonists, vinblastine, vindoline or fer-rocene derivatives) [2-8], enzyme (e.g. calpain) activator/inhibitor [9-11] with a) cell surface-receptor specific ligand (e.g. Erb-B2) [4], b) cell penetrating oligopeptides [5,7,8] , c) branched chain polymeric polypeptide taken up by class A scavenger receptor[2]. We found that the nature (e.g. size, chemical structure) of the targeting moiety, the number and topology of the attached entities as well as the linkage inserted between the partners have marked influence
on binding/recognition properties, cellular uptake, subcellular distribution and on cellular

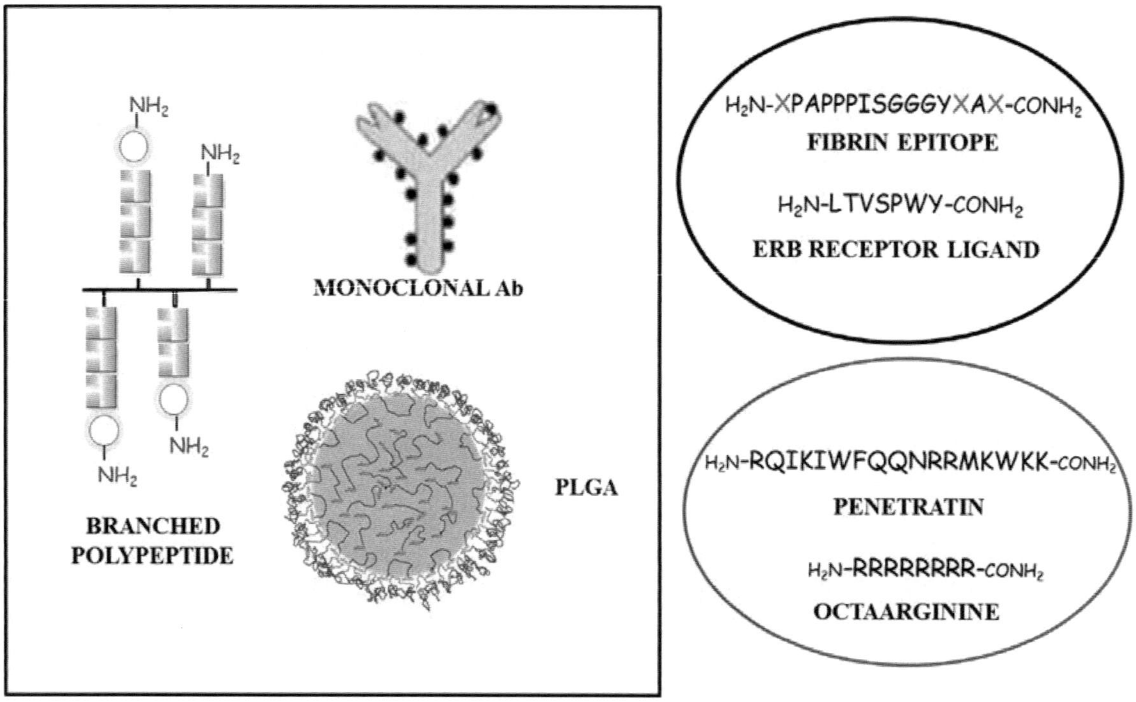

Fig.1 Monoclonal antibodies, branched poly- -amino acids, PLGA nanoparticles and oligopeptides with/without recognition unitresponses (e.g. cytostasis/cytotoxicity, protein expression profile) detected [5-7, 10-11].

Oligo/polypeptides with/without recognition unit are frequently considered as partners for specific targeting of B-/T-cell epitopes or functionalized nanoparticles [12-13]. Post-translational transformation of Arg residues to Cit in relevant proteins could result in structural changes (e.g. charge, conformation) and are connected with autoimmune diseases. We have identified B-cell epitope peptides recognized by anti-Cit protein antibodies (ACPA) to be used for peptide conjugate-based diagnosis of rheumatoid arthritis (RA) in serological assays [14-17]. Recently we have constructed three-party bioconjugates for the selective elimination of autoreactive B-cells producing ACPA (e.g. filaggrin, fibrin). In this, Cit-containing epitope peptide, served as recognition unit, was coupled by covalent linkage with PLGA nanoparticle carriers in multiple copies with uniform orientation and a complement activating peptide from the HIV1 gp120 protein. We have shown that the bifunctional nanoparticles significantly reduced b60-74Cit peptide specific ex vivo ACPA production, by inducing selective, complement dependent lysis of the peptide specific B cells [18-19].

Acknowledgements:
To grants from COST CM1106, MTA-CNR, TÉT_09-1-2010-0010 (RAPEP-09), Hungarian Scientific Research Fund (OTKA) CK 80689, NK 104864, K104385, K104928, NK 105898 and TÁMOP-4.2.2.B-10/1.

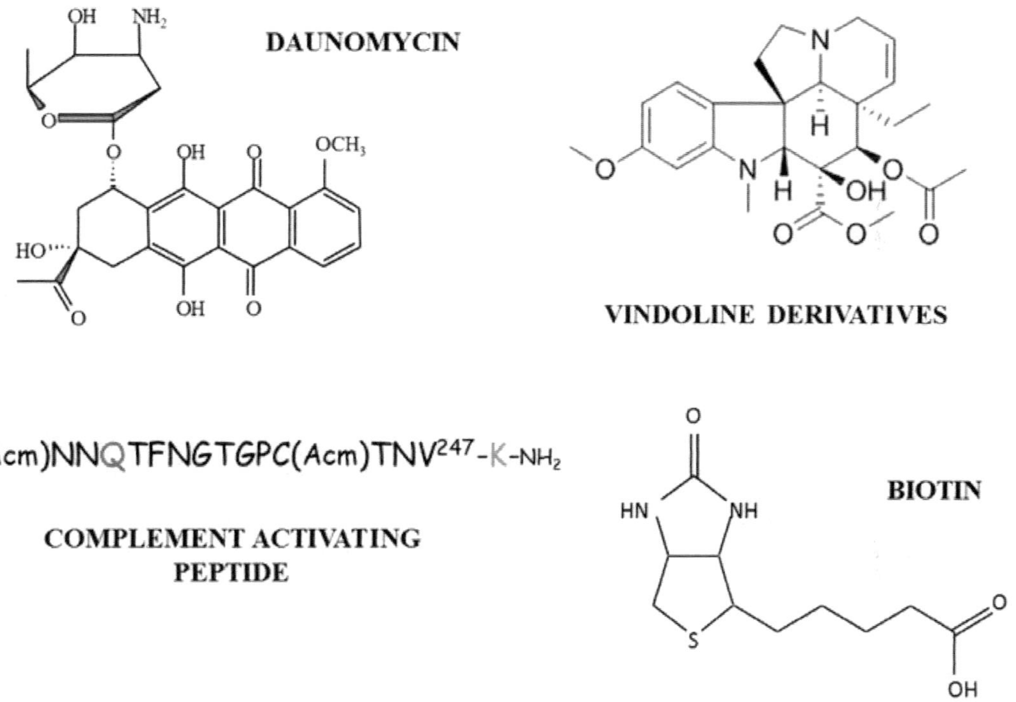

Fig.2 Antitumour compounds, reporter molecule or complement activating peptide from HIV-1 gp 120 protein conjugated with PLGA nanoparticles having fibrin epitope peptide or with oligo/polypeptide with/without recognition unit or with monoclonal antibodies.

References:
01. Mihala, N., Hudecz, F.: Amino acid and peptide bioconjugates. in Amino Acids, Peptides and Proteins (Eds. E. Farkas, M. Ryadnov), RSC, London 37:1-39 (2012)
02. Szabó, R. et al. Bioconjugate Chemistry,19: 1078-1088 (2008)
03. Szabó, R. et al. Biochimica et Biophysica Acta, 1798: 2209-2216 (2010)
04. Orbán, E., et al. Bioconjugate Chemistry, 22: 2154-2165 (2011)
05. Bánóczi, Z., et al. Bioconjugate Chemistry, 21: 1948-1955 (2010)
06. Keglevich, P., Hazai, L. et al. Heterocycles, 87: 2299-2317 (2013)
07. Miklán, Zs., et al. J. Peptide Science, 17: 805-811 (2011)
08. Szabó, I. , Orban, E., Schlosser, G. et al. Eur. J. Med. Chemistry 115: 361-68 (2016)
09. Bánóczi, Z. et al. Bioconjugate Chemistry, 18: 130-137 (2007)

10. Bánóczi, Z. et al. J. Peptide Science, 19: 370-376 (2013)
11. Dókus, L., Tantos, Á., Szabó, I. et al. Eur. J. Med. Chem, 82: 274-280 (2014)
12. Uray, K., Hudecz, F.: Peptide epitopes: Identification and structural modifications of synthetic antigens. in Amino Acids, Peptides and Proteins (Eds. E. Farkas, M. Ryadnov), RSC, London 39: 68-113 (2014)
13.Bösze, Sz.; Hudecz, F.: Proteins and peptides for the immunodiagnosis and therapy of Mycobacterium tuberculosis infections. in Amino Acids, Peptides and Proteins (Eds. M. Ryadnov, F. Hudecz), RSC, London 40: 146-198 (2016)
14. Iobagiu, C., Magyar, A., et al. J. Autoimmunity, 37: 263-272 (2011)
15. Babos, F., Szarka, E., Nagy, Gy. et al. Bioconjugate Chemistry, 24: 817-827 (2013)
16. Ayoglu, B., Szarka, E., Huber, K., Orosz, A. et al. PLOS ONE 9: 1-13, e96403 (2014)
17. Cornillet, M., Sebbag, M., Verrouil, E., et al. Ann. Rheum Dis., 73: 1246-1252 (2014)
18. Szarka, E., Babos, F., Magyar, A. et al. Immunology, 141: 181–191 (2014)
19. Pozsgay, J., Babos, F., Uray, K. et al. Arthritis Research & Therapy, 18: 15-27 (2016)

REGULATED EXOCYTOSIS OF CPP-DELIVERED CARGOES FROM MAST CELLS: A NOVEL CELL-MEDIATED THERAPY?

John Howl[1], Sarah Jones[1]
[1] Molecular Pharmacology Group, Research Institute in Healthcare Science, University of Wolverhampton, Wulfruna Street, Wolverhampton WV1 1LY, United Kingdom

Introduction

Many therapeutic and in vivo applications of cell penetrating peptide (CPP) technologies may be inadvertently compromised by the undesirable ability of polycationic peptides to induce mast cell (MC) degranulation [1,2]. Hence, it is likely that both inert CPP vectors and bioportides (intrinsically bioactive CPPs) may have an enhanced propensity to promote MC secretion either through the general perturbation of plasma membrane integrity and/or by the direct activation of heterotrimeric G proteins [3]. The aims of this study were, therefore, to i) identify inert CPP vectors that did not induce secretion from MCs; ii) utilise CPPs to deliver cargoes into the secretory compartments of MCs, and iii) determine the release kinetics of stored cargoes from MC cells activated by physiological stimuli. Collectively, these studies support the hypothesis that MC degranulation might be exploited to achieve the targeted release of bioactive agents within diseased human tissues [1].

Methods

The release of ß-hexosaminidase, a secretory granule marker, from RBL-2H3 cells is a widely accepted model of regulated exocytosis [1,2]. Hence, we employed the RBL-2H3 model to determine and compare the secretory efficacies of both commonly used inert CPP vectors and a structurally–diverse range of bioportides (Table 1). Confocal microscopy, employing fluorescent CPPs and/or larger protein cargoes, allowed the determination of the precise intracellular localisation of CPP and cargoes within secretory granules or other intracellular sites [1]. Finally, we determined the release of stored cargoes from MCs stimulated with either peptide secretagogues or by antigen-induced aggregation of high affinity IgE receptors [1].

Results

Comparative studies identified two CPP vectors, C105Y [1,4] and Tat [1,5], which readily translocated into RBL-2H3 cells (Table 1). Moreover, these vectors did not induce receptor-independent MC degranulation. When covalently conjugated to the tetramethylrhodamine (TAMRA) fluorophore, TAMRA-C105Y accumulated within acidic secretory lysosomes following efficient cellular translocation (Fig. 1). In contrast, biotinylated-Tat effectively delivered avidin as a non-covalent complex, but the larger protein assumed an intracellular distribution that was not associated with secretory lysosomes [1].
We further analysed the secretion of Tat-delivered cargoes in response to two different stimuli. Both the MC peptide mastoparan and antigen-induced aggregation of high affinity

Table 1. Secretory efficacies of CPP and bioportides. Exocytosis of the lysosomal marker β-hexosaminidase from RBL-2H3 cells was used to determine the secretory efficacies of a range of CPP and bioportides. Efficacies are expressed as the percentage of intracellular ß-hexosaminidase that is exocytosed, minus basal secretory levels following treatment with 10 μM peptide. Data are presented as mean ± S.E.M from 2 independent experiments performed in triplicate (n = 6). ND denotes non-detectable.

Peptide	Sequence	Secretory efficacy
Secretagogues		
Mastoparan	H-INLKALAALAKKIL-NH_2	3.40 ± 0.07
Mitoparan	H-INLKKLAKL(Aib)KKIL-NH_2	10.81 ± 0.37
CPP		
C105Y	H-CSIPPEVKFNKPFVYLI-NH_2	ND
Cyt c^{5-13}	H-KGKKIFIMK-NH_2	ND
Tat	H-GRKKRRQRRRPPQ-NH_2	ND
Penetratin	H-RQIKIWFQNRRMKWKK-NH_2	3.20 ± 0.25
Transportan 10	H-AGYLLGKINLKALAALAKKIL-NH_2	6.09 ± 0.25
Bioportides		
Cyt c^{77-101}	H-GTKMIFVGIKKKEERADLIKKA-NH_2	0.82 ± 0.19
Nosangiotide	H-RKKTFKEVANAVKISA-NH_2	ND
Camptide	H-RKLTTIFPLNWKYRKALSLG-NH_2	2.72 ± 0.41
LRRK2$^{1322-1340}$	H-LQQRLKKAVPYNRMKLMIV-NH_2	0.15 ± 0.17
LRRK2$^{2413-2427}$	H-RVKTLCLQKNTALWI-NH_2	0.33 ± 0.08

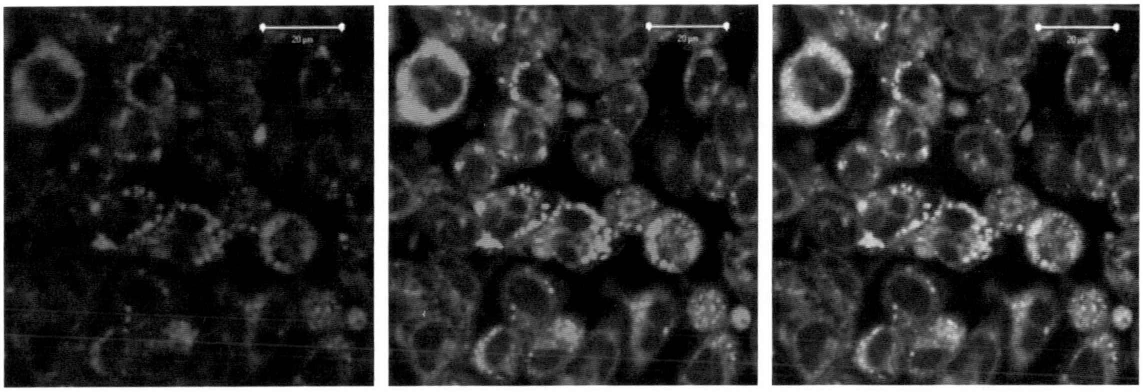

Fig. 1: Non-secretory CPP such as C105Y readily accumulate in secretory lysosomes of RBL-2H3 cells. Cells were treated with TAMRA-conjugated C105Y (3 μM) and LysoSensor™ Green DND-189 (1 μM), the latter to aid in the visualisation of secretory lysosomal structures. Subsequent live cell confocal imaging analysis clearly demonstrates a strong propensity for TAMRA-C105Y to accumulate within acidic secretory granules and is designated here by yellow co-localisation (merge). Scale bars = 20 μm in all panel

IgE receptors, produced a respective concentration- and temporal-dependent exocytosis from structurally-distinct intracellular storage sites [1].

Discussion

It is probable that MC degranulation is a major caveat to the development of both CPPs and bio-portides within a clinical setting. Indeed, numerous animal venoms contain small polybasic peptides that activate MCs upon envenomation. Considering the ubiquitous distribution of MCs it will be difficult to avoid exposing them to peptides delivered by any route of administration. Fortunately, as our more recent studies have indicated [1], a sub-set of inert CPP vectors including C105Y (Fig. 1) can be employed to deliver cargoes into the various secretory compartments of MCs without inducing degranulation. It will be fascinating to repeat such investigations in human MCs. Moreover, stored cargoes, ranging in size from small fluorophores to large proteins, can be released by physiological stimuli. Hence, it is possible that the unique combinatory potential of MCs and non-secretory CPPs could be developed as a novel cell-mediated therapy for the controlled release of bioactive agents at pathological loci.

References

01. Howl J and Jones S. J. Control. Release 202 (2015) 108.
02. Farqhar M, Soomets U, Bates RL, Martin A, Langel Ü and Howl J. Chem. & Biol. 9 (2002) 63.
03. Mousli M, Bueb J-L, Bronner C, Rouot B and Landry YG. Trends Pharmacol. Sci. 11 (1980) 358.
04. Rhee M and Davis P. J. Biol. Chem. 13 (2006) 1233.
05. Vivès E, Brodin P and Lebleu B. J. Biol. Chem. 272 (1997) 16010.

Keisuke Aihara, Kosuke Yamaoka, Naoto Naruse, Tsubasa Inokuma, Akira Shigenaga, Akira Otaka
Japan, Institute of Biomedical Sciences and Graduate School of Pharmaceutical Sciences, Tokushima University, Tokushima, Japan

PP I - 001

Protein chemical synthesis has great potential as a chemical means for elucidating a wide range of protein functions. Native chemical ligation (NCL) is among the most powerful methodologies for the synthesis of proteins. In sequential NCL protocols, a risk of product loss occurring at purification step followed by every round of NCL decreases by the use of one-pot NCLs. Thus, many groups have made extensive efforts toward the creation of one-pot/sequential NCL protocols. In this context, we developed an N-sulfanylethylanilide (SEAlide) peptide as a crypto-thioester for the N–to–C-directive one-pot/sequential three-fragment ligation [1]. In our one-pot protocol, an N-terminal cysteinyl SEAlide peptide initially reacts with a thioester in the absence of phosphate salt to afford the first ligation product with the SEAlide unit remaining intact, and then the resulting SEAlide peptide in the reaction mixture is converted to the corresponding thioester only by addition of phosphate salts, which is then allowed to react with an N-terminal cysteinyl peptide to yield a three-fragment ligated product. However, the use of the SEAlide peptide in a one-pot/sequential ligation involving more than three fragments has yet to be achieved.

With the intention of surmounting this "three-fragment limitation" in the use of the SEAlide peptide for the sequential NCL, we planned to introduce a photo-cleavable protecting group on the free thiol group of the SEAlide moiety. Recently, a 6-nitroveratryl (NV) group was reported to be a useful protecting group in peptide chemistry, because it can be readily removed by UV light under mild conditions without accompanying serious side reactions. Therefore, we attempted to synthesize some small proteins using this photo-caged SEAlide as a photo-tunable crypto-thioester moiety [2]. First, we synthesized the requisite photo-caged SEAlide (1) by the introduction of NV group onto the free thiol of SEAlide as shown in Scheme 1.

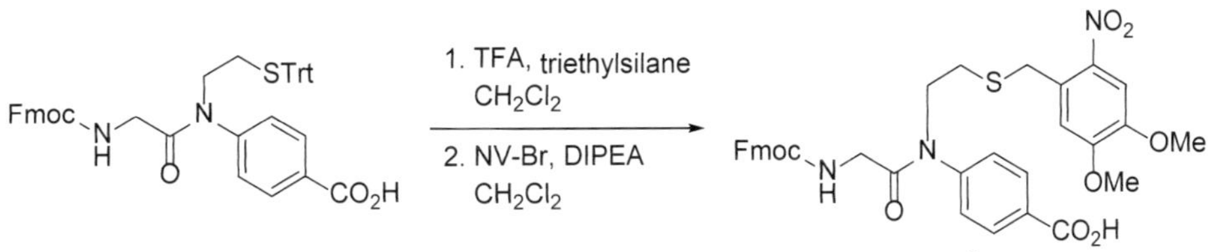

1

Scheme 1. Synthesis of photo-caged SEAlide.

Next, we synthesized reduced form SNX-482 (2) consisting of 41 amino acid residues by one-pot/sequential four-fragment ligation in an N–to–C-directive manner (Scheme 2). Each peptide, including photo-caged or uncaged SEAlide peptide fragments, (3–6) was prepared by standard Fmoc-SPPS. After the first ligation between 3 and 4, the extractive removal of thiophenol from the reaction mixture, followed by UV irradiation for the deprotection of NV group, resulted in the smooth progress of the reaction to yield the desired uncaged SEAlide peptide 7. This obtained reaction mixture containing the ligated peptide was subjected to the subsequent ligation with 5 in a one-pot manner to afford a ligated product 8. UV irradiation to the reaction mixture and ligation with 6 were then performed in a manner similar to the protocol used for the ligation between 5 and 7 to yield reduced form SNX-482 (2).

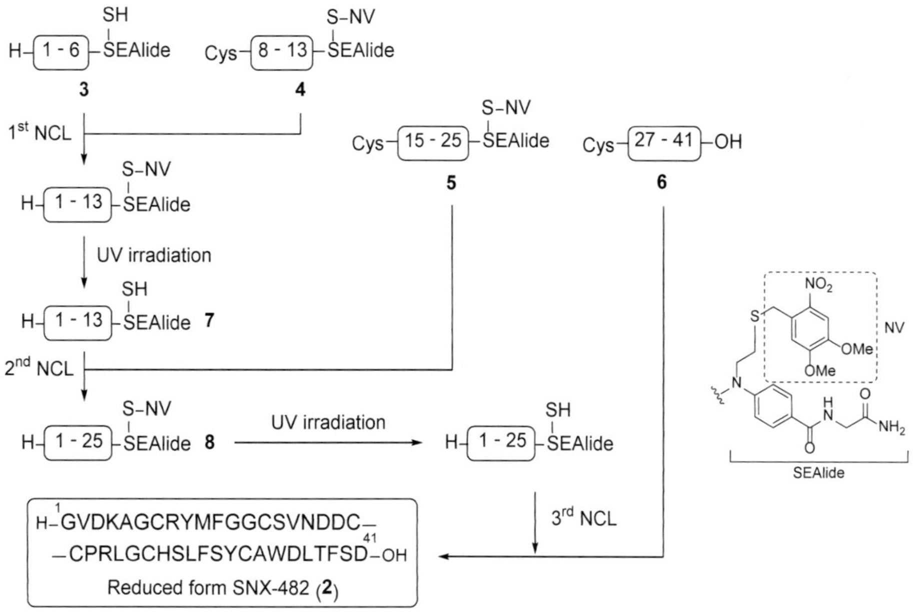

Scheme 2. Synthesis of reduced form SNX-482 using N–to–C directed sequential NCL.

In conclusion, we successfully extended the usefulness of the SEAlide peptide caging its sulfanyl moiety by NV protection, which allowed the reduced form SNX-482 to be synthesized by one-pot/sequential four-fragment ligation in an N–to–C-directive manner.

References

01. K. Sato, A. Shigenaga, K. Tsuji, S. Tsuda, Y. Sumikawa, K. Sakamoto, A. Otaka, ChemBioChem 2011, 12, 1840–1844.

02 K. Aihara, K. Yamaoka, N. Naruse, T. Inokuma, A. Shigenaga, A. Otaka, Org. Lett. 2016, 18, 596–599.

COMBINATORIAL PEPTIDE LIBRARIES MODIFIED BY IONIZATION TAGS ON SOLID SUPPORT FOR INVESTIGATION OF NEW SUBSTRATES OF PROTEASES

Remigiusz Bachor, Alicja Kluczyk, Zbigniew Szewczuk
Faculty of Chemistry, Wrocław University, 50-383 Wrocław, Poland
e-mail remigiusz.bachor@chem.un.wroc.pl

PP I - 002

Proteases are an important class of enzymes that regulate essentially all signaling pathways and biological transformations. The misregulation of peptide and protein proteolysis may cause serious health disorders, therefore the activity of some proteolytic enzymes may be treated as a specific biomarker for many diseases. The one bead-one compound (OBOC) peptide combinatorial libraries are widely used in the investigation of new biologically active compounds, whereas electrospray mass spectrometry (ESI-MS) is currently the method of choice for the direct identification of compounds. However, the necessity of analysis of trace amount of peptide obtained from a single resin bead (about 10^{-15} mole) is insufficient for reliable sequence analysis. Previously we demonstrated that the application of ionization tags in the form of quaternary ammonium (QA) groups increases the ionization efficiency of peptides allowing their ultrasensitive analysis by tandem mass spectrometry (MS/MS) and facilitates their sequencing [1, 2]. Here we present the applicability of QA groups in the analysis of compounds obtained from single resin beads of OBOC peptide library by mass spectrometry.

The main goal of this work was to analyze the utility of QA groups in the analysis of OBOC peptide combinatorial library by mass spectrometry. The synthesis was performed on the TentaGel HL-NH_2 resin and the peptides were connected to the solid support by linker containing methionine residue (selectively cleavable by cyanogen bromide (CNBr), lysine with ε-amino group protected by Mtt, amino acid Peg spacer (9-aza-3,6,12,15-tetraoxa-10-on-heptadecanoic acid) and glycine residue (Fig. 1). Based on the known specificity of caspase 3, caspase 7 and cathepsin G, model libraries were synthesized according to standard Fmoc strategy, using split and mix method. For caspases 3 and 7 both Asp and Ala were introduced in P1 position; Val, Pro, Ser and Thr in P2 position, Glu in P3 and Asp in P4 position. For cathepsin G Val and Asp residues were introduced in P1 position; Leu and Pro in P2 position; Gln, Glu and Leu in P3 and Ala in P4 position. The N-terminal amino acid residues were acetylated and the fixed charge tag in the form of 4-aza-1-azoniabicyclo[2.2.2]octylammonium acetyl group was synthesized at the ε-amino group of the lysine residue according to the procedure described by us previously [3]. Then the side chain protecting groups were removed under acidic conditions and then 3 mg of the obtained QA-OBOC libraries were incubated with a protease for 30 minutes. The enzymatic reaction was terminated by the addition of formic acid (5 μl) and then resin beads were washed with water (3 × 1 min), acetonitrile (3 × 1 min) and methanol (3 × 1 min). To identify the resin beads bearing hydrolyzed sequences the ninhydrin test was

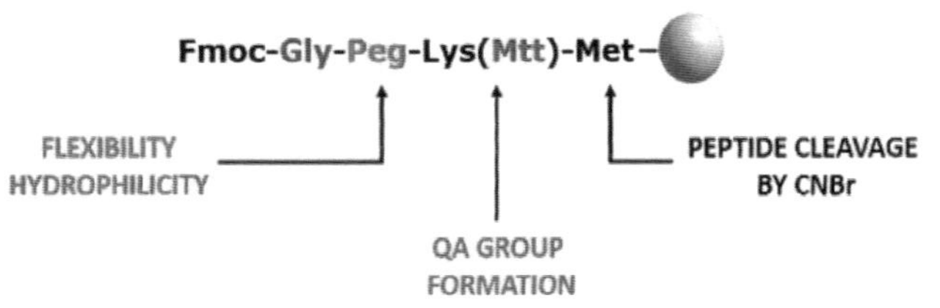

Fig. 1. Schematic presentation of the designed linker.

performed. The enzymatic hydrolysis of peptide bonds results in formation of free amino groups, which are known to yield a purple color (Ruhemann's purple) in the presence of ninhydrin. This allowed for positive beads identification by observation of color change, and enabled selection of beads containing substrates of used enzymes. Colorless or yellow beads suggest complete acylation of all primary amino groups (Fig. 2).

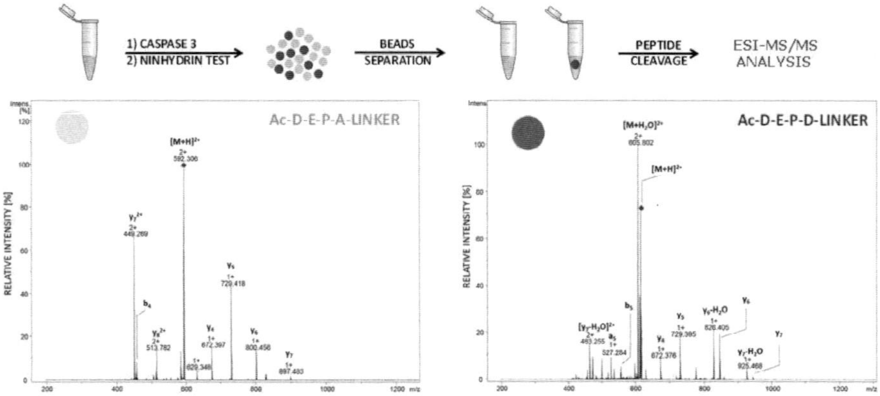

Fig. 2. Schematic presentation of hits identification and the ESI-MS/MS spectra for compound obtained from a single yellow and a single blue resin bead after incubation with caspase 3.

The positive (purple) and negative (yellow) hits were separated manually following by washing with water (3 × 1 min), 70% HCOOH (3 × 1 min), and each selected bead was treated individually with 0.25 M CNBr in 70% formic acid for 18 h. The peptides obtained from single resin beads were analyzed by ESI-MS/MS. In the obtained mass spectra signals corresponding to the 2+ charged ions of undigested peptides were presented, which may suggest that only small amount of the peptide was digested on the resin, although there is enough released primary amino groups to be monitored by ninhydrin test. Only sequences containing aspartic acid in P1 position in the libraries synthesized for caspases and valine residue in the case of cathepsin G substrates were identified on the blue resin beads (positive hits), which is in good accordance with known substrate specificity of used proteases. The proposed methodology was proved to be useful in the investigations of new substrates of important proteases like caspases and cathepsins.

This work was supported by a grant No. UMO-2015/17/D/ST5/01329 from the National Science Centre, Poland.

References

01. R. Bachor, M. Cydzik, M. Rudowska, A. Kluczyk, P. Stefanowicz, Z. Szewczuk. Mol. Divers., 16, 613 (2012).
02. R. Bachor, A. Kluczyk, P. Stefanowicz, Z. Szewczuk. Mol. Divers., 17, 605 (2013).
03. M. Cydzik, M. Rudowska, P. Stefanowicz, Z. Szewczuk. J. Pept. Sci., 17, 445 (2011).

ZYKR1: A NOVEL SHORT CHAIN PEPTIDE BASED PERIPHERALLY RESTRICTED KAPPA (κ) OPIOID RECEPTORS (KOR) AGONIST

Ranjit C. Desai, Rajesh Bahekar*, Vijay Prajapati, Rajendra Chopade, Mukul R. Jain, Vishwanath Pawar, Kanaiyalal Prajapati, Rakesh Patel, Bhavin Sonara, Upendra Bhatnagar, Debdutta Bandyopadhyay, Bhadresh R. Rami, Nuggehally R. Srinivas, Harilal Patel, Poonam Giri and Pankaj R. Patel.
Zydus Research Centre, Cadila Healthcare Ltd., Ahmedabad, Gujarat, India.
*Correspondence: rajeshbahekar@zyduscadila.com; #ZRC communication No: 489

PP I - 003

Abstract

In recent years, considerable attention has been bestowed towards the development of peripherally restricted, selective κ-agonists as potent and efficacious analgesics, devoid of CNS side effects. Unlike δ and μ receptors agonists, peripherally restricted κ-opioid receptor agonists are unlikely to elicit physical dependence, respiratory depression, urinary retention, euphoria and constipation. Thus, peripherally restricted κ-opioid agonists represent an important therapeutic target for the treatment of neuropathic pain, visceral pain, irritable bowel syndrome (IBS) and post-operative pain.

Considering the therapeutic need of peripherally restricted κ-opioid agonists, we identified a short-chain peptide (ZYKR1), as a novel kappa (κ) opioid receptor (KOR) agonist. ZYKR1 showed KOR agonistic activity in cAMP assay (CHO cells transfected with human KOR) with an EC_{50} of 32 pM and more than $3X10^5$ fold selectivity over μ & δ-OR (EC_{50}: > 10 μM). In vivo efficacy of ZYKR1 was evaluated in various animal models (acetic acid induced visceral pain mouse model, ED_{50}: 34 mg/kg,iv; OvHx rat model (post-operative pain), ED_{50}: 32 mg/kg,iv and C48/80 induced pruritus mouse model, ED_{50}: 103 mg/kg,iv).

CNS effects (motor impairment) of ZYKR1 were estimated in the mouse rotarod model with SD_0 (non-sedative dose) of 1 mg/kg,iv, indicated 30 fold CNS safety index. ZYKR1 showed low potential for hypotension, hypernatremia, emesis and respiratory depression, when tested in rats and dogs, at > 5X of ED_{50} doses. ZYKR1 does not exhibit any potential for in vitro hERG blockade (IC_{50} > 100 μM).

In a 4 week repeated dose toxicity study in rats, ZYKR1 showed no major adverse effects, demonstrated dose linear increase in exposure and NOAEL was found to be 30 mg/kg, iv (882 fold of ED_{50} dose).

In conclusion, we have identified, ZYKR1, as a novel, potent, selective and peripherally restricted KOR agonist with promising efficacy in models of visceral, post-operative pain and pruritus, without any apparent toxicity.

Introduction

There are three types of opioid receptors (Mu (μ), Kappa (κ) and Delta (δ)), found to be expressed in both the CNS and in the periphery. In past two decades, several attempts has been made to development peripherally restricted, selective κ-agonists as potent and efficacious analgesics, devoid of CNS side effects[1-2], for the treatment of neuropathic, visceral, post-operative pain and irritable bowel syndrome (IBS)[3].

Because of the significant safety benefits of peripherally restricted κ-opioid agonists, our program was specifically geared towards identification of peripherally restricted short-chain peptides, to improve the therapeutic index of the centrally acting κ-opioid class of drugs. In this context, we discovered ZYKR1 as a novel, potent, selective and peripherally restricted kappa (κ) opioid receptor (KOR) agonist. Synthesis of ZYKR1 was carried out using solution phase peptide chemistry (HPLC purity >99%). KOR agonistic activity of ZYKR1 was checked in cAMP assay. In vivo pharmacological effects of ZYKR1 were evaluated in various animal models.

Methods

In vitro, κ, μ & δ-OR agonistic activity of ZYKR1 was assessed using cAMP assay (CHO cells transfected with human κ, μ & δ- opioid receptors (OR)). All the animal experiments were carried out as per the 'Zydus Research Centre animal ethical committee' approval. In vivo efficacy of ZYKR1 was evaluated in: a) acetic acid induced visceral pain mouse model; b) OvHx rat model (post-operative pain) and c) C48/80 induced pruritus mouse model. In vitro, ex-vivo and in vivo data of ZYKR1 shown in Figure 1a-f.

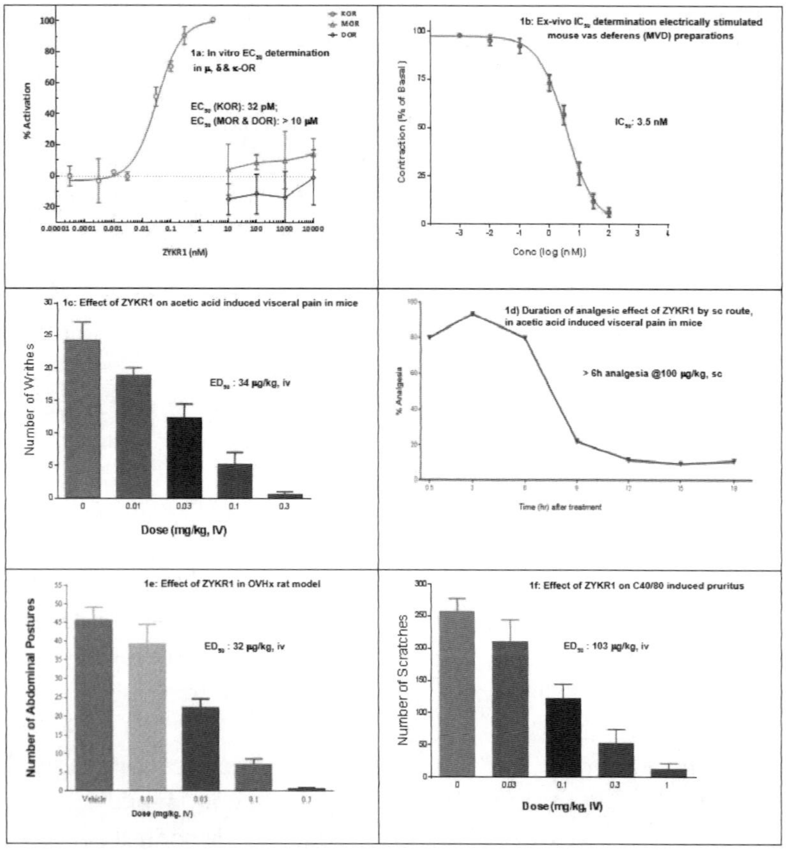

Figure 1a-f:
1a) In vitro, KOR agonistic activity of ZYKR1 was assessed using cAMP based functional assay, in CHO cells transfected with hKOR (cAMP estimated by cAMP direct ELISA kit (Arbor Assays, Cat # K019-H5)). ZYKR1was tested for μ and δ-opioid receptors agonistic activities, CHO cells transfected with human μ & δ-OR (MOR and DOR). ZYKR1 showed selective KOR agonistic activity in cAMP assay, with an EC_{50} of 32 pM and more than $3X10^5$ fold selectivity over μ & δ-OR (EC_{50}: > 10 μM).
1b) Ex-vivo, KOR agonistic activity of ZYKR1 was tested on the electrically stimulated mouse vas deferens (MVD) preparations and IC_{50} (3.5 nM) was determined.
1c) Effect of ZYKR1 on acetic acid induced visceral pain in mice: Vehicle or ZYKR1 administered intravenously in female ICR mice. After 5 min of treatment, 10 ml/kg of 0.6 % v/v glacial acetic acid injected intraperitoneally. Number of writhing responses by the animal in 15 minutes, following acetic acid injection was counted. ZYKR1 showed dose-dependent analgesic effect in acetic acid induced visceral pain mouse model, ED_{50}: 34 mg/kg,iv; N = 5-6, values expressed as mean ± SEM.
1d) ZYKR1 duration of analgesia was tested by subcutaneous (sc) route of administration (dose 100 mg/kg, sc), in acetic acid induced visceral pain in mice (> 6 hrs analgesia).
1e) Effect of ZYKR1 in rat models of overiohysterectomy: Animals were anaesthetized with isoflurane (4 % for induction, 2% for maintenance of anaesthesia) and O_2 mixture. Ovariohysterectomy was performed via a midline abdominal incision (2 cm in length) in the linea alba. The cervix was ligated with 3-0 silk. The ovaries and the uterus were then removed. Incision was closed in two layers. After recovery from anaesthesia, animals were treated with ZYKR1 by iv route. Animals were observed for visceral episodes (Postures) for 30 minutes. ZYKR1 showed dose-dependent analgesic effect in rat models of overiohysterectomy, ED_{50}: 32 mg/kg,iv; N = 6-7, values expressed as mean ± SEM.
1f) Effect of ZYKR1 in ICR mice on C48/80 induced pruritus model. Male ICR mice were injected ZYKR1 by iv route. After 5 min, mice were treated with Compound 48/80 (100 μg/50 μl, sc) into rostral part of back. Numbers of scratches were measured. ZYKR1 showed dose-dependent antipruritic activity in C48/80 induced pruritus model, ED_{50}: 103 mg/kg,iv; N = 6-7, values expressed as mean ± SEM. ED_{50} dose calculated using GraphPad Prism.
Ex-vivo, KOR agonistic activity of ZYKR1 was tested on the electrically stimulated mouse vas deferens (MVD) preparations and IC_{50} was found to be 3.5 nM. ZYKR1 showed <1% brain to plasma ratio, when tested in mice at 1 mg/kg, iv dose. In acetic acid induced visceral pain mouse model, ZYKR1 showed extended duration of analgesia (> 6hrs effect), when tested by subcutaneous route of administration (dose 100 mg/kg, sc). ZYKR1 showed good exposure across species (Table 1). CNS effects (motor impairment) of ZYKR1 were estimated in the mouse rotarod model with SD_0 (non-sedative dose) of 1 mg/kg,iv, indicated 30 fold CNS safety index over ED_{50} dose. ZYKR1 showed low potential for hypotension, hypernatremia, emesis and respiratory depression, when tested in rats and dogs at > 5X of ED_{50} doses. ZYKR1 does not exhibit any potential for in vitro hERG blockade (IC_{50} > 100 μM).
In a 4 week repeated dose toxicity study in rats (doses tested: 7.5, 15 & 30 mg/kg, iv), ZYKR1 showed no major adverse effects, demonstrated dose linear increase in exposure and NOAEL was

found to be 30 mg/kg, iv (882 fold of ED_{50} dose). In conclusion, we have identified, ZYKR1, as a novel, potent, selective and peripherally restricted KOR agonist with promising efficacy in models of visceral, post-operative pain and pruritus, without any apparent toxicity.

Table1:

Intravenous exposure (PK profile) of ZYKR1 in mouse & rats (@ED_{90} dose) & in dogs (@ SD_0 dose)

Parameters	**ZYKR1**		
Species	Mouse	Rat	Dog*
IV Dose (mg/kg)	0.144 (ED_{90})	0.144 (ED_{90})	0.125 (SD_0)
C0 (µg/ml)	0.50	0.68 ± 0.07	1.07+0.24
AUC (0-t) (µg.h/mL)	0.14	0.20±0.01	0.80±0.10
Vss (L/kg)	0.28	0.24 ± 0.03	0.17 ± 0.04
CL (mL/min/kg)	16.62	11.62 ± 0.29	2.58 ± 0.30
T½, iv (h)	0.21	0.30 ± 0.05	0.91 ± 0.27
MRT (h)	0.28	0.35 ± 0.03	1.09 ± 0.39
* Dog did not showed sedation effect at 0.125 mg/kg, iv dose of ZYKR1			

Summary

ZYKR1 is a novel, potent and selective κ-opioid receptor agonist, showed in vitro EC_{50} in pM range, found to be highly selective over μ & δ-OR, with No-CYP & respiratory depression. ZYKR1 showed efficacy in various pain and pruritus animal models, with ED_{50} in microgram range and 30 fold CNS safety index over ED_{50} dose. In pharmacokinetic studies, ZYKR1 showed good plasma exposure in mice, rats and dogs. Existing opioid analgesics such as Tramadol, Fentanyl or Pentazocine exhibits adverse effects, such as nausea vomiting, constipation, drowsiness, addiction, DDI (CYP), itchiness & respiratory depression, while ZYKR1, a selective, potent and peripherally restricted KOR agonist, was found to be devoid of these adverse effects in preclinical animal models. ZYKR1 demonstrated good safety profile in repeated dose toxicity study (in wistar rats), with no major adverse events, dose linear increase in exposure and NOAEL was found to be 882 fold of ED_{50} dose. ZYKR1 is currently undergoing IND enabling studies, for the treatment of visceral and post-operative pain.

References

01. Vadivelu N, et al., Peripheral opioid receptor agonists for analgesia: a comprehensive review, J. Opioid. Manag., (2011), 7, 55.
02. Barber A. et al., Novel developments with selective, non-peptidic kappa-opioid receptor agonists, Exp. Opin. Invest. D rugs, (1997), 6, 1351.
03. Barber A. et al., A pharmacological profile of the novel, peripherally-selective κ-opioid receptor agonist, EMD 61753., Br. J. Pharmacol., (1994) 113,1317.

EFFICIENT SYNTHESIS, DERIVATISATION AND CONFORMATIONAL ANALYSIS OF AMINO-BENZOTRIAZOLODIAZOCINONE SCAFFOLDS VIA TANDEM UGI-HUISGEN REACTION

T. M. A. Barlow, M. Jida, K. Guillemyn, Vicky Caveliers, D. Tourwé and S. Ballet
Research Group of Organic Chemistry, Departments of Chemistry and Engineering Sciences,
Vrije Universiteit Brussel, Pleinlaan 2, Elsene 1050, Belgium

PP I - 004

Introduction

Conformationally constrained amino acids have found use in the synthesis of peptidomimetics with widely different applications. Following on from previous work [1], we further developed a simple, catalyst-free procedure employing an Ugi-4CR between different ortho-azidoanilines, isocyanides, aldehydes and Boc-propargylglycine, followed by a thermal azide–alkyne Huisgen cycloaddition to generate a 16-member library of constrained (diastereomeric) dipeptides with up to six points of diversification in good yields. A range of functional groups – methyl, methoxy, bromo, acetyl – at positions R_1-R_4 was tolerated in these reactions, as shown in Figure 1. The products were obtained in good to excellent yield (up to 82%). When R_2, R_3 or R_6 was a bromine atom, we were able to functionalise these positions with aryl groups via an efficient Suzuki-Miyaura cross-coupling reaction. The concomitant formation of two diastereoisomers cannot be circumvented, but their separation is facile by chromatography. After separation, we demonstrated that the (S,S)-diastereoisomer adopts a β-turn conformation encompassing an intramolecular hydrogen bond, unlike the (S,R)-diastereoisomer which adopts an extended conformation. Both of these observations matched our in silico modelling.

Results and Discussion

Azidoanilines of type 2 (8 examples) were synthesised according to a known procedure [2] and purified by flash column chromatography with slow gradients (typically 1-3% EtOAc in hexanes). The isolated yields of these products were reduced by concomitant formation of diazides (which were not isolated). The differences in substitution patterns of functionalised azidoanilines had no observable difference on yield or diastereoselectivity in these reactions.

When used in an Ugi reaction with Boc-propargylglycine, R_4-bearing aldehydes and R_5-bearing isocyanides, full conversion was observed after overnight reaction at room temperature to acyclic intermediate (not shown), with some products showing spontaneous cyclisations to products of type 3. Subsequent heating at 70 °C for 24 h was sufficient to achieve full conversion to the cyclic product for all but one case (R_1 = Me) where 36 h in dioxane at 100 °C was required. A range of functional groups were tolerated, most strikingly when R_2 and R_3 = Ac, where the aldehyde was reactive enough to react chemoselectively over the ketones present in the reaction mixtures.

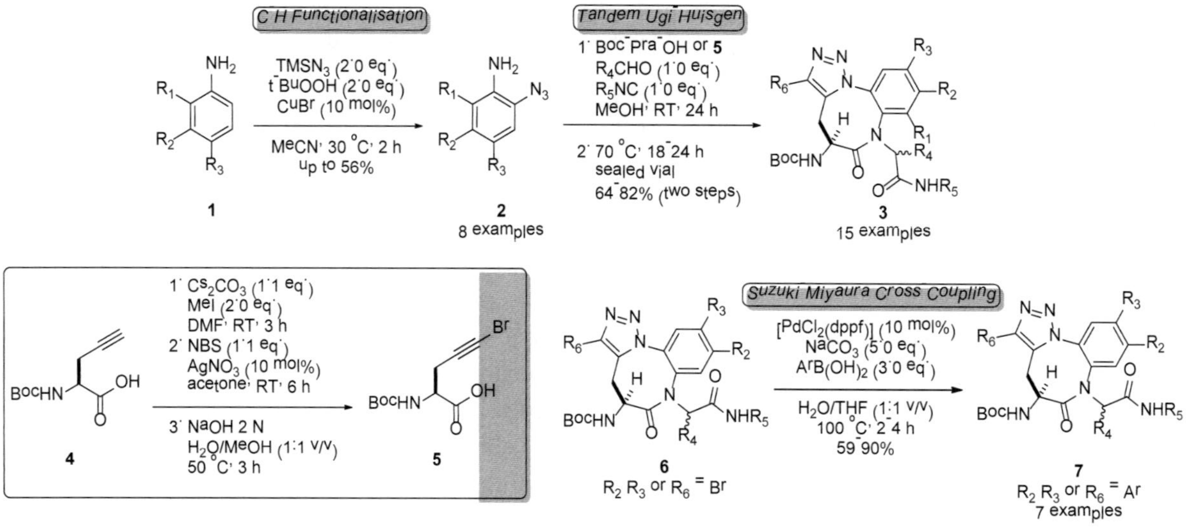

Figure 1 – synthesis of benzazocinone derivatives 3/6 and their arylated derivatives 7

Boc-ω-bromo-propargylglycine 5 was synthesised according to a literature procedure [3]. Three examples of 6 where R_2, R_3 or R_6 = Br were then subjected to Suzuki-Miyaura cross-coupling reactions using electronically neutral (Ph), rich (3'-furyl) and deficient (4'-(CF_3)Ph) boronic acids, and additionally with 4'-pyridyl boronic acid (only at the R_6 position). Despite the basic conditions, no epimerisation was observed by HPLC/^{1}H-NMR.

Molecular modelling (MMFF94x force field) was then used to probe the lowest energy conformations of these molecules. We studied the diastereoisomers of 8 (an iteration of 3 where R_{1-3} and R_6 = H, R_4 = Me and R_5 = tBu) and 9 (3 with R_{1-3} and R_6 = H, R_4 = Bn and R_5 = tBu) individually.

The (S,R)-diastereoisomer of each was shown in silico to adopt an extended conformation whereas the (S,S)-diastereoisomers adopted a turn-like conformation with formation of an intermolecular hydrogen bond. The diastereoisomers of two products were separated by preparative HPLC and then heated in DMSO-d_6 (temperature range 298 to 343K in 5K increments). Thermal coefficients ($\Delta\delta/\Delta T$) for (S,S)-8 and (S,S)-9 were calculated and shown to be consistent in both cases with hydrogen bond formation (thermal coefficient values between 0 and -4 ppb/K are generally accepted for solvent shielded amide protons involved in intramolecular hydrogen bonds [4]).

PP I - 004

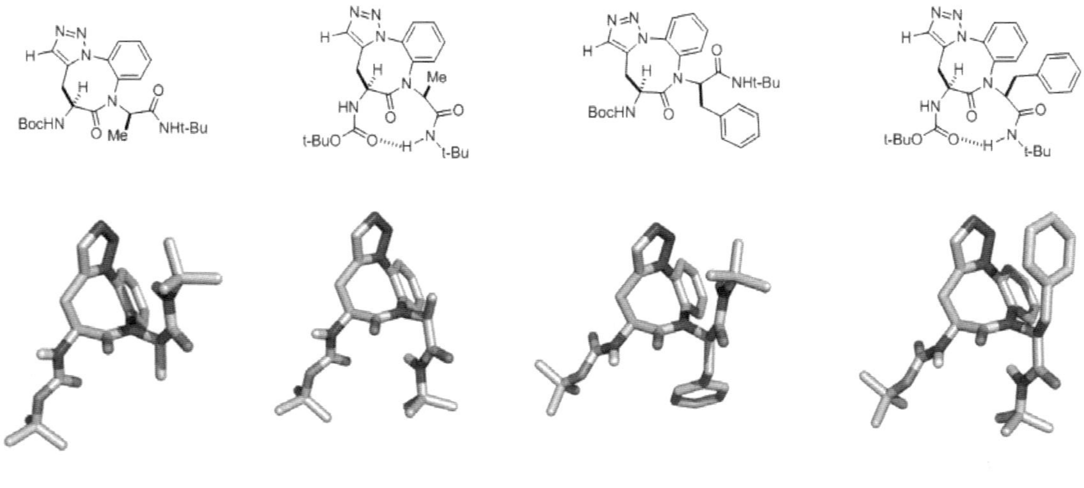

(*S,R*)-**8**	(*S,S*)-**8**	(*S,R*)-**9**	(*S,S*)-**9**
$\Delta\delta/\Delta T = -3.7$	$\Delta\delta/\Delta T = -0.7$	$\Delta\delta/\Delta T = -6.9$	$\Delta\delta/\Delta T = -0.5$

Table 1 - Results of molecular modeling and with the ^{1}H-NMR-derived thermal coefficient studies

Conclusions

An expedient synthesis was developed for new constrained amino acid including facile cyclisation to triazole-fused benzazocanes in good to excellent yields (up to 82%). This scaffold presents up to six points of diversification. Brominated intermediates can undergo further derivatisation through Suzuki-Miyaura cross-coupling reactions. These scaffolds are diastereomeric; the (S,S)-diastereoisomers forms a turn-like structure, as demonstrated by ^{1}H- and thermal NMR studies. Work is ongoing to incorporate this structure into bioactive peptide sequences.

Acknowledgements

We thank Flanders Innovation & Entrepreneurship (VLAIO), as well as the Strategic Research Program – Growth funding of the Vrije Universiteit Brussel for financial support.

References:

Part of this work was published in: Org. Biomol. Chem. (2016), 14, 4669-4677

01. Barlow, T. M. A., Jida, M., Tourwé, D., Ballet, S. Org. Biomol. Chem., 2014, 12, 6986–6989
02. Tang, C., Jiao, N. J. Am. Chem. Soc., 2012, 134, 18294-18297
03. IJsselstijn, M., Kaiser, J., van Delft, F. L., Schoenmaker, H. E., Rutjes, F. P. J. T. Amino Acids, (2003), 24, 263-266

04 Cierpicki, T., Otlewski, J. J. Biol. NMR., (2001), 21, 249-261 ; Halab, D., Lubell, W. D. J. Org. Chem., (1999), 64, 3312-3321

REEXAMINATION OF 2,4-DIMETHOXYPHENYL BASED CYSTEINE PSEUDOPROLINES

Raymond Behrendt[1], Peter White[2]
[1]Merck & Cie, Im Laternenacker 5, 8200 Schaffhausen, Switzerland,
[2]Merck Chemicals, Padge Road, Beeston NG9 2JR, United Kingdom

PP I - 005

Introduction

The interest in polycystine peptides as potential therapeutics has driven developments in methods for their synthesis and more specifically a search for new protecting groups for the cysteine sulfhydryl group. Alternatives to the standard trityl protecting group, such as Dpm, MBom [1] and Thp [2], have been evaluated recently. We were interested to investigate if the incorporation of cysteine as a thiazolidine heterocycle would confer the same advantages of the corresponding serine/threonine dimethyloxazolidines during Fmoc SPPS, namely elimination of epimerization and aggregation [3]. For our studies, we chose to protect cysteine as a 2-(2,4-dimethoxyphenyl)thiazolidine 1 as such a structure more easily regenerates cysteine than the already described 2,2-dimethylthiazolidines (Scheme1) [4].

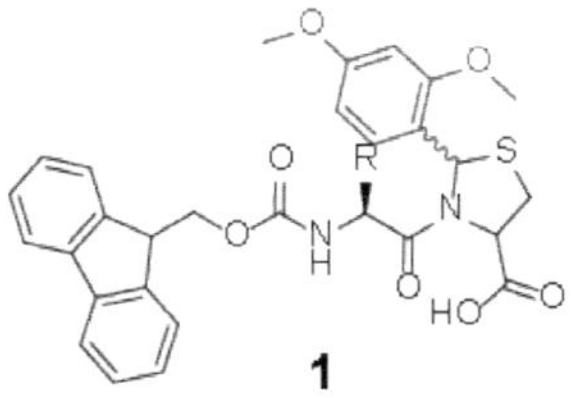

1

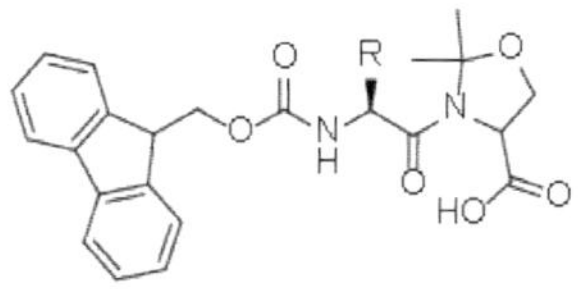

serine pseudoproline

Scheme 1: Structure of the 2-(2,4-dimethoxyphenyl) thiazolidine dipeptides reexamined in this study in comparison to the well-known serine pseudoprolines.

Results and Discussion

Epimerization: The ruminant EGF (36-45) peptide was prepared using Fmoc-Cys(Trt)-OH or Fmoc-Lys(Boc)-Cys(psiDmp,Hpro)-OH 1 coupled with basic TBTU/DIPEA activation. The use of Fmoc-Cys(Trt)-OH resulted in 10 times more D-Cys than Fmoc-Lys(Boc)-Cys(psiDmp,Hpro)-OH (Figure 1) [5].

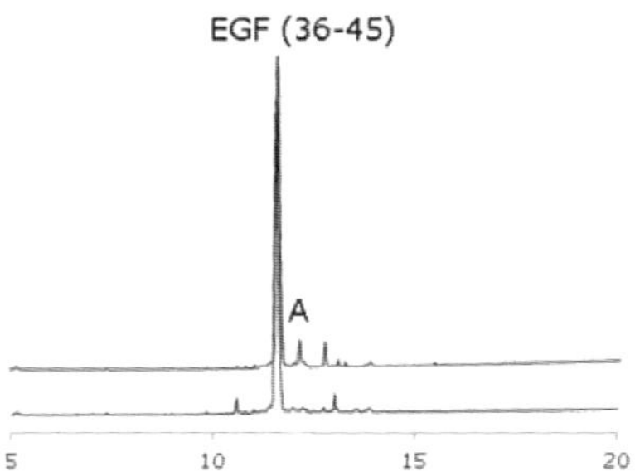

Cysteine compound	EGF (36-45) [area%]	D/L-cysteine
Cys(Trt)	81	3.9
1	81	0.4

Figure 1: Purity and D-cysteine content of the ruminant EGF (36-45) peptide crudes. Left: HPLC of peptide crudes using 1 (lower trace) and Fmoc-Cys(Trt)-OH (upper trace), A: (D-Cys[37])-EGF (36-45).

Aggregation

The difficult peptide sequence of influenza virus hemagglutinin [6] was prepared using either Fmoc-Ser(tBu)-OH, Fmoc-Ala-Ser(psiMe,Mepro)-OH or Fmoc-Ala-Cys(psiDmp,Hpro)-OH (HPLC profiles in Figure 2). As expected, the peptide prepared using Fmoc-Ser(tBu)-OH was highly heterogeneous. In contrast, the purities of the analogs prepared with pseudoproline building blocks were excellent.

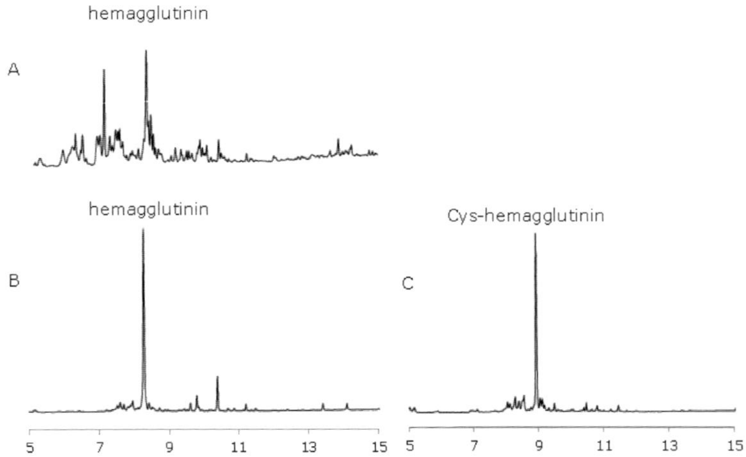

Figure 2: HPLC profiles of influenza virus hemagglutinin using A: Ser(tBu), B: Ala-Ser(psiMe,MePro), C: compound 1.

In conclusion, Fmoc-Xaa-Cys(psiDmp,Hpro)-OH pseudoproline dipeptides are excellent tools for the synthesis of cysteine-containing peptides. In contrast to Fmoc-Cys(Trt)-OH, they can be coupled without epimerization under basic conditions. Furthermore, these reagents should prove useful tools for convergent synthesis strategies as protected peptide fragments containing C-terminal cysteine pseudoproline should couple without epimerization. Finally, they appear to be equally effective as Ser/Thr pseudoprolines in disrupting aggregation during peptide assembly.

To avoid the formation of DMB related by-products during final cleavage from the resin and deprotection of side-chains, TFA cocktails incorporating EDT and TIPS should be used (data not shown).

References

01. Hibino H, Miki Y, Nishiuchi Y. J. Pept. Sci. 2014, 20, 30.
02. Ramos-Tomillero IN, Rodríguez H, Albericio F. Org. Lett. 2015, 17, 1680.
03. Woehr T, Wahl F, NefziA, Rohwedder B, Sato T, Sun X, Mutter M. J. Am. Chem. Soc., 1996, 118,9218-9227.
04. Postma TM & Albericio F. Org. Lett., 2014 16, 1772-1775.
05. Han Y, Albericio F, Barany G. J. Org. Chem. 1997, 62, 4307.
06. Sampson WR, Patsiouras H, Ede NJ, J. Pept. Sci., 1999, 5, 403–409.

MINIMIZING ASPARTIMIDE FORMATION IN FMOC SPPS: FMOC-ASP(OBNO)-OH

Raymond Behrendt[1], Peter White[2]
[1]Merck & Cie, Im Laternenacker 5, 8200 Schaffhausen, Switzerland,
[2]Merck Chemicals, Padge Road, Beeston NG9 2JR, United Kingdom

Introduction

We recently introduced Fmoc-Asp(OBno)-OH 1 (scheme 1), an aspartyl derivative bearing the tributylcarbinol ester at the γ- carboxyl group [2]. The use of this derivative was found to be a simple and effective solution to the aspartimide problem in Fmoc SPPS.

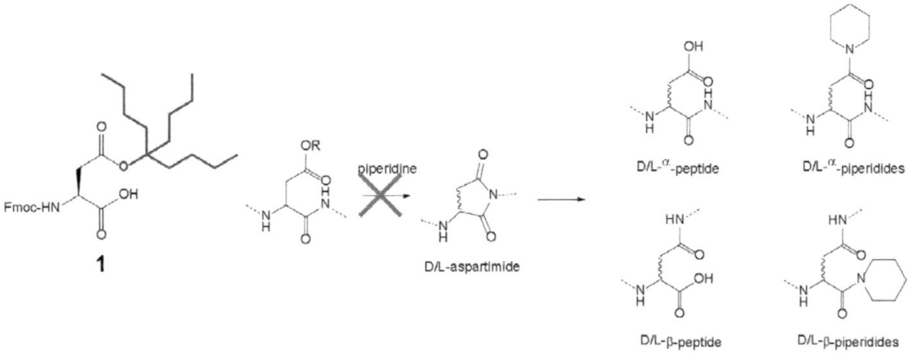

Scheme 1: Fmoc-Asp(OBno)-OH 1 overcoming aspartimide formation and its related impurities.

Results and Discussion

Using the classic scorpion toxin II peptide (VKDXYI, where X=G, N or R) 1 was evaluated by treating the peptidyl resin with 20% piperidine in DMF for 200 min to simulate approximately 100 x 2 min deprotection cycles. For X=N and R, the use of 1 reduced aspartimide formation to almost undetectable amounts (Table 1). In the case of the most problematic case where X=G, aspartimide formation was reduced to only 0.1%/cycle, which is within the purity limits of commercially available N-a-Fmoc amino acids.

Furthermore, virtually no aspartimide related by-products were formed using compound 1 in combination with 0.1 M Oxyma Pure [3] in 20% piperidine/DMF for Fmoc removal (Figure 1, Table 1). In constrast, when Asp(OtBu) was used, numerous Oxyma- Pure related by-products were formed, indicating that OxymaPure can react with the generated aspartimides.

Aspartimides are chirally labile [1], which is reflected in the high D-aspartate values observed with the scorpion toxin II peptides prepared with Asp(OtBu) and Asp(OMpe) (Table1). D-aspartyl peptides are often hidden contaminants of purified peptides due to them having identical molecular mass and almost identical physico-chemical properties [4] to the native sequence. Therefore, the use of Fmoc-Asp(OBno)-OH is highly recommended for peptide manufacturing processes.

Peptide	D-Asp [%]			Asu/2min Fmoc removal cycle [%]		
	tBu	Mpe	Bno	tBu	Mpe	Bno
VKDGYI	18.4	6.0	1.4	2.23	0.77	0.14
VKDGYI[a]	n.d.	n.d.	n.d.	0.67	0.24	0.04
VKDNYI	9.1	4.2	0.9	1.65	0.49	0.06
VKDRYI	25.1	11.0	1.4	1.24	0.40	0.06

Table 1: Epimerization of the aspartyl residue after treating peptidyl resins with 20% piperidine in DMF for 200min or [a]0.1M OxymaPure in 20% piperidine in DMF for 200min (100 cycles) and the corresponding aspartimide formation (Asu) per 2min Fmoc removal cycle calculated for a first order decay ($N=N_0 \cdot e^{-kt}$ -> $k=-ln(N)/t$; $N_0 = 1$; t = number of cycles (100); N = area% of target peptide).

As a further test, the 33mer (Gly2)-GLP-2 was prepared using either Fmoc-Asp(OtBu)-OH or Fmoc-Asp(OBno)-OH. Using compound 1 delivered a crude which contained negligible aspartimide related impurities, increasing the content of target peptide by 25% compared to the standard derivative Fmoc-Asp(OtBu)-OH. Fmoc SPPS quality was exceptionally good and Fmoc-Asp(OBno)-OH enabled the stepwise SPPS, without the formation of des-Asp peptides, additional truncates or Bno alkylation products.

In conclusion, b-tributylmethyl ester protection of aspartic acid provides excellent protection from aspartimide formation, even for peptides containing the most problematic Asp-Gly sequence. Furthermore, despite the bulkiness of OBno, Fmoc-Asp(OBno)-OH appears to couple without difficulties. No evidence was found of alkylation by the tributylmethyl cation during TFA-mediated cleavage. Thus, the routine use of Fmoc-Asp(OBno)-OH provides a simple and robust solution to the problem of aspartimide formation in Fmoc SPPS.

References

01. Subirós-Funosas R, El-Faham A, Albericio F. Tetrahedron 2011, 67, 8595.
02. a) Behrendt R, Huber S, Martí R, White P. J. Pept. Sci. 2015, 21, 680. b) Behrendt R, Huber S, White P. J. Pept. Sci. 2016, 22, 92.
03. Subirós-Funosas R, El-Faham A, Albericio F. Peptide Sci. 2012, 98, 89.
04. Michels T, Doelling R, Haberkorn U, Mier W. Org. Lett. 2012, 14, 5218.

HYDROXYQUINOLYL CYSTEINE DERIVATIVES: PEPTIDE SYNTHESIS AND SIDE PRODUCTS FORMATION

Anna V. Pokhvoshcheva, Maria V. Leko, Marina Yu. Dorosh, Sergey V. Burov
Institute of Macromolecular Compounds RAS, Bolshoi pr. 31, St. Petersburg, 199004, Russia

PP I – 007

Introduction

The potent coordinating and metal chelating ability of 8-hydroxyquinoline gave rise a number of studies devoted to synthesis of its derivatives possessing antioxidant activity or fluorescent properties and their conjugates with peptide carriers [1, 2] (Fig. 1). Moreover, these compounds can be useful for the treatment of different metal-related diseases such as Alzheimer's disease, Parkinson's disease, and multiple sclerosis.

$M^{2+} = Fe^{2+}; Cu^{2+}; Zn^{2+}; Co^{2+}; Ni^{2+}; Mn^{2+}$

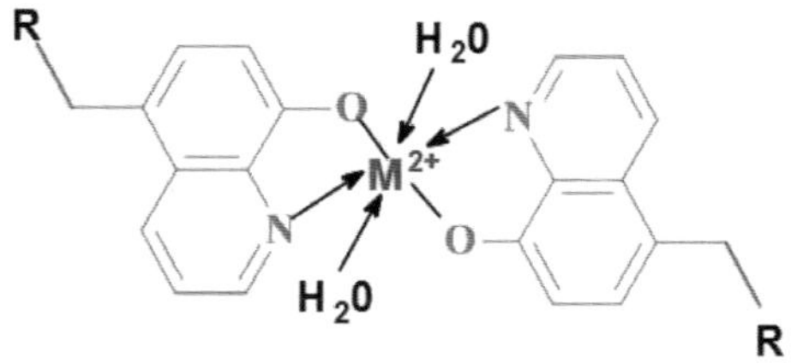

H-Cys(HQ)-Peptide-NH$_2$
Peptide = VIP, SP

Fig. 1. 8-hydroxyquinoline derivatives and their complexes with metal ions.

Therefore, there is a growing demand of efficient synthetic protocols for HQ incorporation into peptide sequence and detailed investigation of possible side products. Here we describe improved procedure for the preparation of BOC and Fmoc derivatives of S-(8-hydroxyquinolin-5-ylmethyl)-L-Cys (Cys(HQ)) and their utility in SPPS.

Results and Discussion

Commonly synthesis of hydroxyquinoline conjugates can be achieved as a result of cysteine incorporation into peptide sequence and subsequent alkylation of free SH group with its 5-chloromethyl or more sophisticated iodoacetamido derivatives. In the first case this approach can be useful for peptide modification both on-resin and in solution [1]. Meantime, application of amino acid derivatives containing HQ moiety in some cases can simplify the process of peptide synthesis.

To the best of our knowledge the only study describing synthesis of Fmoc-Cys(HQ)-OH was done by Youdim et al. [3]; however, the authors did not investigate its utility for peptide synthesis. Moreover, the suggested synthetic scheme implies HPLC purification both intermediate and the final product. Based on the data of Bolognese and co-workers, describing preparation of enantiopure S-(Aminoalkyl)-cysteine [4], we synthesized protected Cys(HQ) derivatives in reasonable yield and purity (Fig. 2).

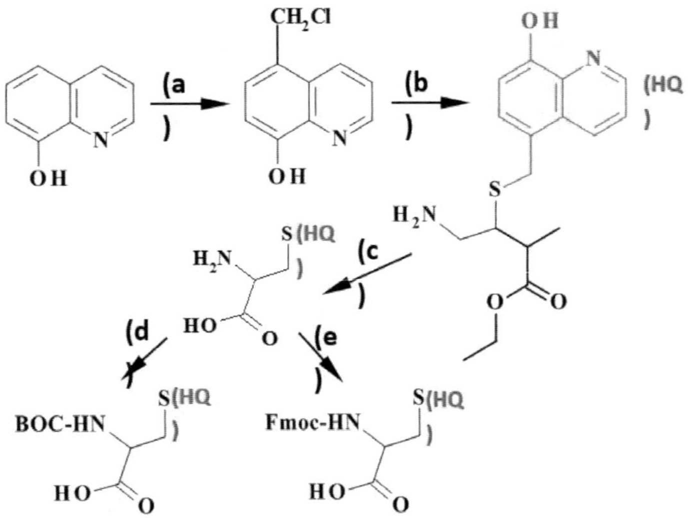

Fig. 2. Synthesis of cysteine derivatives. Reagents and conditions: (a) HCHO/HCl, rt, 8 h; (b) HCl·HCysOEt, Cs_2CO_3, DMF, 16 h; (c) aq. LiOH, MeOH, rt, 2 h; (d) $(BOC)_2O$, THF; (e) Fmoc-OSu, THF

It was shown that Cys(HQ) derivatives are rather stable to oxidation (negligible sulfoxide formation during prolonged storage of Cys(HQ) solution in water). To investigate the utility of synthesized compounds for SPPS we prepared two model peptides using different coupling agents, including DIC/Cl-HOBt, HCTU, HATU and PyBOP.

Standard DIC mediated coupling along with formation of desired product resulted in significant amount (up to 70%) of O-aryl isourea (Fig. 3). Similar side reaction was described previously for Tyr containing peptides and proteins [5]. Formation of O-aryl isourea cannot be suppressed by the excess of Cl-HOBt, while addition of hydroxyquinoline instead of Cl-HOBt completely eliminates the coupling efficiency. Fortunately, O-aryl isourea can be converted to the desired final product by the treatment with 1 N hydroxylamine solution during 24 h.

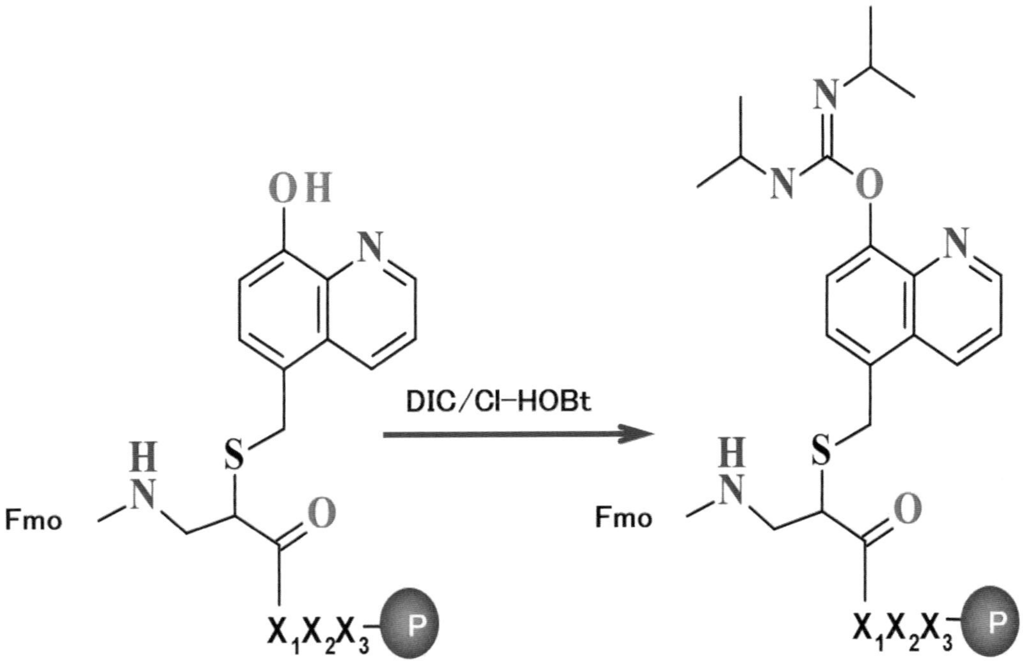

Fig. 3. Side product formation in the course of DIC mediated coupling.

In the case of PyBOP mediated coupling we did not observe any formation of desired product. Surprisingly, application of powerful condensation agents such as HCTU and HATU presumably resulted in mixture of side products including that of amino group guanidinylation. These data are in contradiction to earlier report describing HATU efficiency in the case of unnatural amino acids containing HQ moiety [6].

In conclusion suggested synthetic scheme permits to prepare both BOC and Fmoc derivatives of hydroxyquinolyl cysteine in reasonable yield without HPLC purification. Their incorporation into peptide sequence can be achieved using DIC/Cl-HOBt method followed by hydroxylamine treatment.

References

01. Zheng, H.; Youdim, M.B.H.; Weiner, L.M.; Fridkin, M. J. Peptide Res. 66, 190-203 (2005).
02. Walkup, G.K.; Imperiali, B. J. Org. Chem. 63, 6727-6731 (1998).
03. Youdim, M.B.H.; Fridkin, M.; Zheng, H.; Warshawsky, R. 2003. PCT/IL2003/000932.
04. Bolognese, A.; Fierro, O.; Guarino, D.; Longobardo, L.; Caputo, R. Eur. J. Org. Chem.,169, 169-173 (2006).
05. Carraway, K.L.; Koshland, D.E.Jr. Biochim. Biophys. Acta. 160, 272-274 (1968).
06. Smith, S.J.; Du, K.; Radford, R.J.; Tezcan, F.A. Chem Sci. 4, 3740-3747 (2013).

MACROCYCLES FORMED BY SUBSTITUTED TRITHIOCYANURIC ACID AS TEMPLATE FOR SELF-ASSEMBLY OF PEPTIDE CHAINS

Marta Cal[1,2], Mateusz Waliczek[1], Grzegorz Wołczanski[1], Piotr Stefanowicz[1]
[1] Faculty of Chemistry, University of Wroclaw, Joliot-Curie 14, 50-383 Wroclaw, Poland
[2] Georg-August University Göttingen, Institute of Organic and Biomolecular Chemistry, Tammannstr. 2, 37077 Göttingen, Germany

Introduction

Disulfide bridges are considered as one of the main factor facilitating the formation of spatial structures of native proteins [1]. However their unique redox properties give the opportunity to use them also for the dynamic combinatorial chemistry. The application of disulfide bond for the chemical ligation [2] as well as for the formation of supramolecular compounds [3] have been already described. The main goal of our research project was to apply the trithiocyanuric acid to form a template for self-assembly of peptide in the novel TASP molecules [4,5].

Results and Discussion

The directed reaction of 1,3,5-trithiocyjanuric acid with peptide containing the bromoacetic acid moiety in N-terminal part was unsuccessful. The main product was the three-substituted compound. Therefore we synthesized a new building block - in short DTCCA which was dedicated for the SPPS.

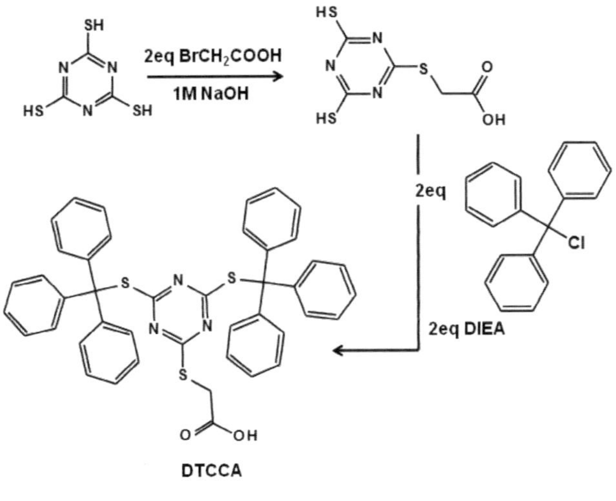

Figure 1. The scheme of synthesis of building block DTCCA.

In the Figure 1 the synthesis of the designed compound is shown. The presented synthetic method contains two steps: the carboxylation and tritylation reaction. After the optimization of the DTCCA synthesis, the yield of the reaction is 10% relative to the pure products, including the two-stage purification. Both reactions (carboxymethylation and tritylation) were performed under nitrogen. The carboxylmethylation reaction was performed in the temp. range from 0 - 8 °C for 2 h. For the following 22 h the reaction was allowed to proceed at the room temperature. The reaction was performed in methyl alcohol using 2 eq. of bromoacetic acid in the 1 M NaOH. The crude product after the ESI-MS analysis was subsequently used for tritylation. This reaction was performed at a room temperature. The trityl chloride and the diisopropylethylamine (DIEA) were used in 2 eq. The reaction was carried out in tetrahydrofuran (THF) for 2 h. Next the solvent was removed and the crude products were dissolved in water and pH was adjusted to 6. Then the final compound was extracted with diethyl ether.
To compare the formation efficiency of the supramolecular systems by two compounds: trithiocy-

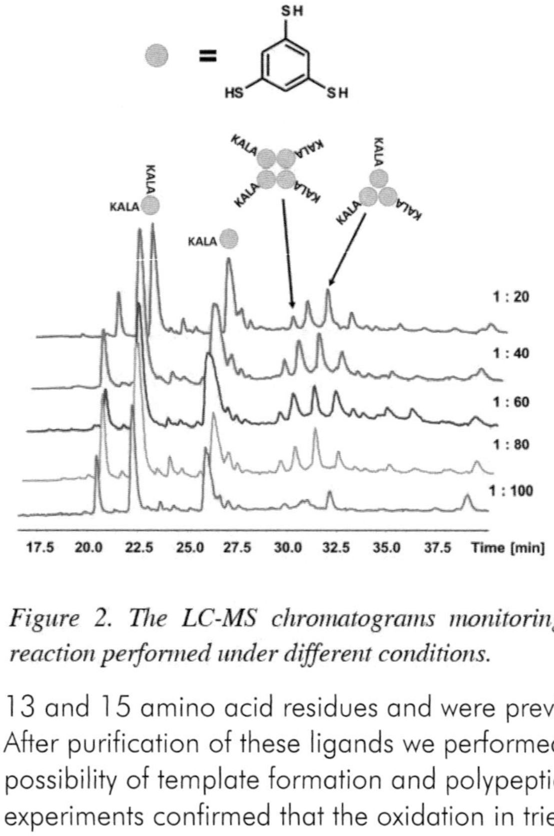

Figure 2. The LC-MS chromatograms monitoring reaction performed under different conditions.

anuric and 1,3,5-trimercaptobenzene, the new synthetic method facilitating the incorporation of 1,3,5-trimercaptobenzene in the polypeptide chains was developed. Introduction of this compound into the peptide chain was obtained by the attachment of bromocetic acid to the peptide followed by the substitution of bromine atom by sulfur. The reaction was performed on model peptide sequence H-KALA-OH on the Rink ChemMatrix® resin and monitored using LC-MS spectrometry (Figure 2). The best ratio of obtained peptide conjugate in monomeric form to the polymeric form (where two polypeptide chains were attached to one 1,3,5-trimercaptobenzene molecules) was obtained for the stoichiometry 1:20 (peptide : thiol moiety). Two derivatives: DTCCA and 1,3,5-trimercaptobenzene were successfully incorporated into the amphipathic peptide sequences ([AD] = H-KALEKALKEALAKLK-OH and [TRI] = H-KALEEKLKALEEK-OH). These peptides contain 13 and 15 amino acid residues and were previously reported to form a tetrahelical bundle [4,5]. After purification of these ligands we performed the oxidation reaction in the order to determine the possibility of template formation and polypeptide chains organization on this template. The ESI-MS experiments confirmed that the oxidation in triethylammonium bicarbonate buffer at pH 8.5 results in the formation of structures with high molecular masses. In the case of the peptide conjugated with 1,3,5-trimercaptobenzene just after 2 days of oxidation the dimer, trimer and tetramer can be found. The preliminary CD studies revealed that the conformational equilibrium for the trimeric systems is dominated by α-helix. In comparison to the peptide conjugate with trithiocyanuric acid the observed products are dominated by the dimeric forms and content of trimeric structures was relatively low. However the oxidation process as well as the obtained structures are still under detailed studies.

Acknowledgement
This work was financially supported by a grant no. UMO-2013/11/N/ST5/01157 from the National Science Center of Poland

References

01. G. Bulaj: Biotechnol. Adv., 23, 87 (2005).
02. N. A. Schnarr, A. J. Kennan: Org. Lett., 7, 395 (2005).
03. J. M. A. Carnall, C. A. Waudby, A. M. Belenguer, M. C. A. Stuart, J. J.-P. Peyralans, S. Otto: Science, 327, 1502 (2010).
04. M. Mutter, S. Vuilleumier: Angew. Chem. Int. Ed. Engl., 28, 535 (1989).
05. M. Cal, Ł. Jaremko, M. Jaremko, P. Stefanowicz: New. J. Chem., 37, 3770 (2013).

PHOSPHOLE AMINO ACIDS PROVIDE FLUORESCENT PROPERTIES TO PEPTIDES

Mathieu Arribat, Benjamin Renaud, Emmanuelle Remond, Florine Cavelier
Institut des Biomolécules Max Mousseron, IBMM, UMR-5247, CNRS, Université de Montpellier, ENSCM
Place Eugène Bataillon 34095 Montpellier cedex 5, FRANCE. florine.cavelier@umontpellier.fr
Keywords: Unnatural amino acids, phosphole, fluorescence, peptides.

PP I - 009

Introduction

Fluorescence microscopy is a powerful technique to visualize three-dimensional imaging of tissues and living cells. This technique, which is highly sensitive and gives a fast response time, is continually renewed by the progresses of instruments and the discovery of new fluorophores and tagging methods.[1,2] Phospholes are more and more exploited for their photophysical properties as potential fluorescent probes for sensor applications at molecular level.[3,4] Until today, phospholyl amino acids were few described, and in all cases isolated as sulfur derivatives, more stable than free P(III)-compounds easily oxidized under reaction conditions.[5,6] Herein, we report the first stereoselective synthesis of phospholyl borane amino esters and their use in peptide synthesis, as well as the preparation of derivatives by oxidation, sulfuration or quaternization of the P-center.

Results and discussion

The stereoselective synthesis of phospholyl borane (L)-α-amino esters 4a-e and 5d-e was achieved by substitution of β or γ-iodo amino esters 3a-e with phospholide anions, then the reaction mixture was quenched with BH_3.DMS (Scheme 1). Stable and easily handle phospholyl borane amino esters 4a-e and 5d-e were isolated in 50 to 78% yield as stable compounds. The enantiomeric purity of all phospholyl borane amino esters was checked by HPLC on chiral column per comparison with a corresponding racemic sample.

The phospholyl borane amino ester 4d was successfully transformed into the free phosphole 6 by decomplexation of the borane with DABCO. In addition, the phospholyl(sulfide) 7 and phospholyl(oxide) 8 were obtained by in situ reaction with sulfur or tert-butyl hydroperoxide in the presence of DABCO respectively. Quaternization with methyl iodide led to the corresponding dibenzophospholinium salt 9 in 70% isolated yield (Scheme 1).

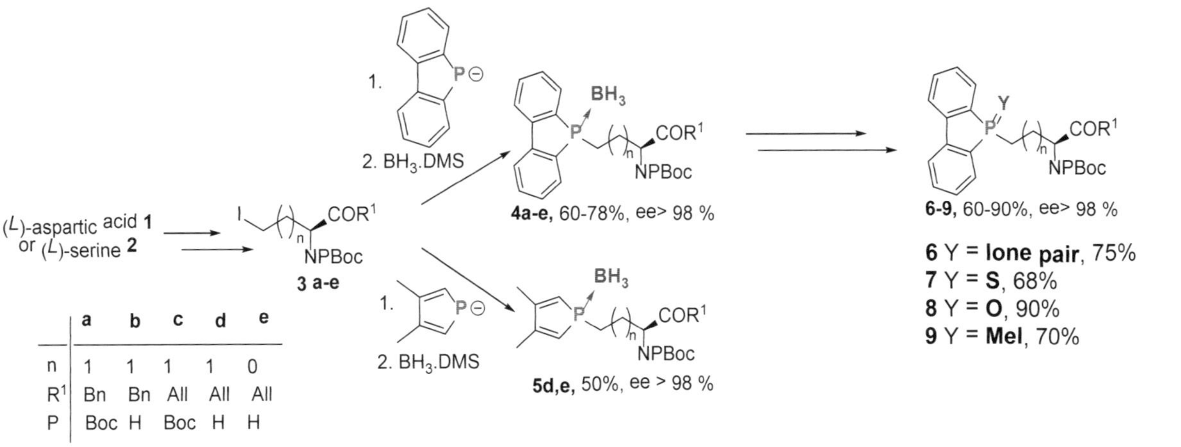

	a	b	c	d	e
n	1	1	1	1	0
R[1]	Bn	Bn	All	All	All
P	Boc	H	Boc	H	H

Scheme 1. Modular synthesis of phosphole borane amino esters

Subsequent deprotections of the dibenzophospholyl borane amino ester 4d were achieved to afford the corresponding enantiopure amine hydrochloride salt 10 or free phospholyl carboxylic acid 13 in good yields (Scheme 2). The borane complex 10 (or 13) was used in peptide coupling using IBCF with protected (L)-alanine derivative 11 (or 14), to afford stereoselectively the dibenzophospholyl borane dipeptide 12 (or 15) in yield up to 50% (Scheme 2).

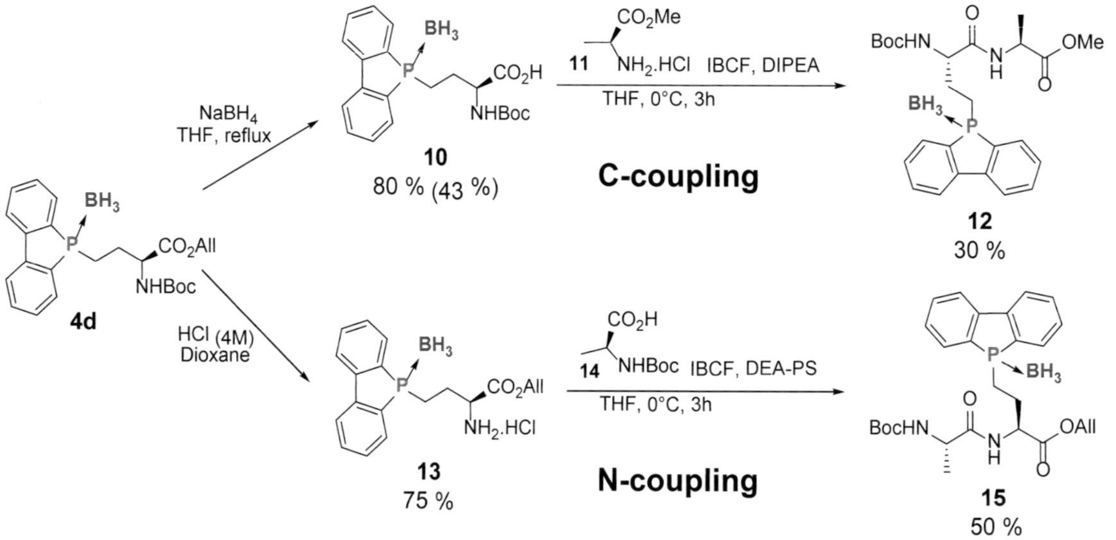

Scheme 2. Synthesis of dibenzophospholyl borane dipeptides

The photopysical properties of phospholy amino esters 6-9 were determined by fluorescent spectroscopy in CH_2Cl_2.

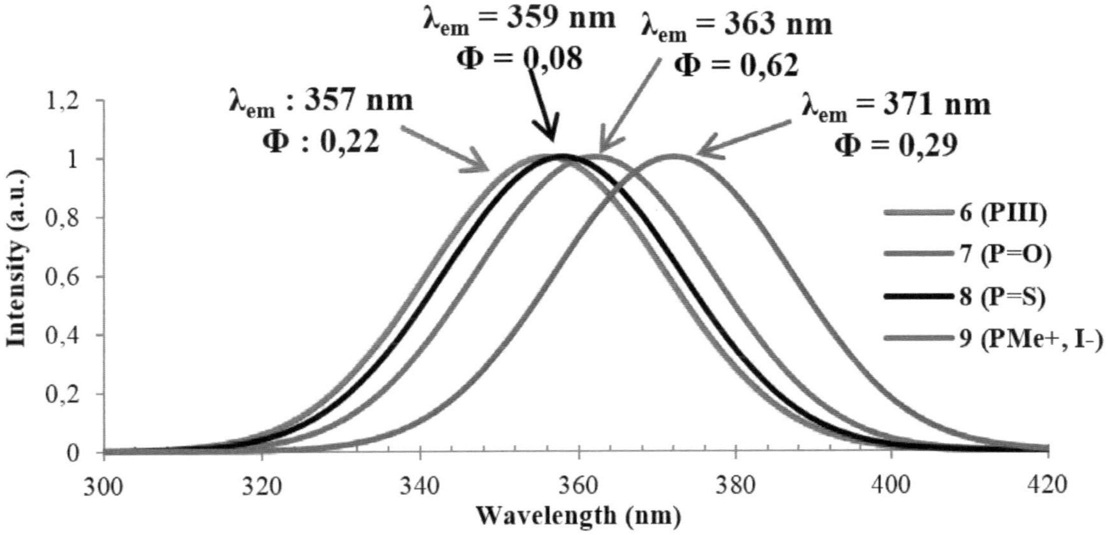

Figure 1: Emission spectra of 6-9 in CH_2Cl_2.

In summary, new phospholyl borane amino acid and peptide derivatives were stereoselectively synthesized by P-C bond formation. The nucleophilic substitution involves a phospholide anion with β or γ-iodo amino esters as electrophilic building block, followed by in situ trapping phosphorus atom with a borane group. Phospholyl borane complexes are key precursors for the preparation of phospholinium salts, oxide and sulfur derivatives, while the free phosphole precursors are highly oxidizable.

Phospholyl amino esters exhibit fluorescent emission between 357 nm to 371 nm in up to 62 % of quantum yield. Consequently, these amino acids are of great interest for the development of fluorescent peptides.

Acknowledgements

The authors thank Montpellier University for the grant of Mathieu Arribat.

References

01. Haugland, R. P. Handbook of Fluorescence Probes and Research Products, 9th Ed, 2002.
02. Krueger, A. T.; Imperiali, B. ChemBioChem, 2013, 14, 788-799,
03. Stolar, M.; Baumgartner, T. Chem. Asian J., 2014, 9, 1212-1225.
04. Duffy, M. P.; Delaunay, W.; Bouit, P. A.; Hissler, M., Chem. Soc. Rev., 2016, 45, 5296-5310.
05. Van Zutphen, S.; Margarit, V. J.; Mora, G.; Le Floch, P., Tet. Lett., 2007, 48, 2857-2859.
06. Bisaro, F.; Le Floch, P., Synlett, 2010, 2010, 3081-3085.

INCORPORATION OF TRIFLUOROMETHYLATED AND S-TRIFLUOROMETHYLATED AMINO ACIDS INTO PEPTIDES AND QUANTIFICATION OF THEIR HYDROPHOBICITY

Charlène Gadais, Emmanuelle Devillers, Nathalie Saraiva Rosa, Vincent Gasparik, Julien Pytkowicz, Evelyne Chelain, Thierry Brigaud.
LCB, EA 4505, University of Cergy-Pontoise, 5, Mail Gay-Lussac, Neuville-sur-Oise, 95031 Cergy-Pontoise Cedex, France.

Introduction

Hydrophobicity of lateral chains in peptides is known to have a critical impact on stabilization of protein, protein–protein interactions, peptide–receptor binding…[1] Due to particular properties of fluorine, trifluoromethylated amino acids (TfmAAs) are excellent candidates to locally increase the hydrophobicity of peptide. If the hydrophobicity of Fmoc-protected fluorinated amino acids has already been investigated by Koksch et al.,[2] to our knowledge, the magnitude of the hydrophobicity increase due to the incorporation of TfmAAs into peptides has never been evaluated so far. Thus, to quantify the hydrophobicity induced by TfmAAs, we have designed specific tripeptides (H-AA-Ala-Leu-OH and H-Ala-AA-Leu-OH) and studied them using a reliable RP-HPLC method. We have recently reported the incorporation of TfmAla in short peptide sequences.[3] Here, we extend our study to S-trifluoromethylated AAs, since SCF_3 group exhibits a higher Hansch parameter (π = 1.44) than CF_3 (π = 0.88).

Results and discussion

We achieved the synthesis of S-trifluoromethylcysteine (TfmCys) and trifluoromethionine (TFM) starting from orthogonally Boc/Bn protected cystine and homocystine via a radical trifluoromethylation (scheme 1).[4] Then, adequate deprotection allowed the incorporation of TfmCys in central position via mixed anhydride activation method, while its incorporation at the N-terminal position could be easily achieved by SPPS under micro-wave irradiation starting from a pre-loaded Fmoc-Leu-Wang resin.

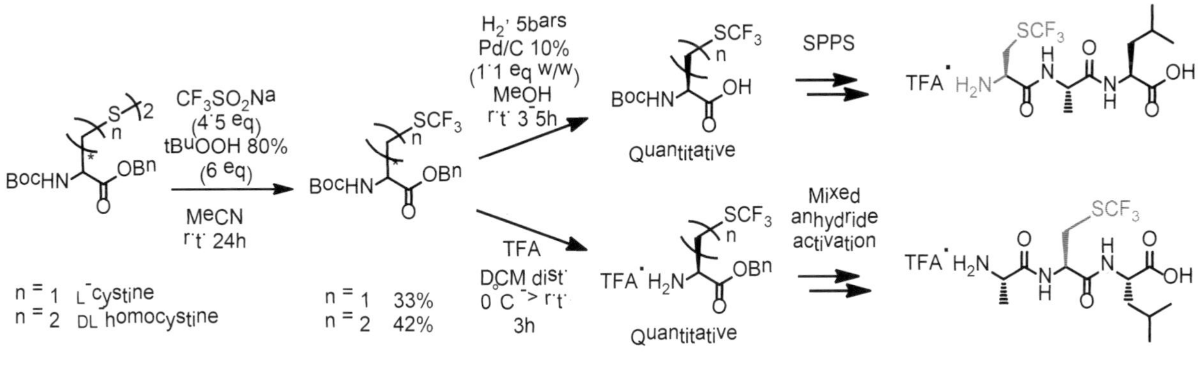

Scheme 1 Synthesis of TfmCys and TFM and incorporation of TfmCys in tripeptides

An alternative strategy based on a late trifluoromethylation has also been developped. The Togni's reagent proved to be efficient on thiol derivatives such as cysteine and homocysteine.[5] The SCF_3 containing tripeptides were built by electrophilic trifluoromethylation of disulfide bridged dimers. After reduction of the sulfur-sulfur bond, the generated thiol reacts in situ with Togni's reagent to lead to the desired tripeptide (scheme 2).

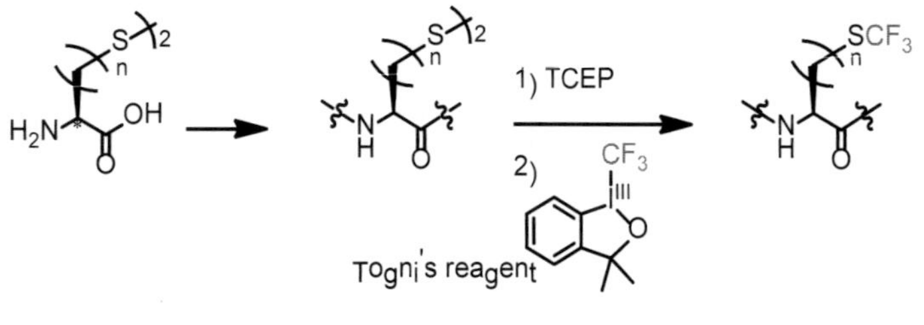

Scheme 2 "Late-trifluoromethylation" strategy

We chose to quantify the hydrophobicity of the peptides by determination of their φ_0, an index derived from the RP-HPLC peptides retention time.. The higher is the φ_0 the more hydrophobic is the compound. This procedure was initially disclosed by Kovacs et al.[6] and exhibited accuracy, reproducibility, and the possibility to work on milligram-scale. We adapted this method to our trifluoromethylated peptides and we studied the effect of both the position of the variant AA and the pH.. We observed that the trifluoromethylated AAs: TfmAla, TfmCys and TFM, displayed globally a higher φ_0 than natural hydrophobic AA such as Ile (figure 1).

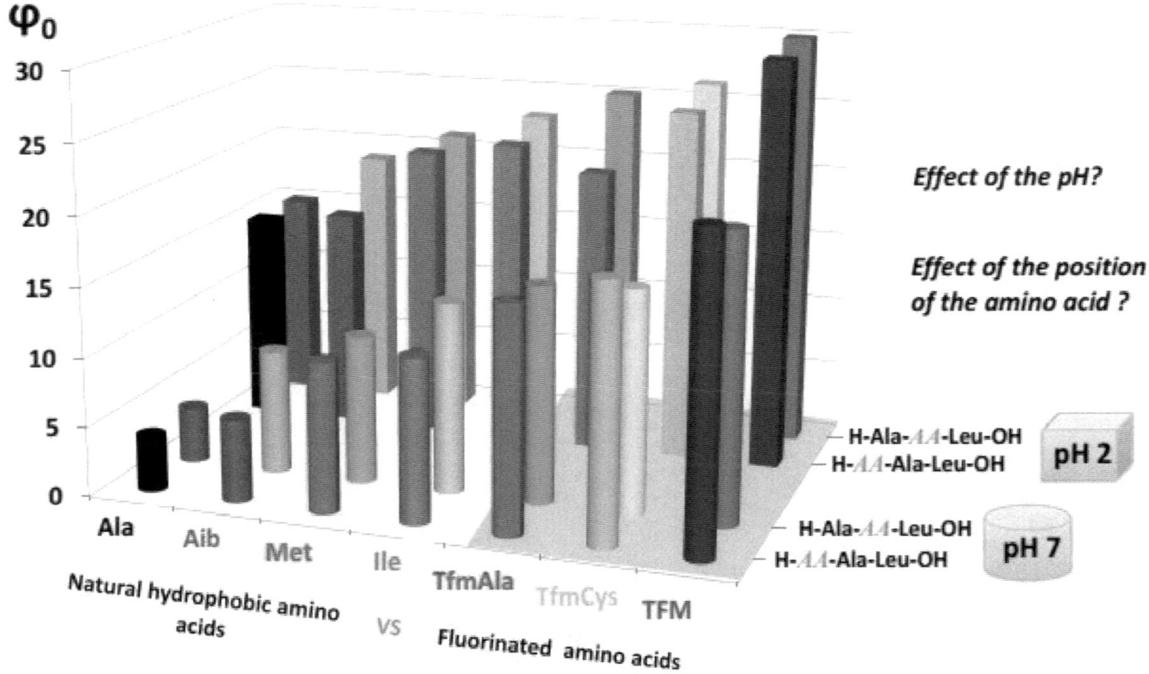

Figure 1 Hydrophobicity index φ_0 of studied tripeptides

So the selective incorporation of TfmAA significantly increase the hydrophobicity of peptidic sequences giving us a remarkable tool for rational design of biological active peptides.

References

01. Dalvia, V. H.; Peter J. Rossky, P. J. PNAS 2010, 107, 13603.
02. Gerling, U. I.; Salwiczek, M.; Cadicamo, C. D.; Erdbrink, H.; Czekelius, C.; Grage, S. L.; Wadhwani, P.; Ulrich, A. S.; Behrends, M.; Haufe, G.; Koksch, B. Chem. Sci. 2014, 5, 819.
03. Devillers, E.; Pytkowicz, J.; Chelain, E.; Brigaud, T., Amino Acids 2016, 48, 1457-1468.
04. Langlois, B.; Montègre, D.; Roidot, N., J. Fluorine Chem..1994, 68, 63-66.
05. Kieltsch, I.; Eisenberger, P.; Togni, A., Angew. Chem. Int. Ed. 2007, 46, 754-757.
06. Kovacs, J. M.; Mant, C. T.; Hodges, R. S. Biopolymers (Pept. Sci.) 2006, 84, 283.

AN EFFICIENT STRATEGY FOR THE SYNTHESIS OF INSULIN DERIVATIVES VIA 'INVERTED' MINI-PROINSULIN PRECURSORS

Alexandra Anastasiou[1], Dimitrios Gatos[1*] and Kleomenis Barlos[2]
[1]Department of Chemistry, University of Patras, 26500 Patras, Greece
[2]CBL-Patras, Patras, Greece

PP I - 014

Introduction

Insulin and its derivatives are the most important drugs for the treatment of diabetes with annual sales of over 20 billions and with steadily increased market. Despite of numerous efforts, the chemical and economically feasible route to insulin has not yet been developed. The methods which have been applied to date include the random mixing of the linear A and B chains and their air oxidation, the mixing of the sulfonated A and B-chains, the site-directed building of the three disulfide bonds and the biomimetic folding of single-chain precursors[1,2]. A-C-B 'inverted'-proinsulins with a connecting C-peptide, which consists of at least 8 amino acids, are known to yield also insulin derivatives. It is a common sense among peptide chemists and biologist that the C-peptide of a natural or inverse proinsulin must be of a minimum length of several amino acids, in order to have the required flexibility to fold correctly to the natural mature proinsulin or reverse proinsulin.
Very surprisingly, we found that C-peptides which contain in their sequence only few amino acids, even only one, for example the A-Arg-B inverse proinsulin and their corresponding protected or partially protected derivatives act as excellent mature proinsulin precursors.

Results and Discussion

To facilitate the chemical synthesis of proinsulins by the solid-phase method we followed a strategy which was based on the following simple common knowledge:
Because the peptide must not contain difficult peptide synthesis regions, that means not to form β-turns and β-sheets during the on resin peptide chain elongation, we synthesized proinsulins not only in the natural B-C-A order but also in the A-C-B order, which has been proven in our hands to be much more easy to synthesize than the corresponding B-C-A proinsulins. Best results were achieved in cases where the C-peptides contained the β-sheets and β-turns disorganizing residues Pro, Hyp or pseudoprolines. Insertion of such residues in the C-peptide allows the effective synthesis of inverse proinsulins.
To improve their solubility in solvents used for their purification we inserted in the C-peptide Pro, Hyp, basic amino acids, such as Arg or Lys or acidic and hydrophilic amino acids, such as Glu or Ser.
To obtain proinsulins of high purity smaller protected peptides were condensed in solution or on solid-phase. In order to avoid racemisation the condensations were performed using as C-terminal amino acid of the fragments the amino acids Gly, Pro, β-Ala or an amino acid which contains an oligo or poly-glycol part in its structure, for example the $-NH-(CH_2CH_2O)_n-CO-$ structural element. To be able to remove the C-peptide, basic amino acids at the C-terminus of the A-chain and the amino terminus of the B-chain were positioned.
All syntheses were performed using the Fmoc/tBu-protection scheme and the 2-chlorotrityl resin as the solid support. For the protection of the Cys-residues the Trt or the Trt/Acm protecting groups

were used. The folding of the obtained inverse 'super mini' proinsulins was performed either 'randomly'or 'directed'. The removal of the C-peptide was performed as usually by trypsin and/or carboxypeptidase B. Below we describe, as an example, the synthesis of an insulin derivative via an ACB-mini-proinsulin precursor using for the protection of the side-chains of the Cys residues Acm and Trt groups.

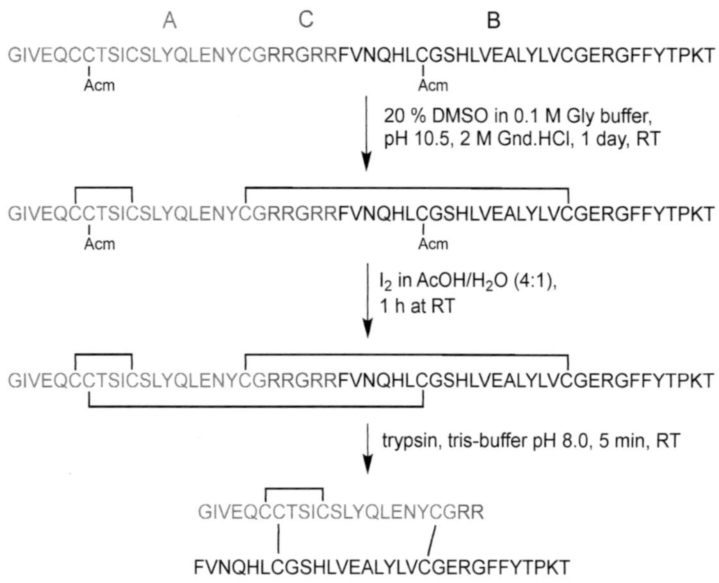

Figure 1. Synthesis of [Gly^{A21}, $Arg^{A22, A23}$] insulin via an ACB-mini-proinsulin precursor.

Acknowledgements

Authors acknowledge CBL-Patras S.A. for financial support.

References:

01. Sohma, Y. and Kent, S. B. J. Am. Chem. Soc. 2009, 131, 16313–16318.
02. Tofteng, A. P.; Jensen, K. J.; Schaffer, L.; Hoeg-Jensen, T. Chem- BioChem 2008, 9, 2989-2996.

Da'san M. M. Jaradat[1]
[1] Department of Chemistry, Faculty of Science, Al-Balqa' Applied University, Al-Salt 19117, Jordan
dasan.jaradat@bau.edu.jo, jaradatdasan@gmail.com

PP I - 020

In this paper, peptide 4 which contains two disulfide bridges was synthesized by orthogonal protection approach. This peptide represents a mimetic of natural antimicrobial peptides Magainin and Defensins and it showed antimicrobial activity against a number of bacterial strains including Pseudomonas aeruginosa, Salmonella typhimurium and some other strains.

Introduction

A number of natural peptides, that exhibit antimicrobial activity, have been isolated from nearly all groups of organisms. Most antimicrobial peptides (AMPs) are membrane active, composed of less than 100 amino acid residues and have a net positive charge.[1,2] AMPs can be classified into two major groups; the first group consists of cysteine-containing peptides including single or multiple disulfide bridges. The second group consists of linear molecules which either tend to adapt the secondary structure α-helix or they are enriched with certain amino acid residues such as Trp, Arg, Pro, His, and Gly.[3] There are some important families of AMPs, namely, Bombininis, Cathelicidins, Cecropins, Ceratotoxins, Defensins, Dermaseptins, Magainins, and Histatins.[4]

The emergence of bacterial resistance to common chemical-based antibiotics could cause a serious threat to human health.[5] This potential threat has motivated many research groups to be interested in AMPs as an alternative for chemical-based antibiotics. Naturally occurring AMPs have some drawbacks represented in their poor metabolic stability and low oral bioavailability. Some of these limitations can be overcome by synthesizing non-natural peptido-mimetics of natural bioactive peptides. In an attempt to synthesize such mimetics, we focus on the synthesis of mimetics of Magainin and Defensins peptides that contain multiple disulfide bridges. The S—S bridges and circular polypeptide chains confer structural stability.

Results and Discussion

Solid phase peptide synthesis (SPPS) was performed by standard Fmoc-couplings on a Rink amide resin, in which cysteine residues were orthogonally protected; indeed, two cysteine residues were trityl (Trt) protected and the other two cysteine residues were acetamidomethyl (Acm) protected in peptide 1 (Scheme 1). Peptide 1 was cleaved from the resin and partially deprotected by a solution containing 90% trifluoroacetic acid (TFA), water, and a scavenger, to deliver peptide 2 which contained two unprotected cysteine residues, whereas the Acm protected cysteine residues remained intact. Next, oxidation and formation of a disulfide bridge were achieved by dissolving peptide 2 in a solution of DMSO / H_2O (1:2) and stirring for 30 hours to produce peptide 3 which was confirmed by LC-MS/MS.

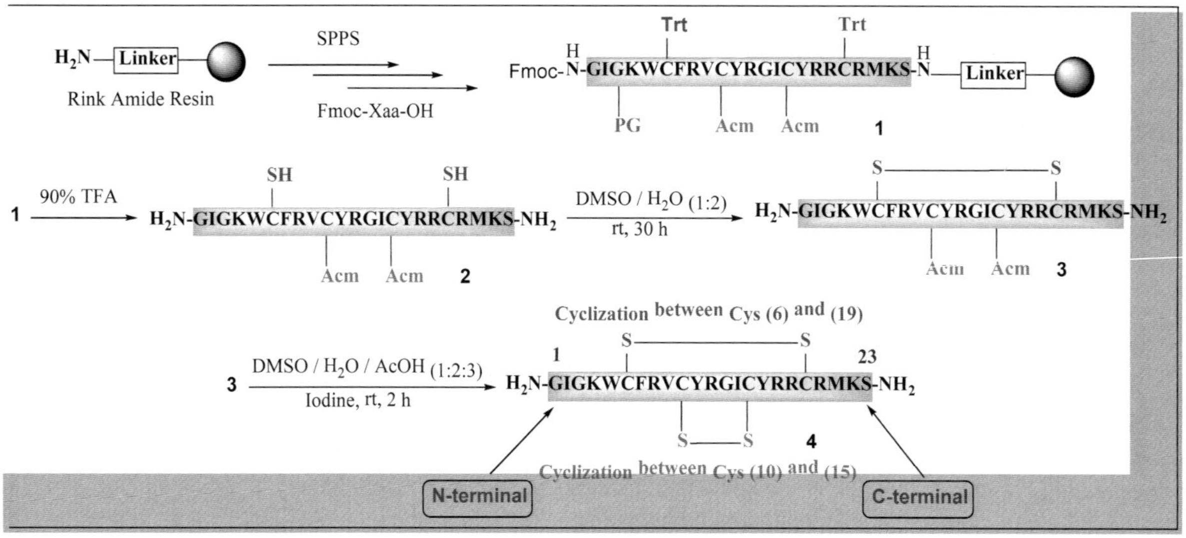

Scheme 1: Synthesis of fully protected immobilized peptide 1, cyclized partially protected cleaved peptide 3 containing one disulfide bridge, and cyclized deprotected peptide 4 containing multiple disulfide bridges. Acm = Acetamidomethyl, Trt = Trityl, Fmoc = 9-Fluorenylmethoxycarbonyl, TFA = trifluoroacetic acid, DMSO = Dimethyl sulfoxide, AcOH = Acetic acid, SPPS = Solid phase peptide synthesis, PG = Protecting groups of other side chains.

In order to remove the Acm protecting groups, peptide 3 was treated with a solution of DMSO / H_2O / acetic acid (1:2:3) followed by iodine, and the reaction mixture was stirred for two hours.[6] Removal of the Acm groups and complete conversion to the two disulfide peptide 4 was confirmed by LC-MS/MS. Excess solvent was evaporated in vacuo, leaving yellow oil. The residue was diluted with water and charged onto a preparative HPLC column.[7] After Lyophilization, pure peptide 4 was isolated in a yield of 51% (Figure 1). Ellman's test was performed to confirm the absence of sulfhydryl groups.

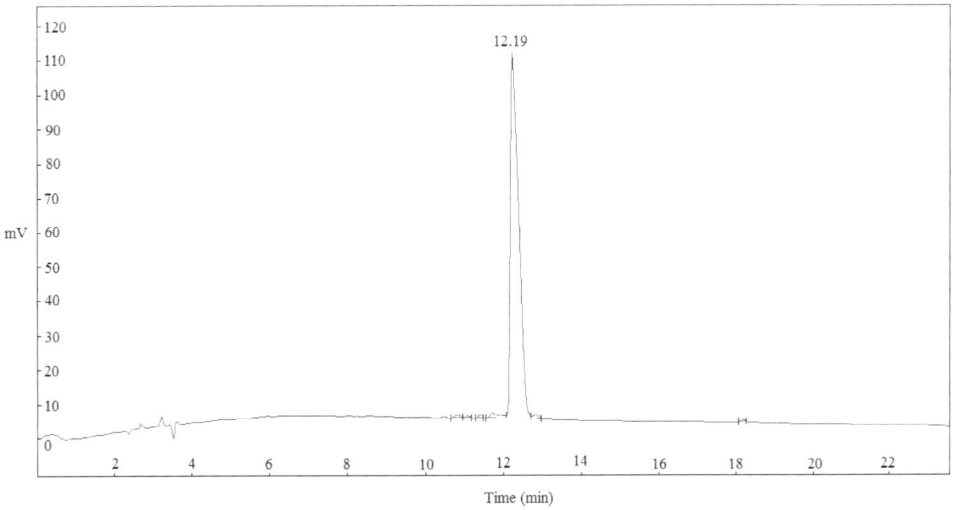

Figure 1: HPLC profile of cyclized peptide 4.

Peptide 4 showed antimicrobial activity against a number of bacterial strains including E. coli O157:H7, Staphylococcus aurous, Pseudomonas aeruginosa and Salmonella typhimurium and others. This antimicrobial activity can be attributed to the presence of the disulfide bridges and the positively charged amino acid residues such as Lysine and Arginine.

Conclusion

In summary, we have shown that a synthetic antimicrobial peptide containing two disulfide bridges was achieved by standard Fmoc-based solid-phase peptide synthesis utilizing orthogonal protecting groups for cysteine residues. The key step in this strategy involves the utilization of cysteine residues protected with Trt and Acm. Trt groups will be removed during the cleavage step with TFA whereas Acm will remain intact. Then, a disulfide bridge can be formed between the two free sulfhydryl groups. Next, Acm groups are removed and oxidation is achieved by iodine to form the second disulfide bridge. Synthesis of peptides containing triple and tetra disulfide bridges can be done using cysteines protected with tert-butyl (tBu) and 4-methylbenzyl (MeBzl) in addition to those described in our strategy.

Materials and Experimental Methods

General: All reagents, amino acids, and solvents were purchased from commercial suppliers and used without further purification. Peptide Synthesis: Solid phase peptide synthesis was carried out by APPTEC Focus Xi peptide synthesizer using standard amide coupling conditions HBTU/HOBt utilizing Rink Amide resin. Characterization: Analytical HPLC spectra were recorded by Shimadzu LC-10AT vp using phenomenex C18 column (5 μm, 150 x 4.6 mm), mass spectra were recorded by Agilent 1200 – API 4000 LC–MS/MS system. Purification of peptides was performed by preparative KNAUER HPLC system using phenomenex C18 column (10 μm, 250 x 21.2 mm). Antimicrobial Activity: The bioassays were carried out as previously described.[8]

Acknowledgments
The author acknowledges financial support from the Jordan Scientific Research Support Fund (SRSF) within the grant (Project MPH/2/03/2012).

References

01. M. Zasloff, Nature, 415, 389-395 (2002).
02. D. Andreu, L. Rivas, Biopolymers, 47, 415-433 (1998).
03. D. A. Devine, Mol. Immunol., 40, 431-443 (2003).
04. S. H. Marshall, G. Arenas, Electronic Journal of Biotechnology, 6, 271-284 (2003).
05. N. A. AL-Haj, N. I. Mashan, M. N. Shamsudin et al. Am. J. Pharm. & Toxicol., 5(2), 95-102 (2010).
06. A. Cuthbertson, B. Indrevoll, Org. Lett., 5(6), 2955-2957 (2003).
07. D. M. M. Jaradat, H. Hamouda, C. P. R. Hackenberger, Eur. J. Org. Chem., 5004-5009 (2010).
08. C. Lòpez-Abarrategui et al. The FASEB Journal, 29(8), 3315-3325 (2016).

Alicja Kluczyk, Alina Dambinova, Remigiusz Bachor, Bartosz Setner, Monika Kijewska, Monika Biernat, Piotr Stefanowicz, Zbigniew Szewczuk
Faculty of Chemistry, Wrocław University, 50-383 Wrocław, Poland
e-mail alicja.kluczyk@chem.un.wroc.pl

PP I - 023

OP
PP I
PP II
PP VI
PP VII
PP VIII
PP IX
PP X
PP XI
PP XII
PP XIII
PP XIV

Introduction

In the search for new methods of ultrasensitive peptide analysis by mass spectrometry, we developed a series of peptide modifications, based on quaternary ammonium (QA) moieties [1]. The application of cyclic QA-acetyl groups simplifies sequence analysis and allows attomole level detection of peptides [2]. QA-peptides undergo hydrogen-deuterium exchange (HDX) under mild conditions (1% triethylamine in D_2O), producing isotopically labelled ionization markers [3]. Another category of permanent charge tags is based on substituted triarylphosphonium modification [4]. Such tags are easily synthesized and the collision-induced dissociation of derivatized peptides allows straightforward sequence determination.

Results and Discussion

The aim of our work was to investigate whether the effects of incubation of triarylphosphoniumacetyl-modified peptides in 1% triethylamine in D_2O resemble that of QA-acetyl peptides. The preliminary results suggested that HDX occurs in the case of tris-trimethoxyphenylphosphoniumacetyl-modified peptides (TMPP), whereas in the case of unsubstituted triphenylphosphonium analogues (TPP), a different reaction takes priority. A model TPP-peptide conjugate was synthesized on solid phase and subjected to incubation in 1% triethylamine in D_2O. The reaction was monitored by mass spectrometry (Fig. 1).

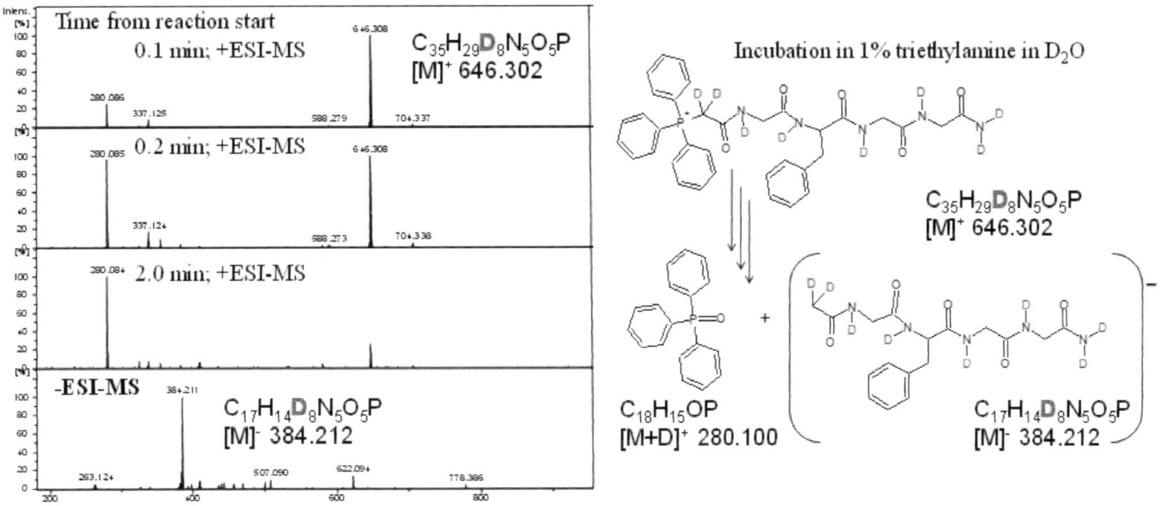

Figure 1. MS monitoring of incubation of TPP-peptide conjugate in 1% triethylaminein D_2O. In negative ion mode MS, residual peptide fragment is observed (the structure is not confirmed).

Triphenylphosphine oxide, which could be identified by MS, is a typical side-product of Wittig reaction [6]. Therefore we decided to test this hypothesis by performing a Wittig reaction on solid phase using Fmoc-based strategy. The details of the final stages of reaction, LC-MS chromatogram of crude cleavage product and isomer assignment by NMR spectra are presented in Fig. 2.

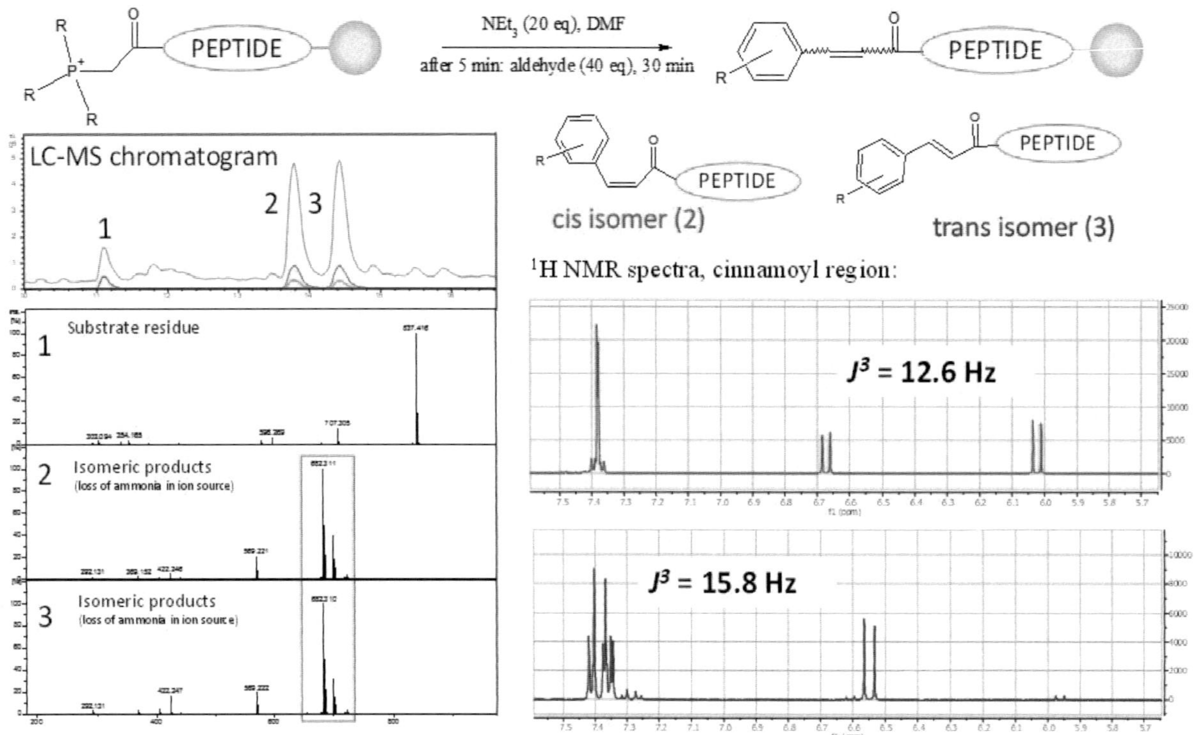

Figure 2. Wittig transformation of triphenylphosphoniumacetyl-peptide conjugate on solid support. The LC-MS analysis revealed two isomeric products, which were identified by ^{1}H and COSY NMR after separation.

In conclusion, the results of HDX next to phosphonium motif depend on the substitution of aromatic rings in triarylphosphine. In TMPP-peptides a regular HDX occurs, whereas in TPP-peptides it is possible to induce Wittig reaction, opening a way to novel unsaturated peptide derivatives.

Acknowledgments

This work was supported by grant UMO-2013/09/B/ST4/00277 from the National Science Centre, Poland.

References

01. M. Cydzik, M. Rudowska, P. Stefanowicz, Z. Szewczuk, J. Pept. Sci., 17, 445 (2011)
02. R. Bachor, P. Mielczarek, M. Rudowska, J. Silberring, Z. Szewczuk, Int. J. Mass Spectrom., 362, 32 (2014)
03. M. Rudowska, D. Wojewska, A. Kluczyk, R. Bachor, P. Stefanowicz, Z. Szewczuk, .J. Am. Soc. Mass Spectrom., 23, 1024 (2012)
04. Z.H. Huang, J. Wu, K.D. Roth, Y. Yang, D.A. Gage, J.T. Watson, Anal. Chem. 69, 137 (1997)
06. S.R. Wilson, J. Perez, A. Pasternak, J. Am. Chem. Soc., 115, 1994 (1993).
07. T. Itaya, T. Iida, S. Shimizu, A. Mizutani, M. Morisue, Y. Sugimoto, M. Tachinaka, Chem. Pharm. Bull., 41, 252 (1993)

Chiaki Komiya, Keisuke Aihara, Tsubasa Inokuma, Akira Shigenaga, Akira Otaka
Tokushima University, Institute of Biomedical Sciences and Graduate School of Pharmaceutical Sciences, Tokushima, Japan

PP I - 024

Spatiotemporal control of peptide/protein function has contributed to the progress of research in various fields of chemical biology. An external stimulus-induced amide bond processing system has received increasing attention, because such a system is potentially useful to control activity of peptides/proteins at a desired time and location [1]. These systems require processing reactions that proceed under mild conditions, but chemical cleavage of amide bonds generally occurs under harsh conditions. In this context, we focused on intein-mediated protein splicing, which includes amide cleavage reaction proceeding under physiological conditions [2]. In this system, a conformationally restricted asparagine (Asn) at a cleavage site is activated as nucleophile by neighboring basic residue. Consequently, the Asn side chain cyclizes to cleave the amide bond via aspartimide formation. Being inspired by the Asn-induced amide bond cleavage reaction of intein-mediated protein splicing, we envisioned development of an Asn-based stimulus-responsive amide processing device that induces amide bond processing without using a whole intein system.

Design of the Asn-based processing device is shown in Figure 1 [3]. The Asn derivative contains a secondary amine unit and geminal dimethyl groups to mimic the intein-mediated processing reaction. We expected that the amine unit would work as an intramolecular base to activate the amide nitrogen on the Asn side chain, and the geminal dimethyl groups would fix conformation of the Asn derivative by Thorpe-Ingold effect to accelerate the amide cleavage of main chain. Furthermore, we introduced a stimulus-removable protecting group onto the appended amine to achieve a stimulus-triggered processing system.

In this study, an o-nitrobenzyloxycarbonyl (oNBnoc) group was used for N-protection to achieve UV-responsive amide cleavage. First, Fmoc-protected Asn derivative 1 was synthesized and incorporated into model peptide 2 (H-YGGFLXSGF-NH_2; X = Asn derivative) using standard Fmoc-based solid phase peptide synthesis. We then examined a UV-irradiation experiment. UV irradiation (365 nm) to peptide 2 in 6 M Gn·HCl and 0.2 M phosphate buffer at pH 7.4 induced the complete removal of oNBnoc protection. After incubation of the resulting solution at 37 °C for 24 h, processing reaction of the deprotected intermediate proceeded in about 80% conversion. On the other hand, when peptide 2 was incubated without photo-irradiation, the processing reaction was not observed. These results clearly indicate that the developed Asn derivative possessing the photo-removable N-protection functions as a UV-responsive amide processing device.

In summary, we developed an intein-inspired UV-responsive Asn derivative that induces UV-triggered amide bond cleavage. In principle, the newly developed Asn scaffold could respond to any other stimulus by simply varying the protecting group according to employed stimulus. Development of other stimulus-responsive Asn derivatives and their application to control peptide/protein activity are in progress.

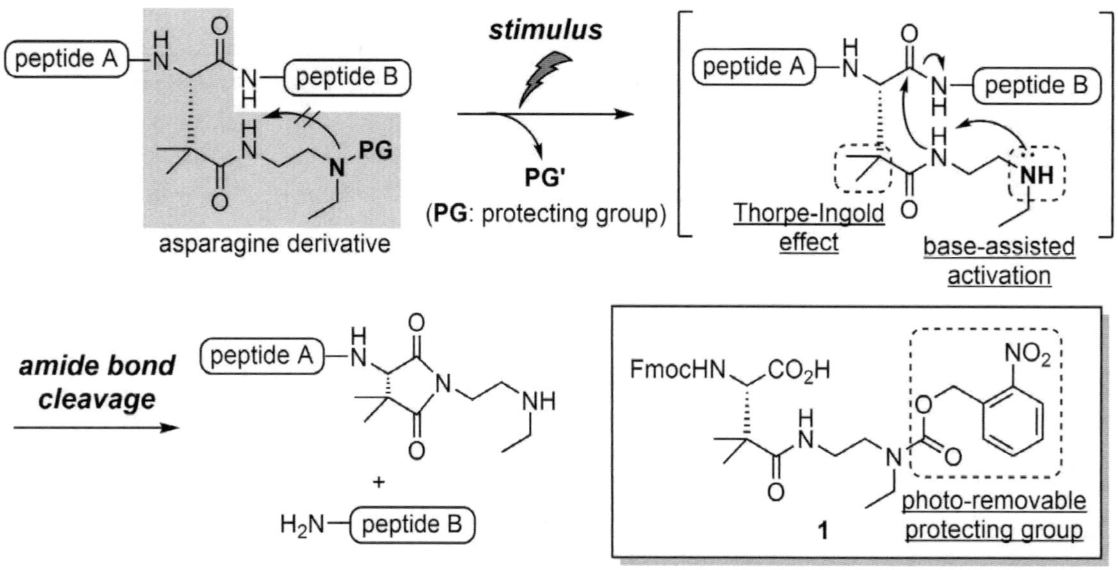

Fig. 1. Design of stimulus-responsive amide processing device.

References

09. (a) A. Shigenaga, D. Tsuji, N. Nishioka, S. Tsuda, K. Itoh, A. Otaka, ChemBioChem 2007, 8, 1929-1931.
10. (b) A. Shigenaga, J. Pharm. Soc. Jpn. 2012, 132, 1075–1082.
    Z. Liu, S. Frutos, M. J. Bick, M. Vila-Perelló, G. T. Debelouchina, S. A. Darst, T. W. Muir, Proc. Natl. Acad. Sci. U. S. A. 2014, 111, 8422–8427.
11. C. Komiya, K. Aihara, K. Morishita, H. Ding, T. Inokuma, A. Shigenaga, A. Otaka, J. Org. Chem. 2016, 81, 699–707.

CHARACTERIZATION OF THE SPYTAG – SPYCATCHER INTERACTION

Jonas Ludwig[1], Nicolas Ulm[1], Marlene Pröschel[2], Heinrich Sticht[3], Uwe Sonnewald[2], Jutta Eichler[1]
[1]Department of Chemistry and Pharmacy, University of Erlangen-Nürnberg, Schuhstr. 19, 91052 Erlangen, jonas.ludwig@fau,de
[2]Department of Biology, University of Erlangen-Nürnberg, Staudtstr. 5, 91056 Erlangen
[3]Institute of Biochemistry, University of Erlangen-Nürnberg, Fahrstr. 17, 91052 Erlangen

Protein structures are stabilized by secondary structure elements, as well as by covalent bonds, including disulfide bridges between cysteine residues, and isopeptide bonds between asparagine/aspartate and lysine residues[1]. Using the fibronectin binding protein (FbaB) of Streptococcus pyogenes (Spy), which contains an aspartate-lysine isopeptide bond, Howarth et al. demonstrated that this bond is spontaneously formed not only within the protein (intramolecular), but also between two separate fragments of the protein, which each contain one of the involved aspartate and lysine residues, respectively, (intermolecular)[2]. These two fragments, termed SpyTag (a 13 amino acid peptide) and SpyCatcher (the truncated protein), can thus serve as adapter molecules for the generation of new protein complexes and architectures, as well as for biotechnology applications that involve stable attachment of proteins to surfaces or other materials.

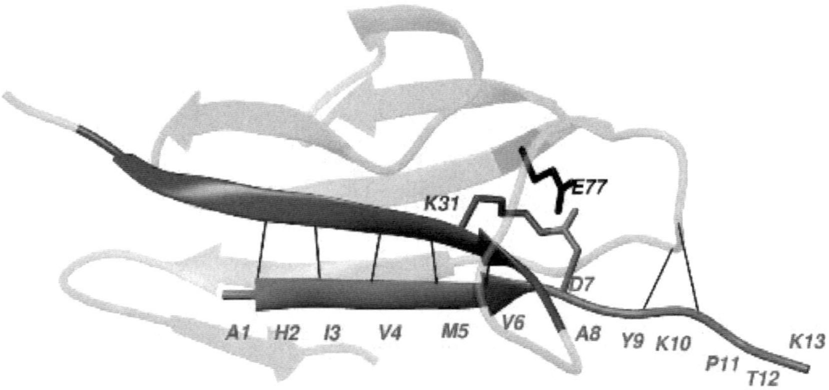

Figure 2: Crystal structure of the covalent SpyTag (red) - SpyCatcher (green/blue) complex with an isopeptide bond between D7 and K31, whose formation is catalyzed by E77. (pdb: 4MLI) .3

In order to explore the SpyTag – SpyCatcher interaction in more detail, we have generated an array of SpyTag substitution variants, including alanine, proline and D-amino acid scans of the peptide, and analyzed their interaction with the SpyCatcher. In the ELISA binding assay, the SpyCatcher was coated to microtiterplates, and then incubated with biotinylated SpyTag peptides. Non-covalently bound peptides were removed using guanidinium hydrochloride, and bound peptides were detected via an anti-biotin-HRP conjugate.

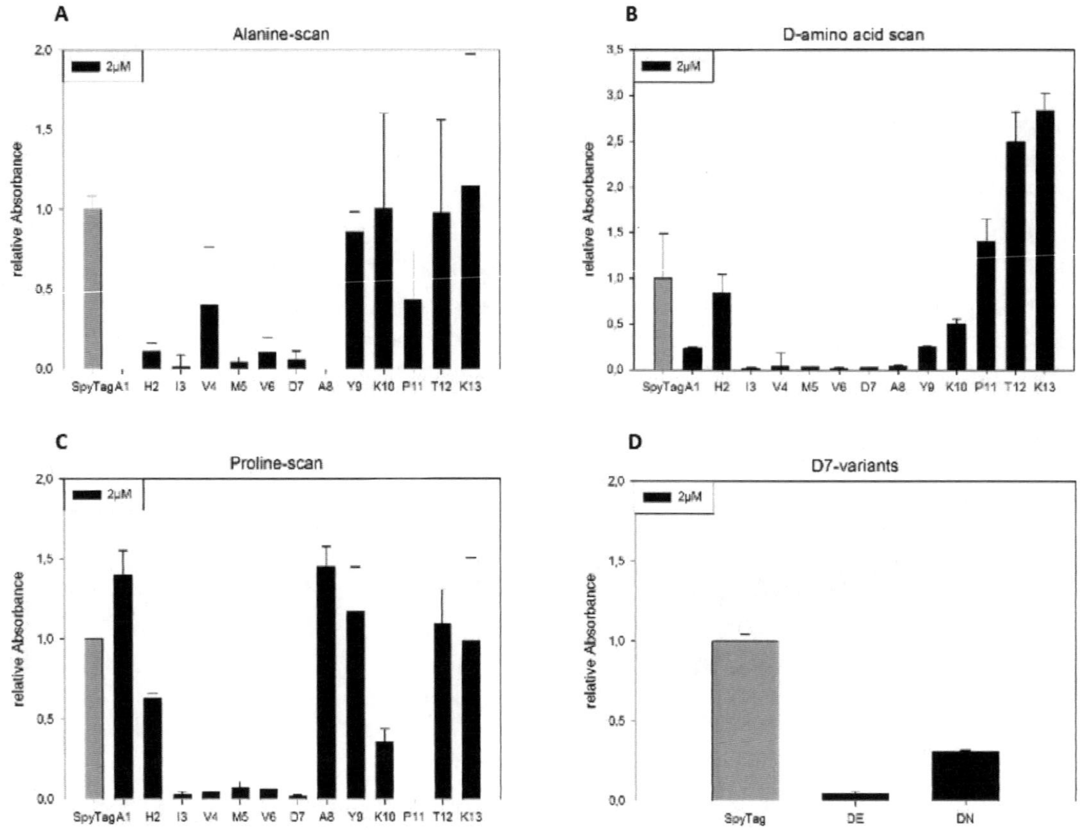

Figure 2: Covalent interaction of SpyTag variants with the SpyCatcher protein. A) Alanine scan, B) D-amino acid scan, C) Proline scan and D) D7N and D7E variants of SpyTag.

As shown in Figure 2A-C, essentially all amino acid side chains in the N-terminal part of the SpyTag, which interacts with the K31 containing β-strand of the SpyCatcher, are important for isopeptide bond formation, as replacing them with alanine, the respective D-amino acid and proline, respectively, greatly hampers the formation of the covalent SpaTag-SpyCatcher complex. Amino acid exchanges in the C-terminal part, on the other hand, are better tolerated, or even beneficial for the interaction, providing potential for future optimization. Furthermore, substitution of D7 in the Spy-Tag, which is involved in the isopeptide bond with K31 of the SpyCatcher, with asparagine (D7N) or glutamic acid (D7E), greatly impedes, or even completely abrogates isopeptide bond formation (Figure 2D), possibly due to incorrect positioning within the catalytic triad formed by D7, K31 and E77. These data provide new insight into the molecular determinants of the SpyTag – SpyCatcher interaction, which may be used to further optimize this highly useful molecular adapter system.

References

01. Kang H.J.; Baker E.N., Trends Biochem. Sci., 2010, 36(4), 229-237.
02. Zakeri, B; Fierer, J.O.; Celik, E.; Chittock, E.C.; Schwarz-Linek, U.; Moy, V.T.; Howarth, M., Proc. Natl. Acad. Sci. USA, 2012, 109(12), E690–E697.
03. Li, L.; Fierer, J.O.; Rapoport, T.A.; Howarth, M. J.Mol.Biol. 2014, 426(2), 309- 317.

SECOND-GENERATION SYNTHETIC STRATEGY OF GM2-ACTIVATOR PROTEIN (GM2AP) ANALOGUES APPLICABLE TO THE PREPARATION OF A PROTEIN LIBRARY

Takahiro Nakamura, Akira Shigenaga, Naoto Naruse, Tsubasa Inokuma, Kohji Itoh, Akira Otaka
Tokushima University, Graduate School of Pharmaceutical Sciences, Tokushima, Japan

Question

One of the most useful methods for elucidation of functions of proteins has utilized chemically synthesized proteins that contain artificial units such as a fluorescence probe. Among various strategies for the synthesis of such proteins, native chemical ligation (NCL) with the use of peptide thioesters and N-terminal cysteinyl peptides has been widely employed.[1] Nowadays, a number of NCL-mediated chemical syntheses of small- and medium-sized proteins consisting of more than 100 amino acids have been reported.[2] Recently, we achieved the NCL-based chemical synthesis of 162-residue GM2-activator protein (GM2AP), which is a lysosomal glycoprotein involved in the hydrolysis of GM2 to GM3 by β-hexosaminidase A (Hex A) through the formation of Hex A-GM2AP complex (first-generation synthetic strategy).[3] Notably, the synthesized GM2AP exhibited hydrolytic activity of GM2 to GM3 in the presence of Hex A similarly to inherent proteins. In this context, we explored GM2AP analogues having greater activity than that of parent protein.

Methods:

Computational analysis of the Hex A-GM2AP complex predicted that position of Thr69 residue in GM2AP molecule could be critically important to the formation of the stable protein complex. This prediction encouraged us to replace Thr69 with some other suitable amino acid residues, because we assumed that such replacement would lead to enhancement of the hydrolytic activity of GM2AP through formation of more stable complex. To verify this hypothesis, however, we needed the development of a new strategy for synthesis of GM2AP, because our first protocol required laborious and time-consuming preparation of the 43-residue fragment (Fr. 2) that contains the replacement position at its C-terminus. Thus, we envisioned that the use of an easily accessible short peptide fragment containing the replacement would overcome difficulties mentioned above, and divided 43-residue Fr. 2 into smaller 36- and 7-residue fragments (Fr. 2N and Fr. 2C, respectively), with latter Fr. 2C involving the replacement position (second-generation synthetic strategy) (Fig. 1). This division required a new ligation between Pro67 thioester and Cys68; however, NCL reactions using prolyl thioesters had been reported to be difficult to carry out.[4] In this regard, we have recently disclosed that prolyl thioesters could participate in NCL by tuning the concentration of thiol catalyst and the reaction temperature.[5] Therefore, we speculated that this division was applicable to our second synthetic protocol for the preparation of GM2AP analogues containing various amino acid substitutions.

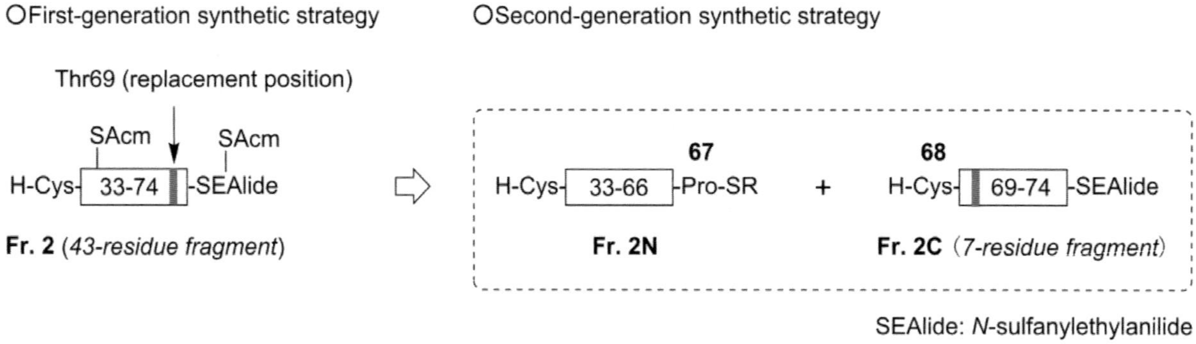

Fig. 1. Second-generation synthetic strategy of GM2AP analogues.

Results

Based our new strategy, we initially prepared N- (67-residue), middle (7-residue) and C-segment (88-residue) from six peptide fragments. We then performed two different kinetically controlled ligation reactions using these three segments which were based on the controllable reactivity of prolyl thioesters and our previously reported SEAlide methods.[6] Consequently, we successfully obtained five types of GM2AP analogues with substitution for Thr69, and indicated that they all had the hydrolytic activity of GM2 to GM3 through formation of Hex A-GM2AP.[7]

Conclusion

We developed a new platform for the second-generation synthesis of GM2AP analogues bearing various amino acid substitutions for Thr69. In our new protocol, the use of a readily accessible short peptide in combination of two types of kinetically controlled ligation enabled the facile incorporation of these replacements into whole protein and the construction of GM2AP protein library.

References

01. S. B. H. Kent. et al. Science 1994, 266, 766-779.
02. S. B. H. Kent. Chem. Soc. Rev. 2009, 38, 338-351.
03. a) A. Otaka. et. al. Angew. Chem. Int. Ed. 2013, 52, 7855-7859; b) A. Otaka. et. al. Chem. Commun. 2015, 51, 9946-9948.
04. S. B. H. Kent. et. al. Chem. Commun. 2011, 47, 2342-2344.
05. A. Otaka. et. al. Chem.Commun. 2014, 50, 58-60.
06. A. Otaka. et. al. Top. Curr. Chem. 2015, 363, 33-56.
07. A. Otaka. et. al. ChemBioChem 2016, in press. (doi: 10.1002/cbic.201600400)

EVALUATION OF CYS RACEMIZATION DURING SOLID PHASE PEPTIDE SYNTHESIS UNDER MICROWAVE IRRADIATION

Anna Pantelia[1], Mqtthaia Ieronymaki[1], Maria Eleni Androutsou[2], Dimitrios Gatos[1], Theodore Tselios[1]
[1]University of Patras, Department of Chemistry, Rion Patras, 26504, Greece
[2]Eldrug S.A., Pharmaceutical Company, 26504, Platani, Greece

Introduction

Peptides are involved in many biological processes and pathways and are used for the rational design of potent molecules. Due to their low stability in proteolysis, peptides have been less popular candidates for therapeutic purposes. The configuration of each amino acid (except for Gly) influences the properties and biological activity of the synthetic peptides. In addition, specific microwave (MW) irradiation conditions are required for the synthesis of "difficult" peptide sequences to avoid undesired byproducts and achieve negligible racemization and high yield synthesis.
MW irradiation in solid phase peptide synthesis (SPPS) is based on the intensity of the thermal effects and not the intensity of the electromagnetic field. The polar peptide backbone continuously aligns with the alternating electromagnetic field, resulting in decrease in steric hindrance and reduction in chain aggregation that may accelerate the reactions during peptide synthesis.
In this study, the SPPS of peptides was achieved on the acid labile CLTR-Cl resin. The advantages of the CLTR-Cl resin are: (i) the fast reaction and the reduced racemization of the first amino acid attached to the resin via esterification, (ii) the prevention of the formation of diketopiperazine (DKP) during the synthesis of the first dipeptide and (iii) the cleavage of the synthesized protected peptide from the resin under mild acidic conditions without affecting the side chain protecting groups.

Results and Discussion

The Fmoc/tBu synthetic methodology was used for the synthesis of the model peptides using a combination of conventional SPPS and MW irradiation. SPPS was conducted for the coupling of the first four amino acid residues in the desired sequence at RT. The coupling reaction and Fmoc deprotection of D/L Cys was achieved under MW irradiation. The MW assisted coupling reactions were performed at two temperatures, 60°C and 75°C in the presence of Fmoc protected amino acids dissolved in DMF, while an excess of HOBt/DIC dissolved in DMF was used as coupling reagents. Effective HPLC analytical methodology was developed with the capability of efficient separating the L-Cys containing peptide from its enantiomeric D-isomer (Fig. 1).

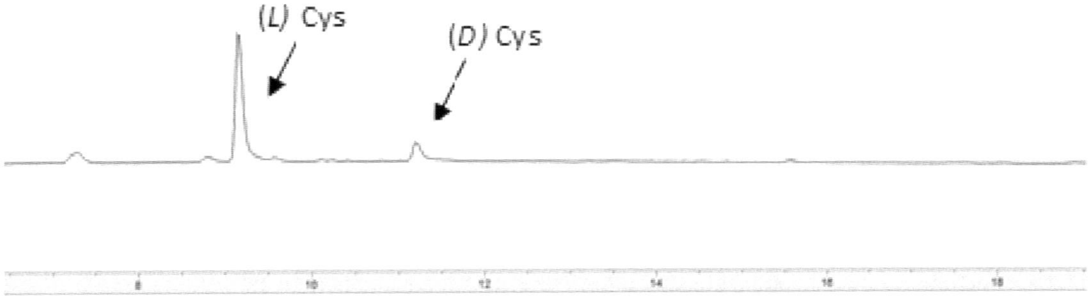

Figure 1

RP-HPLC chromatogram of model peptide H-(L)Cys-Leu-Phe-Ala-Gly-OH at 214nm. The coupling of Fmoc-(L)Cys(Trt)-OH and Fmoc deprotection were achieved using MW-SPPS (60°C, 34W). The main peak refers to the desirable peptide, containing L-Cys while only the 3% refers to the corresponding peptide with D-Cys.

RP-HPLC Conditions:

i) Column: HICHROM, Lichrosorb RP18-5 (250x4.6mm),

ii) Solvents: H_2O (0.08% TFA), ACN (0.08%TFA),

iii) Gradient elution: from 20% ACN to 80% ACN over 20min at RT

iv) Detection: UV at 214nm

Conclusion

The scope of this work was the evaluation of Cys racemization during SPPS under MW irradiation. The results from the RP-HPLC chromatogram of model peptides show that no significant racemization (<3%) was observed at 60°C. In contrary, at 75°C the rate of racemization was increased at 10%. Furthermore, the detectable racemization of Cys using the MW irradiation at 60°C was almost similar to that of the model peptides obtained from conventional SPPS (data not shown). This work demonstrates that the CLTR-Cl resin combined with MW-SPPS and the Fmoc/tBu methodology can be used for the synthesis of protected peptides with minimal level of racemization

Acknowledgment

We thank the organizing committee of 34th EPS for Travel Grand to AP. We are grateful to Eldrug SA for providing access to the CEM Liberty automated MW peptide synthesizer.

References

01. Chien, S.T.; Chiou, S.H.I.; Wang K.T. J. Chin. Chem. Soc., 1991,38, 85-91
02. Yu, H.M.; Chen, S.T.; Wang K.T. J. Org. Chem.,1992 57(18), 4781-4784).
03. Pedersen, A.L.; Tofteng, A.P.; Malik, L.; Jensen K.J. Chem. Soc. Rev., 2012, 41, 1826-1844.
04. Kappe, C.O. Acc. Chem. Res., 2013, 46, 1579-1587.
05. Barlos, K.; Gatos, D.; Schafer, W. Angew. Chem. Int. Ed Engl., 1991, 30, 590-593.
06. Friligou, I.; Papadimitriou, E.; Gatos, D.: Matsoukas, J.; Tselios, T. Amino Acids, 2011, 40(5), 1431-1440.
07. Ieronymaki, M.; Androutsou, M.E.; Pantelia, A.; Friligou, I.; Crisp, M.; High, K.; Penkman, K.; Gatos, D.; Tselios, T. Biopolymers, 2015, 104(5), 506-514.

INVESTIGATING RACEMIZATION IN HIS COUPLINGS IN SPPS

Jan Pawlas[1]
[1]PolyPeptide Laboratories AB, Limhamnsvägen 108, PO BOX 30089, 20061 Limhamn, Sweden, jap@polypeptide.com

PP I - 034

Introduction

The exceptional propensity of histidine to undergo side-chain assisted racemization renders His couplings a challenging endeavor in peptide chemistry[1]. Methods in art include using racemization resistant His derivatives[2] or coupling reagents[3] as well as coupling His at lower temperatures than used for other amino acids[4]. Nevertheless, cumbersome removal of protecting groups, inadequate chemoselectivities as well as low reaction rates still hamper efficient synthesis of His containing peptides.

Results and Discussion

Table 1. Fmoc-His(X)-OH vs coupling additives in GHF SPPS

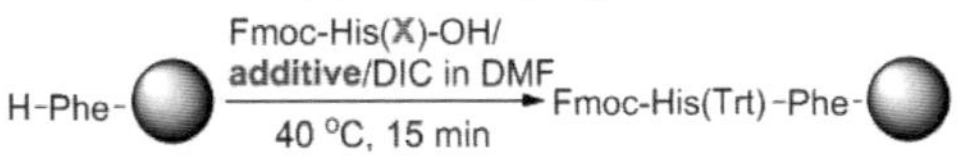

entry	additive	X	% D-His	% des-His	% endo-His
1	HOBtxH_2O	πMBom	**0.1**	0.1	0.2
2	Oxyma	πMBom	**0.2**	0.1	0.1
3	Oxyma-B	πMBom	**0.1**	0.1	0.1
4	HONB	πMBom	**0.1**	0.2	2.3
5	HOBtxH_2O	Boc	**0.1**	0.1	0.5
6	Oxyma	Boc	**0.1**	0.1	0.1
7	Oxyma-B	Boc	**0.1**	0.1	0.1
8	HONB	Boc	**0.1**	0.5	3.1
9	HOBtxH_2O	Trt	**1.1**	0.1	0.4
10	Oxyma	Trt	**3.6**	0.1	0.2
11	Oxyma-B	Trt	**1.0**	0.3	0.2
12	HONB	Trt	**0.2**	1.1	2.5
13	HOBtxH_2O	ClT	**1.0**	0.2	0.1
14	Oxyma	ClT	**3.5**	0.3	0.1
15	Oxyma-B	ClT	**0.4**	0.3	0.2
16	HONB	ClT	**0.3**	5.2	1.2

We set out to develop methods to couple His with minimal racemization regardless of temperature and without the need to hinge on difficult-to-deprotect His derivatives. To this end we screened various His derivatives vs coupling reagents to find AA/reagent combinations for further assessment. An example of such screen is shown in Table 1 in which four His(X) compounds vs four coupling reagents were used in a SPPS of GHF trimer[5]. The racemization resistant His(pMBom)[6] (entries 1 - 4) as well as His(Boc) (entries 5 - 8) both exhibited negligible racemization albeit scarce availability/tedious PG removal for the former and limited stability to piperidine for the latter[7] hinders the use of these His derivatives. Using His(Trt) (entries 9 - 12) racemization was unsatisfactory for all of the coupling additives except for HONB[8], which gave minimal D-His but poor chemoselectivity. It is worth noting that for Oxyma-B[9], racemization with His(ClT) was significantly lower than with His(Trt) (entries 11 vs 15) [10].We next examined solvent effects in a His(Trt)/HONB/DIC coupling (Table 2), which revealed that for a number of DMF/solvent mixtures a chemoselectivity improvement over neat DMF could be attained without increasing racemization (entries 2, 4, 6 - 8) [11]. Among the solvents which showed improvement over DMF we opted to further examine DMF/MeCN and DMF/EtOAc for their good resin swelling properties as well as suitable EHS characteristics (Table 3). Thus, we determined that i) altering DMF/MeCN (entries 1 - 3) and His/HONB ratios (entries 1, 4 - 5) from 1:1 is not beneficial ii) the amount of DIC used vs His/HONB appears to be of less importance (entries 1 vs 6) iii) addition of bases gives no improvement (entries 7 - 8). Next, in a MeCN vs EtOAc comparison all EtOAc runs gave slightly lower D-His contents than the corresponding MeCN experiments did (entries 9 - 16). With the exception of the shortest reaction times

Table 2. Solvent effects in GHF SPPS

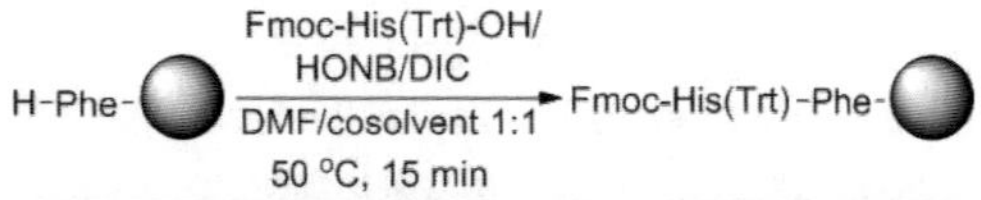

entry	cosolvent	% D-His	% des-His	% endo-His
1	none	0.3	1.1	1.2
2	MeCN	0.2	0.4	0.9
3	THF	0.3	19.0	0.4
4	EtOAc	0.2	0.5	0.8
5	DMSO	0.3	1.0	1.7
6	heptane	0.2	0.3	1.2
7	DCE	0.2	0.3	0.6
8	CPME	0.1	0.8	0.9

Table 3. Assessment of His/HONB/DIC couplings in DMF/MeCN and DMF/EtOAc as solvents in GHF SPPS[1]

entry	cosolvent	DMF/cosolvent	His/HONB	coupling time (min)	% D-His	% des-His	% endo-His
1	MeCN	1:1	1:1	1 x 15	0.3	0.6	0.9
2	MeCN	3:1	1:1	1 x 15	0.3	1.3	1.1
3	MeCN	1:3	1:1	1 x 15	0.6	1.0	1.3
4	MeCN	1:1	2:3	1 x 15	0.9	0.4	1.1
5	MeCN	1:1	2:1	1 x 15	27.9	2.9	0.6
6[2]	MeCN	1:1	1:1	1 x 15	0.3	0.7	1.0
7[3]	MeCN	1:1	1:1	1 x 15	0.3	0.4	1.3
8[4]	MeCN	1:1	1:1	1 x 15	1.4	0.7	1.1
9	EtOAc	1:1	1:1	1 x 15	0.2	0.4	0.5
10	MeCN	1:1	1:1	1 x 15	0.3	0.2	0.9
11	EtOAc	1:1	1:1	1 x 30	0.2	0.1	0.5
12	MeCN	1:1	1:1	1 x 30	0.3	0.1	0.6
13	EtOAc	1:1	1:1	2 x 15	0.2	0.1	0.9
14	MeCN	1:1	1:1	2 x 15	0.3	0.1	1.8
15	EtOAc	1:1	1:1	2 x 30	0.2	0.1	1.1
16	MeCN	1:1	1:1	2 x 30	0.3	0.1	1.6

[1] Carried out at 50 °C using Fmoc-His(Trt)-OH/HONB/DIC; [2] Amount of DIC reduced twofold; [3] 20 mol% DIEA added; [4] 20 mol% NMM added

(entries 9 - 10), coupling rates for EtOAc vs MeCN as cosolvents were quite comparable. Having found useful conditions for low racemization His couplings we set out to suppress the content of endo-His impurity (~ 0.5 – 3.0% using HONB/DIC). As HONB/DIC couplings proceed at relatively high pH (pH ~ 6.5), we attempted to suppress endo-His by addition of acids (Table 4). The addition of TsOHxH_2O did result in a significant endo-His reduction although increasing the amount of the strong acid also led to the increase in the content of D and des-His impurities (entries 1 - 4). Among the other additives which we tested (entries 5 - 7) HOBt hydrate gave the best result. Replacing DIC with EDCxHCl suppressed endo-His however both D and des-His were increased (entry 8). Next, we tested scalability of our His coupling conditions by carrying out multigram syntheses of the test GHF trimer. These experiments afforded crude peptides of minimal His racemization and low des- and endo-His content (Table 5)[12]. Finally, we examined His couplings with the sterically highly encumbered H-Aib-Phe resin. Using a His(Trt)/HONB/DIC protocol afforded a slightly lower conversion (~ 96%) than in the SPPS of GHF, however upon examining conditions from Table 1 we found that His(Boc)[13]/Oxyma/DIC couplings give negligible racemization as well as low contents of des and endo His impurities (Scheme 1).

In summary, we have developed protocols that enable efficient couplings of easy-to-deprotect His derivatives at elevated temperatures essentially without racemization [14].

Table 4. Effect of pH on His/HONB/DIC coupling in GHF SPPS[1]

entry	pH modulator	carbodiimide	% D-His	% des-His	% endo-His
1	none	DIC	0.2	0.2	1.9
2	5 mol% $TsOHxH_2O$	DIC	0.4	0.2	0.5
3	15 mol% $TsOHxH_2O$	DIC	1.0	1.2	0.3
4	30 mol% $TsOHxH_2O$	DIC	1.4	2.9	0.3
5	15 mol% CSA	DIC	0.6	1.0	0.3
6	15 mol% $HOBtxH_2O$	DIC	0.3	0.2	0.4
7	15 mol% Oxyma-B	DIC	0.6	0.2	0.5
8	none	EDCxHCl	0.5	1.6	1.2

[1] Carried out at 50 °C for 15 min using Fmoc-His(Trt)-OH/HONB/DIC in DMF/EtOAc (1:1)

Table 5. Gram-scale His/HONB/DIC couplings in GHF SPPS[1]

temperature	time	**% D-His**	% des-His	% endo-His
40 °C	40 min	**0.1**	0.1	0.3
50 °C	30 min	**0.2**	0.1	0.3

[1] Carried out using 3 g of a H-Phe resin and Fmoc-His(Trt)-OH/HONB/DIC/15mol%$HOBtxH_2O$ in DMF/EtOAc (1:1)

Scheme 1. Assessment of His coupling in SPPS of H-His-Aib-Phe-NH_2

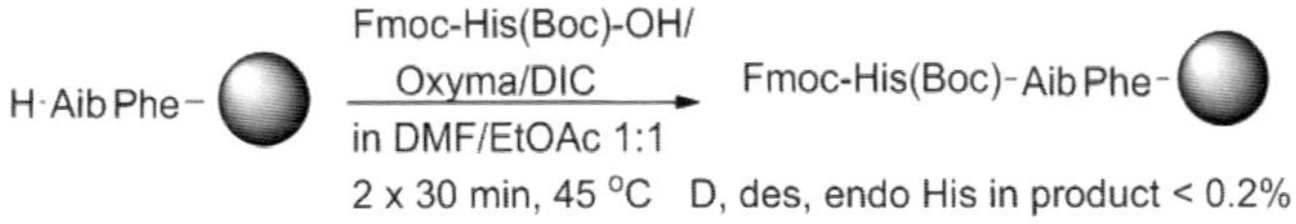

References

01. Isidro-Llobet, A.; Álvarez, M.; Albericio, F. Chem. Rev. 2009, 109, 2455.
02. Hibino, H.; Nischiuchi, Y. Tetrahedron Lett. 2011, 52, 4947.
03. El-Faham, A.; Albericio, F. Chem. Rev. 2011, 111, 6557.
04. Bacsa, B.; B sze, S.; Kappe, C. O. J. Org. Chem., 2010, 75, 2103.
05. Van Den Nest, W.; Yuval, S.; Albericio, F. J. Pept. Sci. 2001, 7, 115.
06. Hibino, H.; Miki, Y.; Nischiuchi, Y. J. Pept. Sci. 2012, 18, 763.
07. Sieber, P.; Riniker, B. Tetrahedron Lett. 1987, 28, 6031.
08. Fujino, M.; Kobayashi, S.; Obayashi, M.; Fukuda, T.; Shinagawa, S.; Nishimura, O. Chem. Pharm. Bull. 1974, 22, 1857.
09. Jad, Y. E.; Khattab, S. N.; de la Torre, B. G.; Govender, T.; Kruger, H. G.; El-Faham, A.; Albericio, F. Org. Biomol.Chem. 2014, 12, 8379.

10. We determined that Fmoc-His(ClT)-OH deprotection (using TFA/TIS/H_2O, 90:5:5) was markedly slower ($t_{1/2}$ > 2 h) than deblocking Fmoc-His(Trt)-OH ($t_{1/2}$ < 10 min). For kinetics of removal of side chain bound His derivatives from differently substituted trityl resins see: Eleftheriou, S.; Gatos, D.; Panagopoulos, A.; Stathopoulos, S.; Barlos, K. Tetrahedron Lett. 1999, 40, 2825.
11. For the crude GHF obtained in Table 2, entry 6 experiment D-His content by chiral GC-MS analysis was 0.13 %.
12. Identities of D, des and endo His impurities respectively were confirmed by spiking with authentic samples of these peptides. Conditions for HPLC analyses: detection at 220 nm, XSelect CSH130 C18 2.5m 4.6x150mm stationary phase, TFA/H_2O/MeCN mobile phases.
13. Due to the limited stability of His(Boc) to piperidine (ref 7) it is advisable to use less nucleophilic bases in Fmoc deprotections see e.g. Ralhan, K,; KrishnaKumar, V. G.; Gupta, S. RSC Adv. 2015, 5, 104417.
14. Experimental: Knorr and tricyclic amide linkers respectively coupled to PS/1%DVB AM resins were used throughout. Piperidine was used for deFmocings except for His(Boc) containing resins (DBU). Unless noted Oxyma/DIC was used for couplings. In His couplings, His/coupling additive/DIC ratio of 1:1:3 was used (50% of DIC added t=0 and t=50% of coupling time, respectively). Resin cleavages were done using TFA/TIS/H_2O (90:5:5) for 1 h followed by ether precipitation. $MeONH_2$xHCl was added to cleavages of His(pM-Bom) containing resins.

PREPARATION OF DEUTERATED ANALOGS OF PEPTIDES WITH THE 5-AZONIASPIRO[4.4]NONYL-CARBONYL IONIZATION TAG

Bartosz Setner, Magdalena Wierzbicka, Marek Lisowski, Zbigniew Szewczuk
Faculty of Chemistry, University of Wroclaw, F. Joliot Curie 14, Wroclaw, Poland

PP I - 042

Betaine derivatives of peptides allow efficient analysis of their trace amounts by electrospray-mass spectrometry.[1] The synthesis of isotopologues of analytes, which can be used for quantitative analysis in mass spectrometry by isotopic dilution [2], is an important goal of modern analytical methods. The deuterons, which are introduced at the α-carbon of betaine derivatives, are resistant to the back-exchange in acidic aqueous solution.[3] 5-Azoniaspiro[4.4]nonyl-carbonyl (ASN^+CO) and benzazoniaspiro[4.4]nonyl-carbonyl ($BASN^+CO$) [4] are examples of cyclic quaternary ammonium salts, formed from the proline residue, which increase the ionization efficiency. Our goal was to examine the hydrogen-deuterium exchange (HDX) of the azoniaspiro systems and its application to isotopic tagging for peptides and proteins analysis. We also investigated the racemization of the derivatized proline residue.

Model peptide H-MQIFVKT-OH as well as its two derivatives, ASN^+CO-MQIFVKT-OH and $BASN^+$-CO-MQIFVKT-OH, were synthesized using a standard Fmoc procedure and the quaternarization of the N-terminal proline residue was performed according to the method developed by Setner et al.[4] (Fig.1). Analysis of the hydrogen-deuterium exchange in 1% TEA/D_2O solution followed by the back-exchange in H_2O was performed, using electrospray mass spectrometry.

*The samples of ASN^+CO-MQIFVKT-OH were incubated for different periods in 1% TEA/D_2O, followed by lyophilization and redissolving in H_2O. Next, the samples of ASN^+ (d_1)CO-MQIFVKT-OH were incubated in 1% TEA/H_2O, followed by lyophilisation and redissolving in H_2O (Fig. 2). During the HDX all labile hydrogens exchange to deuterons and undergo DHX after lyophilisation and redissolving in water but the deuteron at α-C of the azoniaspiro system does not back-exchange in such conditions. The HDX of the α-C hydrogen in 1% TEA/D_2O is completed within 30 minutes but the DHX in 1% TEA/H_2O is slower and takes over 3 hours. The ionization enhancers facilitate the peptides sequencing since the N-terminal fragmentation non-protonated *a and *b ions series are present in the MS/MS spectra of peptides. The coelution of the ASN conjugate (m/z = 509.294 for $[M+H]^{2+}$) and its isotopologue (m/z = 509.797 for $[M+D]^{2+}$) was confirmed by LC-MS analysis (Fig. 3). The racemization of the proline derivatives was observed during the tested HDX conditions (1% TEA).*

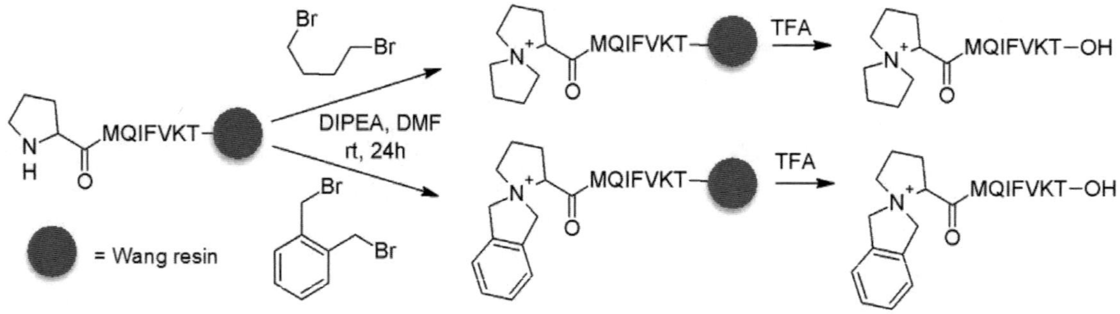

Fig. 1 The scheme of synthesis of ASN+CO and BASN+CO peptide derivatives

In conclusion, the 5-azoniaspiro[4.4]nonylcarbonyl and benzoazoniaspiro[4.4]nonylcarbonyl ionization enhancers facilitate the peptides sequencing by MS/MS analysis. The hydrogen at the C-α of the derivatized proline residue undergoes HDX in 1% TEA in D_2O and it does not back-exchange in water solution. We developed an inexpensive method of isotopic labelling of the 5-azoniaspiro[4.4] nonylcarbonyl group which may be useful for the qualitative analysis of peptides by mass spectrometry using the isotopic dilution method. The developed ionization tag is the first known example of a proline derivative that undergoes racemization in the presence of 1% TEA.

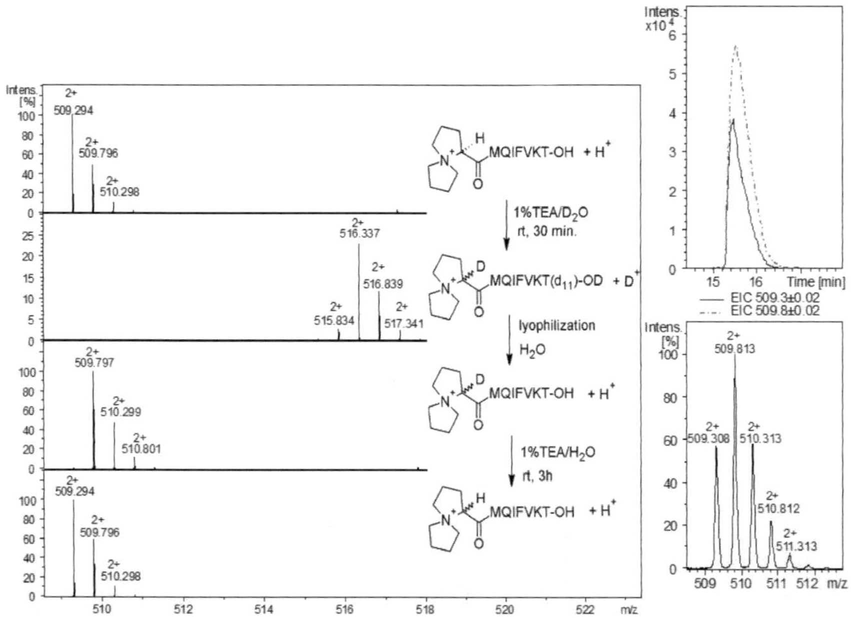

This work was supported by a grant UMO-2013/09/B/ST4/00277 from the National Science Centre, Poland.

References

01. P. Stefanowicz, A. Kluczyk, Z. Szewczuk, Amino Acids, Pept. Proteins, 2016, 40, 36-74.
02. M. Rudowska, D. Wojewska, A. Kluczyk, R. Bachor, P. Stefanowicz, Z. Szewczuk, J. Am. Soc. Mass Spectrom., 2012, 23, 1024-1028.
03. R. Bachor, M. Rudowska, A. Kluczyk, P. Stefanowicz, Z. Szewczuk, J. Mass Spectrom. 2014, 49, 529-536
04. B. Setner, M. Rudowska, D. Wojewska, A. Kluczyk, P. Stefanowicz, Z. Szewczuk, J. Pept. Sci. 2014, 20, S64-S65.

SYNTHESIS AND TRANSFORMATIONS OF 1,3-DIYNE CONTAINING TETRAPEPTIDES

Steven Verlinden, Steven Ballet, Guido Verniest
Research Group of Organic Chemistry, Vrije Universiteit Brussel (VUB), Pleinlaan 2, B-1050, Brussels, Belgium

PP I - 046

Introduction

Alkyne-alkyne coupling reactions are currently being evaluated in peptide chemistry as promising tools to cyclise or functionalise peptidic backbones.[1-4] Cyclic peptides containing a 1,3-diyne moiety can be prepared via either the derivatization of peptide side chains with bis-functionalized 1,3-diynes[5] or via Glaser-Hay type oxidative alkyne-alkyne coupling reactions of α,ω-dialkynylated peptides.[1,2] To date there are no examples in literature known where post-cyclisation transformations of the 1,3-diyne linker in a macrocyclic peptide have been applied. Interestingly, starting from one parent structure the reaction of such diynes with nucleophiles could lead to a set of different cyclic peptides that contain a heterocyclic tether (e.g. furans, thiophenes, pyrazoles, isoxazoles), which in turn could be of value in SAR studies.

Results and Discussion

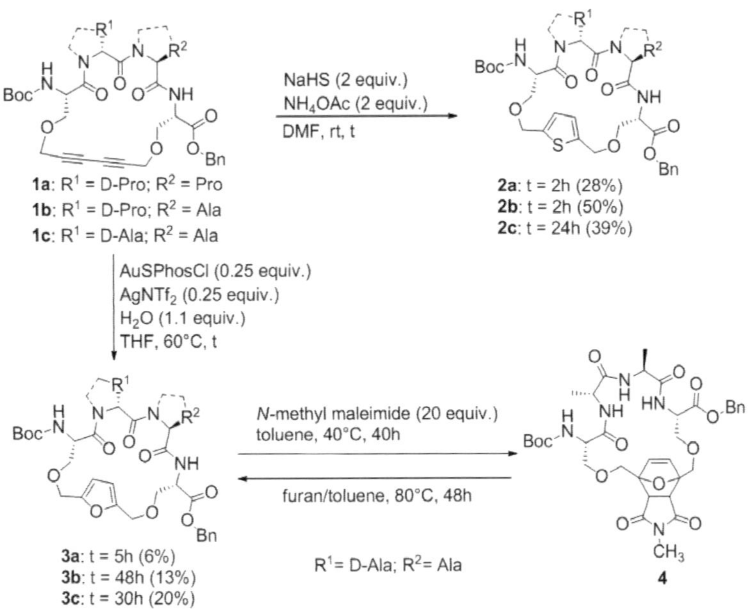

Scheme 1. Synthesis of thiophene **2a-c** and furan derivatives **3a-c** and cycloadduct **4**.

In a first evaluation to transform macrocyclic diynes to the corresponding thiophenes (Scheme 1), macrocyclic peptides 1a-c were treated with 2 equivalents of NaHS and NH_4OAc in DMF at room temperature and resulted in a clean conversion (70-92%) towards thiophene derivatives 2a-c in moderate yields (28-50%). In a next transformation of macrocyclic 1,3-diynes 1a-c, water was evaluated as nucleophile. Without activation of the alkyne moiety no reaction took place even at elevated temperatures. However, when $SPhosAuNTf_2$ was used as an alkyne activating catalyst, the desired furans 3a-c were obtained, even though more side products were formed, as compared to the formation of thiophenes 2a-c. The introduction of furans into macrocyclic peptide structures, via transformation of the 1,3-diyne linker, can expand the current toolbox for reversible furan-based peptide labelling techniques. Having peptide-bridged furans 4 in hands, Diels-Alder reactions with maleimides were therefore evaluated in a model study. After heating macrocycle 3c in toluene at 40°C in the presence of an excess of N-methyl maleimide, cycloadducts 4 (both diastereomers) were obtained after 40h in a nearly peak-to-peak HPLC conversion. Subsequently, the obtained and isolated compounds 4 were heated to 80°C to afford the retro-Diels-Alder adduct. This conversion provides a proof-of-principle for the reversible linkage of macrocycle-bridged furans of type 3c without affecting the peptidic linker, and opens a gateway for labelling studies at a site distant from the potential recognition domain within peptide macrocycles.

Scheme 2. Synthesis of pyrazole derivatives **5**, **6** and isoxazoles **7**. [a]isolated as mixture of two regioisomers (1:1 ratio); [b]isolated as mixture of three regioisomers (3:1:1 ratio); [c]isolated as mixture of two regioisomers (2:1 ratio)

All the above described transformations give rise to symmetrically substituted heterocycles, whereas the use of hydrazines or hydroxylamines can result in different regioisomers depending on whether position 1 or 4 of the 1,3-diynes is first attacked by these bisnucleophiles. To evaluate if the nature and conformation of the macrocycles could direct the incoming nucleophile and give regioselective reactions with bisnucleophiles, compound 1c was treated with hydrazine, hydroxylamine and N-methylhydrazine (Scheme 2). In these cases no or low regioselectivity was observed and mixtures of pyrazole regioisomers 5 and 6, and isoxazoles 7 were obtained. Pyrazoles 5 and isoxazoles 7 could not be separated and were obtained as a mixture of two regioisomers in a 1:1 and 2:1 ratio respectively, as shown by NMR. The reaction of N-methyl hydrazine with 1c gave rise to a mixture of three regioisomers in a 3:1:1 ratio. The reason for the observed lack of selectivity is most probably due to the quasi-symmetric nature of the 1,3-diyne moiety in substrate 1c. This could change drastically when 1-aryl-4-alkyl-1,3-diynes are used since an electronic bias (aryl conjugation) is present in such cases.

To verify this, the synthesis of macrocyclic substrates 9a-c was envisaged (Scheme 3). Instead of using two propargylated serine residues, one serine residue was exchanged with an ethynylated

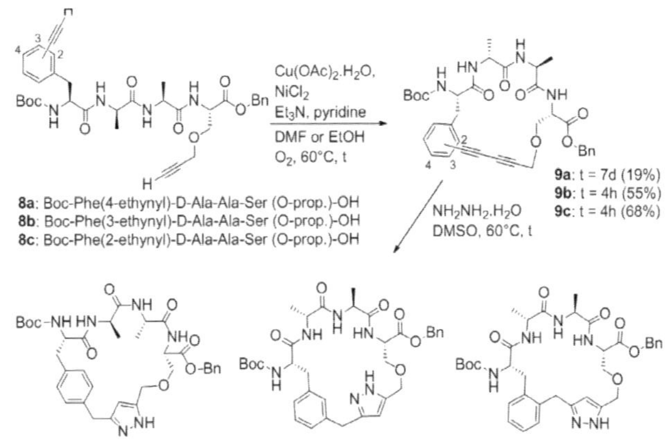

Scheme 3. Synthesis of cyclic peptides 9a-c and pyrazoles 10a-c. [i]peptide 10a was not purified

phenylalanine residue. These ethynylated phenylalanine residues were obtained after Sonogashira coupling of 2-, 3- and 4-iodinated Boc-Phe-OMe with TMS-acetylene, followed by a simultaneous silyl deprotection and saponification with NaOH. The ethynylated phenylalanine derivatives were then used as a first amino acid in a solution phase peptide synthesis strategy. Coupling of ethynylated Boc-Phe-OH with H-D-Ala-OMe using HATU in DMF and subsequent saponification afforded the first dipeptides, which were then coupled with H-Ala-Ser(O-propargyl)-OBn using DIC/HOAt in DMF, and resulted in linear tetrapeptides 8a-c in good yield (67-82%). Next, these were then cyclised to corresponding peptides 9a-c using the optimized Glaser-Hay conditions.[1] Arylated diynes 9a-c were then treated with hydrazine in DMSO for 15h at 60°C, which fortunately resulted in the formation of single regioisomers 10a-c as observed by HPLC and NMR. This is in agreement with the expected reactivity where the first nucleophilic attack occurs at the 4-position of a 1-aryl-1,3-diyne.

In conclusion, the synthesis of various different peptide macrocycles with heterocycle-bearing tethers (thiophene 2, furan 3, cycloadduct 4, pyrazole 5, N-methyl pyrazole 6 and isoxazole 7) was realized using mild reaction conditions. In addition, it is shown that a non-symmetric arylated 1,3-diyne linker (as in 9a-c) gives rise to the regioselective formation of the corresponding pyrazoles (10a-c) upon treatment with hydrazine. Further application of these methods to larger peptide sequences and the potential for reversible linking of such macrocycles via Diels-Alder reactions is currently being evaluated and will be reported in due course.

Acknowledgments

S.V., S.B. and G.V. acknowledge the Vrije Universiteit Brussel (VUB) for financial support.

References

01. S. Verlinden et al., Org. Biomol. Chem., 2015, 13, 9398-9404.
02. N. Auberger et al., Bioorg. Med. Chem., 2014, 22, 6924-6932.
03. J.C. Maza et al., Chem. Comm., 2016, 52, 88-91.
04. J.S. Lampkowski et al., Angew. Chem. Int. Ed., 2015, 54, 9343-9346.
05. K. Fujimoto et al., Chem. Eur. J., 2008, 14, 857-863.

PALLADIUM-CATALYSED DERIVATISATION OF PEPTIDES IN AN AQUEOUS ENVIRONMENT

T. Willemse[1,2], H. W.T. Van Vlijmen[3], W. Schepens[3], B. U. W. Maes[2], S. Ballet[1]
[1]Departments of Chemistry and Bio-engineering Sciences, Vrije Universiteit Brussel, Pleinlaan 2, B-1050 Brussels.
[2]Organic Synthesis, University of Antwerp, Groenenborgerlaan 171, B-2020 Antwerp.
[3]Janssen Pharmaceutica, CREATe, Turnhoutseweg 30, B-2340 Beerse.

PP I - 050

Introduction

Peptide therapeutics received significant interest in the past decades. However, limitations of these potential drugs include fast clearance rates, low bioavailability and high conformational flexibility.[1] Medicinal chemists have provided several solutions to overcome the problems associated to their intrinsic properties, including cyclisation and the replacement of natural amino acids. In this respect, bioorthogonal palladium-catalysed reactions were introduced as a macrocyclisation alternative to classical lactam- or disulfide bridges, and to generate a wide array of derivatised amino acid/peptide analogues.[2] Within peptide sequences the late-stage derivatisation of halophenylalanine and halotryptophan residues,[3] amino acids that play a significant role in peptide-protein recognition, allow the establishment of SAR-studies and provide valuable information to improve peptide lead compounds. During initial studies on unprotected dipeptide fragments we have shown that high functional group tolerance was achieved by using a ferrocene-based phosphine catalyst 1, $PdCl_2$(dppf), in mixed aqueous conditions (iPrOH/H_2O).[4] Based on these results, we focused our efforts on the preparation of tetrapeptide analogues of H-Dmt-D-Arg-Phe-Gly-NH_2, with potential antinociceptive activity. The commercially available Fmoc-protected iodinated phenylalanine building blocks were inserted and optimisation of Suzuki-Miyaura cross-coupling allowed access to a diverse set of opioid ligands. Furthermore, in view of cyclic peptides, the incorporation of orthogonally protected biaryl-containing dipeptide building blocks was examined, which would present cyclic peptides after a final lactamisation.

1a

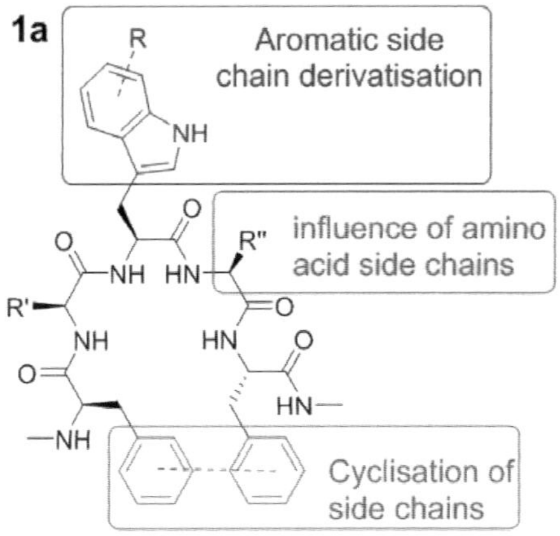

1b

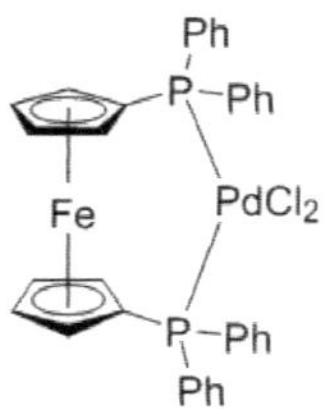

1
$PdCl_2$(dppf)

Fig 1a: Strategies for improve-ment of aromatic moieties of peptide leads. 1b: PdCl2(dppf) catalyst

Results and discussion

In this work, the derivatisation of a short peptide was envisaged by insertion of a halophenylalanine residue. The tetrapeptide was first derivatised in solution, but due to limited conversion and hydrolysis of the C-terminal amide, our strategy was altered to a full synthesis on solid support. The peptide was prepared following standard Fmoc-based protocols on Rink Amide resin, to give Boc-Dmt-D-Arg(Pbf)-Phe(I)-Gly-NH-Rink Amide resin 2a-c with the iodine on the respective p-, m- or o-position (Fig 2). Based on the swelling properties of the polystyrene-based resin, the initial Suzuki-Miyaura arylations were carried out in pure organic solvent (DMF), but even in the presence of organic bases such as Et_3N low conversions were attained. Gratifyingly, by changing the solvent system to a biphasic mixture of THF and H_2O in 1:1 ratio with $PdCl_2$(dppf) (1) in a catalytic amount (10 mol%) complete conversions to the expected biaryl products were obtained (Fig 2). With this methodology it was possible to introduce a broad range of (hetero)aryl groups on the 2, 3 or 4-position (Fig 2, in blue). Previous results[3] also showed that vinylation could be achieved on the amino acid level. We therefore attempted similar conditions for the vinylation of the peptides 2a-c. However, following the acidic cleavage from the resin only side products were obtained. Repeating the experiment on highly acid-labile Sieber resin (1% TFA in CH_2Cl_2) it was shown that the vinylation reaction did occur, indicating that treatment with 95% TFA and scavengers is responsible for the degradation of the derivatized peptide. To our delight, solution-phase derivatisation with vinyltrifluoroborate afforded the envisaged peptides (Fig 2, purple). The reaction time of the vinylation was limited compared to the arylation (2h vs 24h), thus avoiding partial hydrolysis of the C-terminus. Combination of both solid- and solution-phase derivatisation, it was possible to generate 14 analogues 3a-n that will be tested for biological activity at the opioid GPCRs.

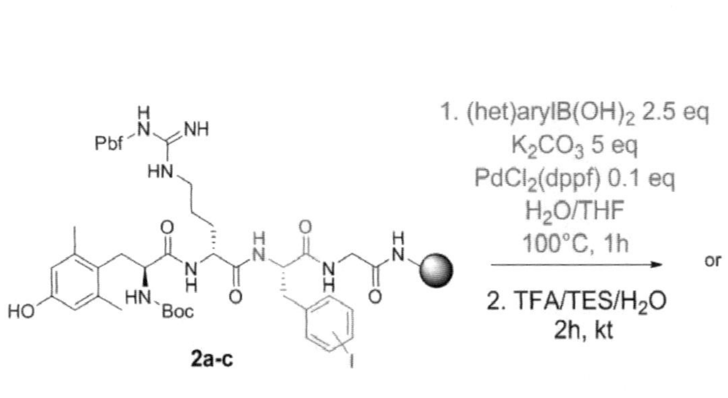

Fig 2: Derivatisation of tetrapeptide H-Dmt-D-Arg-Phe(X)-Gly-NH2 on solid-phase or solution-phase

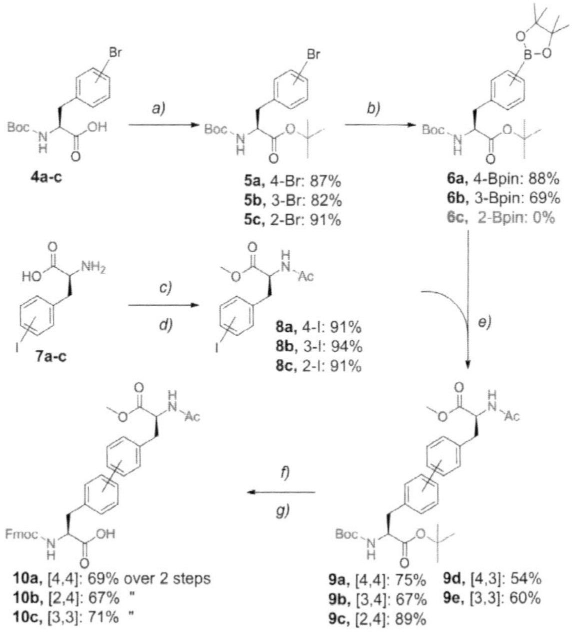

a) 2,2,2-tertbutyltrichloroacetamidate 3 eq, $BF_3.OEt_2$ 0.3 eq, CH_2Cl_2, 2h, rt; *b)* B_2pin_2 1.1 eq, $PdCl_2(dppf)$ 0.06 eq, dppf 0.06 eq, KOAC 3 eq, DMSO,16h, 90°C; *c)* $SOCl_2$ 2.5 eq, MeOH, 16h, 45°C; *d)* Ac_2O 3.3 eq, pyridine, 6h, rt; *e)* $PdCl_2(dppf)$ 0.05 eq, K_2CO_3 5 eq, DMSO, 16h, 80°C; *f)* TFA/DCM 1:1, 3h, rt; *g)* FmocOSu 1.1 eq, $NaHCO_3$ 1.2 eq, THF/dioxane 1:1, 16h, rt

Fig 3: Synthesis of the orthogonally protected biaryl-building block for application in peptide cyclisations

Next, we turned towards cyclic peptides. Biaryl-building blocks that are readily incorporated in Fmoc-based SPPS (see Fig 3, 10a-c) were synthesized. These building blocks were prepared according to a sequential Miyaura-borylation and Suzuki-Miyaura cross-coupling, starting from commercially available halophenylalanine derivatives 4a-b and 7a-c. By alteration of the position of the halo/boryl-substituent it is possible to prepare regioisomers with different biaryl-patterns, potentially leading to a distinct secondary structures or structural motifs after incorporation and lactam formation. This cyclisation can be realised by on-resin hydrolysis of the methyl ester[5] present in 10a-c and subsequent activation of the free carboxylic acid to undergo lactamisation by attack of the peptide's N-terminal amine. This method allows the final lactamisation step to be performed on-resin which should benefit from pseudo-dilution effects. However, it was shown that by use of Fmoc-Phe[(4-(Ac-Phe(4-)-OMe)]-OH (10a), leading to final products with high ring strain,
a mixture of inseparable dimers/oligomers were obtained. Gratifyingly, with the less linear dipeptide Fmoc-Phe[(4-(Ac-Phe(2-)-OMe)]-OH (10b) cyclisation proceeded with high efficiency towards monomeric cyclic products. The influence of the regio-isomers on cyclisation efficiency is currently under investigation. By developing a method for solid- or solution-phase derivatisation for short peptide ligands, combined with the cyclic peptide strategy, we expect to enhance the pharmacological properties of natural peptides.

Acknowledgements

The IWT Flanders and Janssen Pharmaceutica are acknowledged for providing the financial support of T.W. The CGB-CBB initiative is thanked for covering the travel expenses and the EPS registration fee of T.W.

References

01. Vlieghe, P. et al., Drug Discov. Today 15 (2010) 40-56, Fosgerau K. et al., Drug Discov. Today 20 (2015) 122.
02. Boutureira O et al., Chem. Rev. 115 (2015) 2174-2195.
03. Smith, D. R. M., Willemse, T. et al., Org. Lett. 16 (2014) 2622-2625.
04. Willemse, T. et al., Chemcatchem 7 (2015) 2055-2070.
05. Cantel S. et al., J. Peptide Sci.10 (2004), 326-328.

Sequeira, M. A.; Dodero, V. I
Depto de Química- INQUISUR, Universidad Nacional del Sur-CONICET, Bahía Blanca (AR).

Introduction

Numerous self-assembled proteins are physiologically active and others have a pathological role in some diseases. The most significant case involving supramolecular organization changes associated with disease is the formation of amyloid "type" oligomers and fibers implicated in the development of Alzheimer's disease, Parkinson's disease and type II diabetes, among others. [1, 2] Taking into account the increased prevalence of protein aggregates in human diseases, much effort is directed towards the understanding of the molecular mechanisms of self-assembly and to develop agents, which inhibit or modulate the supramolecular assembly. [3, 4] In this context, directed assembly has emerged as a powerful strategy to obtain structures under non-equilibrium conditions. In this case, external forces direct the self-organized process. [5] Short peptides have been widely used as model systems to study biological self-assembly processes due to their structural simplicity. [6] Gazit et.al have defined minimal homo-aromatic dipeptide fragments, which mediate the molecular recognition and self-organization process in minor scale, foregoing polypeptide aggregation. [7] Herein, we present the directed assembly of Ac-Phe-Phe-Cys-NH_2 in water and its supramolecular organization depending on a reductive or oxidative environment.

Experimental Procedure

Synthesis: Manual peptide solid phase synthesis using Fmoc / tBu methodology and a Rink amide resin was performed. Peptide was cleaved from resin, precipitate in cold ether and pellet was suspended in water and lyophilized. ^{1}HNMR (500MHz), MS-ESI, MS-MALDI-Tof and MS-ESI-HR confirm the identity and purity of Ac-Phe-Phe-Cys-NH_2.

Results and Discussion

The directed assembly in water at pH 8.0 was performed from a concentrated solution of Ac-Phe-Phe-Cys-NH_2 in Hexafluoroisopropanol (HFIP) (1,37mM). HFIP is a fluorinated alcohol with acidic character (pKa = 9) which is mostly used to dissolve β-amyloid peptides for their study and analysis. [8] Due to its electrophilic character, it leads up hydrophobic stacking interactions with peptides, being favored by the two CF_3 groups.

In a first experiment, assembly was triggered by adding a Ac-Phe-Phe-Cys-NH_2 HFIP stock solution via syringe into water at pH 8.0 (oxidative conditions, Figure 1). Then, in a second experiment the peptide:HFIP solution was added to water at pH 8.0 with tris(2-carboxyethyl)phosphine (TCEP) (reductive condition, Figure 1). Both self-organized systems were monitored at different time points by Uv-Vis spectroscopy and electron microscopy techniques (SEM and TEM) at different time points.

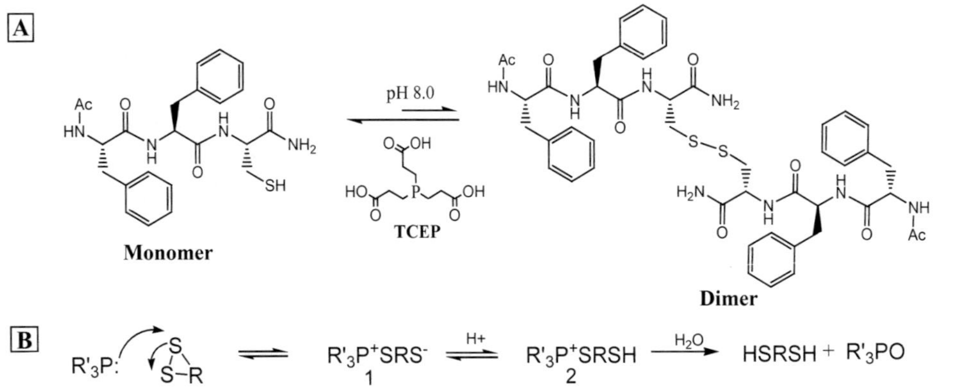

Figure 1. A) Oxidation-Reduction of Ac-Phe-Phe-Cys-NH_2 B) Mechanism of reduction of disulfide bond mediated by TCEP.

Simple time dependent UV-Vis spectroscopy was useful to calculate the aggregation index (AI). The AI index is calculated considering Phenylalanine (Phe) maximum absorbance (λ_{max}: 258 nm, corresponding to ∏→∏* transitions) relative to 350-400 nm region of the spectrum. This A_{350}/A_{258} relation is distinctive of aggregates formation. [9] According to this index the following scale is utilized: between 0 and 2: the system is a solution; between 2 and 5: some aggregates are formed; >5: high aggregation occurs. Under oxidative aqueous conditions, the oligomerization process is not only promoted by hydrophobic effect but also favored by peptide covalent dimerization through disulfide bond (Figure 1A). At t0 the AI is 24.5% and increases to 40.3% after t24 h incubation. No precipitation was observed at all-time scales (10 min. to 24 h). We hypothized that the formation of covalent disulfide dimers which increases the inherent molecular hydrophobicity is the driving force of the high aggregation kinetics. The second derivative analysis of the spectrum provided information of the molecular microenvironment of the Phe. [9] It was observed a slightly blue shift of the minimum absorbance of Phe from 257.8 nm (10 min.) to 257.4 nm (24 hrs.), indicating solvent-exposed chromophores through time. An increase in the polarity of the chromophores microenvironment led us to hypothesize that the disulfide bonds formation induced the exposure of Phe to basic aqueous media. [10]

Under reductive conditions, the initial AI was 15% (10 min) which slightly increase to 22% after 24 hrs. Again, precipitation or pellet formation was not observed. However, a marked red shift of the Phe band was detected from 261.6 to 262.2 nm, showing a decrease in the polarity of Phe microenvironment, which means that water molecules are excluded, and the aggregates are more compact. Under reductive conditions, oligomerization occurs via monomer ∏-stacking interactions, probably polar cysteine groups are more exposed to the solvent, and thus Phe are more buried in the supramolecular structure.

Electron microscopy observation confirms our solution experiments, detecting morphological differences between the superstructures obtained under oxidative or reductive conditions. Under oxidative condition, the clear solution primarily contained spheres of 34 nm diameter average size at t10 min. Then, concurrently with the appearance of elongated assemblies similar to nanofibrils (ranging from 8 to 21 nm of width), spheres became extinct (30min). Finally, after 24 hrs. the fibrils could further be organized into bundled nanofibrils, which were stabilized in the solution by water molecules (Figure 2).

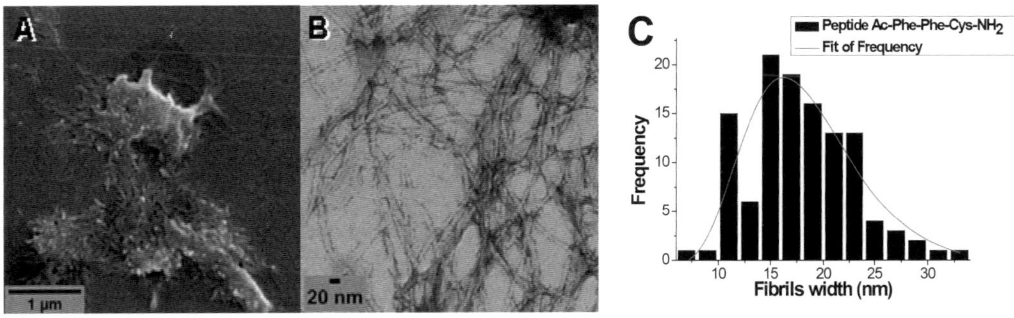

Figure 2. A) SEM microphotograph of Ac-Phe-Phe-Cys-NH_2 fibrils at pH 8.0 after 24hs. B) TEM of fibrils of Ac-Phe-Phe-Cys-NH_2 at pH 8.0 after 24hs. C) Average fibrils width obtained from Panel B

In the reductive condition experiments, we observed different types of nanofibrils instead of spherical oligomers and bundled fibrils. Fibrils grew through time to 62 nm width forming clusters at 24 hrs., as it was observed by SEM. Further experiments are in progress to validate our observations and to propose a mechanism of oligomerization.

Acknowledgements

Supported by Universidad Nacional del Sur, Argentina.

References:

01. D. Eisenberg, M. Jucker, Cell. 2012, 148, 1188.
02. M. G. Herrera. et.al. Soft Matter 2015, 11, 8648.
03. Z. Wu, Y. Yan, J. Huang Langmuir, 2014, 30, 14375.
04. Z. Quirolo, L. Benedini, A. Sequeira, G. Herrera, T. Veuthey, V. Dodero, Curr Top Med Chem., 2014, 14, 730.
05. M. A. Sequeira, et al. RSC Adv., 2016, 6, 108132.
06. L. Adler-Abramovich, E. Gazit, Chem. Soc. Rev., 2014, 43, 6881.
07. M. Reches, E. Gazit, Phys. Biol. 2006, 3 (1), S10-9.
08. L. Malavolta, et al. Protein Science: A Publication of the Protein Society. 2006, 15, 1476.
09. R. Esfandiary, C. R. Middaugh, in Analysis of Aggregates and Particles in Protein Pharmaceuticals (Eds: H-C. Mahler, W. Jiskoot) John Wiley & Sons, Inc. 2012, chapter 8.
10. A. Saunders, G. Young, G. Pielak, Protein Science 1993, 2, 1183.

INJECTABLE PEPTIDE HYDROGELS FOR CONTROLLED DRUG RELEASE

Oyen E.,[1,2,3] Martin C.,[1] Hernot S.,[4] Gardiner J.,[5] Van Mele B.,[2] Madder A.,[6] Hoogenboom R.,[3] Spetea M.,[7] Ballet S.[1]

[1]Vrije Universiteit Brussel, Research Group of Organic Chemistry, Brussels, Belgium | [2]Vrije Universiteit Brussel, Research Group of Physical Chemistry and Polymer Science, Brussels, Belgium

[3]Ghent University, Supramolecular Chemistry Group, Ghent, Belgium | [4]Vrije Universiteit Brussel, In Vivo Cellular and Molecular Imaging, Brussels, Belgium | [5]CSIRO Materials Science and Engineering, Clayton, Australia

[6]Ghent University, Organic and Biomimetic Chemistry Research Group, Ghent, Belgium | [7]University of Innsbruck, Institute of Pharmacy, Innsbruck, Austria

Introduction

Currently, most drugs are directly administered into patients orally or systemically, without any specific formulation, via parenteral routes. Therefore, to get the desired therapeutic effect, high doses are required due to substantial biodegradation of the drug prior to interaction with the biological target. These high doses can however also result in the appearance of adverse effects. To overcome the need of repeated high dose administration, hydrogels have been reported as suitable controlled drug-delivery systems.

Due to their biocompatibility, biodegradability, their low toxicity and their physically cross-linked properties, peptide-based hydrogels represent an important class of injectable hydrogels suitable to be used as matrices for controlled and slow drug release. Amphipathic peptide hydrogelators, with hydrophobic amino acids being alternated with hydrophilic amino acids, represent a subclass. It is conceived that these kinds of peptides form sheets and -sheet bilayers that self-assemble into fibers. Entanglement of the fibers forms the hydrogel network (Figure 1).

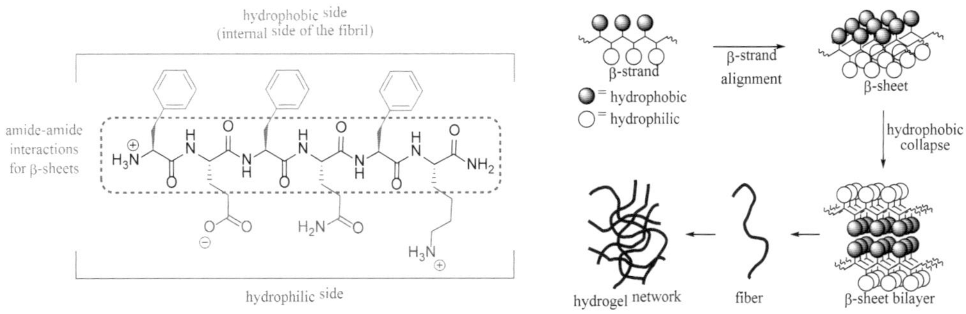

Figure 1. Chemical structure of P1 and the proposed self-assembly process to a hydrogel network, valid for amphipathic peptide hydrogelators

In this work, a new family of hydrogel-forming peptides was designed starting from the short, tunable and amphipathic hexapeptide hydrogelator H-FEFQFK-NH_2 P1 (Figure 1). This peptide showed interesting results in terms of the in vivo release profile of morphine, presenting an effect up to 24 h.[2] Analogues of this sequence were synthesized and all hydrogels were characterized at the macroscopic and microscopic level by dynamic rheometry and transmission electron microscopy (TEM). Their in vivo release profiles were recorded for morphine as a drug, after confirming non-cytotoxicity by an in vitro Live/Dead assay. Opioid administration by subcutaneous injection of the co-formulated hydrogel and subsequent testing in the tail-flick assay (acute pain model) showed sustained antinociceptive effects over longer periods of time (up to 72 h), as compared to drug injections in solution. Within a second in vivo window, the biostability of the hydrogel based on sequence P1 was visualized by nuclear imaging.

Results and Discussion

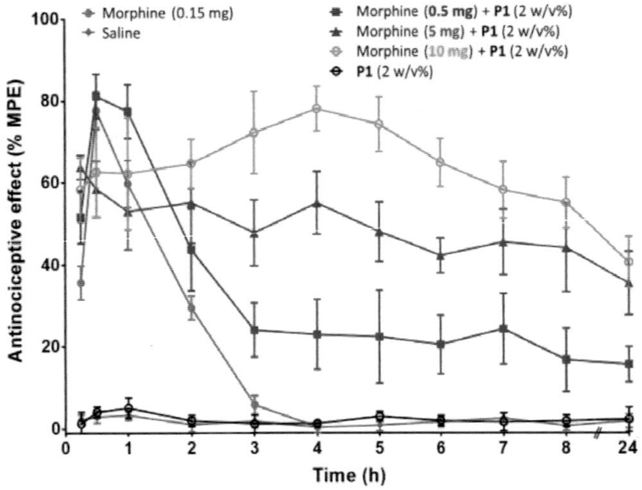

Figure 2. Antinociceptive effects of morphine in the mouse tail-flick test after s.c. administration, applied in solution (red) or co-formulated with hydrogel P1 (2 w/v%). Time course of the antinociceptive response as % of the Maximum Possible Effect (% MPE). Doses of morphine are given per mouse. Data are the mean ± SEM of 5 to 10 mice per group.

In previous work,[2] peptide hydrogelator H-FEFQFK-NH_2 P1 was co-formulated with different amounts of morphine (0.5 mg, 5 mg and 10 mg per 150 μl injections) and subcutaneously injected as a hydrogel (2 w/v% composition in physiological saline) (Figure 2).

Compared to the injection of morphine in solution and compared to control data where only physiological saline or an unloaded gel were injected, a sustained antinociceptive effect over 24 h was observed, with the best effect for a loading of 10 mg morphine. In order to improve the extended release potential, analogues of P1 were synthesized. Several modifications were considered (Table 1), such as side chain changes (within the hydrophobic or the hydrophilic part of the peptide) for a more efficient self-assembly of the fibers, or incorporation of unnatural amino acids like D- or -homo amino acids for an increase of the proteolytic stability. Hydrogel storage moduli (2 w/v% compositions in physio-logical saline) and peptide half-lifes were determined by dynamic rheometry and stability studies in human blood plasma, respectively (Table 1). Note that for sequence P5, no degradation could be determined after 24 h.

Code	Sequence	G' (Pa)	$t_{1/2}$ (min)
P1	H-Phe-Glu-Phe-Gln-Phe-Lys-NH_2	1900	15
P2	H-Phe-Gln-Phe-Gln-Phe-Lys-NH_2	550	3
P3	H-Trp-Glu-Trp-Gln-Trp-Lys-NH_2	4250	21
P4	H-Lys-Phe-Gln-Phe-Glu-Phe-NH_2	950	5
P5	H-DLys-DPhe-DGln-DPhe-DGlu-DPhe-NH_2	1250	-
P6	H-Phe-Glu-β^3*h*Phe-Phe-Gln-Phe-Lys-NH_2	< 50	25

Table 1. Synthesized amphipathic hydrogelators with corresponding storage moduli G' (for 2 w/v% hydrogels in physiological saline) and half-lifes in human blood plasma.

After confirmation of their non-cytotoxicity by in vitro cell assays, morphine was co-formulated within the peptide hydrogels (10 mg/150 μl injections) and the antinociceptive effect was observed using the tail-flick test. All co-formulations resulted in approximately the same antinociceptive effect, where an effect could be monitored up to 72 h, except for P3, were a slight improvement was observed.

While the primary sequence of the peptide gelator and its ability to form stable, long-term fibrillary networks in vivo was hypothesized to be crucial for the final extent of the release, these investigations showed that the stabilization of the peptide hydrogel, by the addition of, for instance, D- or β-amino acids does not give a significant difference in the antinociceptive effect. This observation might probably mean that the release of the drug is more related to: i) the interaction that the drug can make with the peptide hydrogelator and ii) the long-term stability of the hydrogel after s.c. injection indicating that hydrogel erosion is a key determining factor for the release profiles. Therefore, the in vivo release profiles were complemented by: i) TEM images of (ethyl morphine-loaded) gels and ii) in vivo biostability studies by nuclear imaging. The latter study consisted of the s.c. injection of a ^{111}In-radiolabeled gel, after which the radioactive counts at the injection site could be monitored in function of time. Here, the gel degrades over 72 h, with the largest volume reduction occurring within the first 12 h post-injection. The above mentioned drug-loaded TEM images showed an association of ethyl morphine with the fibers of the hydrogel network. As such, the release mechanism seems to be based on the dissolution of the fibers followed by their hydrolysis, where the loaded cargo can possibly increase the stability of the gel and slow down gel erosion (cfr. largest volume reduction within 12 h for unloaded (radiolabeled) gel and antinociceptive effect over 72 h for morphine-loaded gels).

Conclusion

In summary, we have developed a set of amphipathic supramolecular hydrogelators which form hydrogels suited for extended drug release under physiologically relevant conditions. While the various chemical modifications of the hydrogelators resulted in the desired fibers and hydrogel formation, no significant influence on the extended release profiles was noticed, except for P3, which showed a longer effect. Gratifyingly, significant analgesic effects were noticed up to 72 h post-administration, demonstrating the hydrogel's effectiveness as extended-release systems. Our results indicate that, under the examined conditions, a stabilization of the peptide's primary sequence does not result in a prolonged in vivo effect, but that the interaction of the drugs with the peptide hydrogelator and the long-term stability of the hydrogel after s.c. injection might be dominating determinants of in vivo efficacy.

References

01. Dasgupta A., et al. RSC Advances 2013, 3, 9117-9149.
02. Martin C., et al. MedChemComm 2016, 3, 542-549.

Andrea Orlandin[1], Geta Hilma[2], Simona Oancea[2], Fernando Formaggio[1], Paolo Dolcet[3], Cristina Peggion[3]
[1]ICB, Padova Unit, CNR, Department of Chemistry, University of Padova, Padova 35131, Italy;
[2]Department of Biochemistry and Toxicology, University "Lucian Blaga" of Sibiu, Sibiu 550012, Romania;
[3]Department of Chemistry, University of Padova, Padova 35131, Italy; e-mail: fernando.formaggio@unipd.it

Introduction

The need to develop new biocompatible materials for a variety of applications is greatly promoting academic and industrial research. To this aim, we recently directed our interest to the creation of antimicrobial textiles for healthcare environments, characterized by a durable antimicrobial power imparted by peptides [1,2]. Antimicrobial peptides (AMPs) represent a valid alternative to common antibiotics as they interact with the outer membrane of bacteria, with no need to enter the cells. Therefore, bacteria are less prone to develop resistance against AMPs. Moreover, as the chemical nature of AMPs is not modified when they interact with bacteria, we expect our functionalized textiles to maintain their antimicrobial ability even after sterilization and washing cycles.

Results and Discussion

We synthesized a series of short AMPs (Table 1), growing them directly on a cotton fabric, according to common SPPS procedures (Figure 1) [2].
Different synthetic approaches were explored to generate stable chemical bonds between cotton and peptides. Besides the published procedure [2], we implemented a synthetic "green" method based on a chemoselective ligation strategy. In particular, we functionalized the cotton fibres with N-Boc-2,3-epoxypropylamine. After Boc removal and coupling with Boc-aminoxyacetic acid, we ended up with the appropriate functionalization to exploit, for peptide anchoring, the chemoselective ligation via oxime formation.

	Sample	McFarland units*		
		Initial	After 24 hours	After 6 days
control	cotton	0.5	0.5	---
1	Pal-Lys-Ala-D-Ala-Lys-Linker-cotton	0.5	0.5	0.5
2	Pal-His-Ala-D-Ala-His-Linker-cotton	0.5	0.4	0.1
3	Pal-Arg-Ala-Ala-Arg-Linker-cotton	0.5	0.3	0.4
4	Pal-Arg-Ala-Aib-Arg-Linker-cotton	0.5	0.3	0.4
5	Pal-Arg(NO_2)-Ala-Aib-Arg(NO_2)-Linker-cotton	0.5	0.5	0.2
6	[(Pal-Arg-Ala-Aib-Arg)$_4$-Lys$_2$]Lys-Linker-cotton	0.5	0.4	0.2

Table 1. Primary structures of the AMP-cotton conjugates synthesized and antimicrobial activity against Straphylococcus aureus ATCC25923.

To quantify the loading on the cotton fibres, we measured the UV absorbance of the dibenzofulvene (ε = 7800, λ_{max} = 301 nm) released upon removal of the N-protecting Fmoc group. Cotton loadings ranged from 0.6 to 0.8 mmol/g. In addition, we exploited the XPS (X-ray Photoelectron Spectroscopy) technique to determine the atomic elements of our samples. In particular, we detected the presence of nitrogen, not present in pure cotton.
The antibacterial activity against Gram-positive and Gram-negative bacteria strains was checked by turbidity measurements of cells in suspension (inoculum) (Table 1). The standard procedure was appropriately modified for the peptide-cotton samples. Results are expressed as McFarland units, used to approximate the concentration of cells in a suspension. [0.5 McFarland units correspond to a concentration of 1.5×10^8 Colony Formation Unit/ml (CFU/ml)]. The antibacterial effect in these suspensions is revealed by a decrease of McFarland units to 0.2 or 0.1. Peptide-cotton samples 2, 5, and dendrimer 6 were the most active against Staphylococcus aureus.

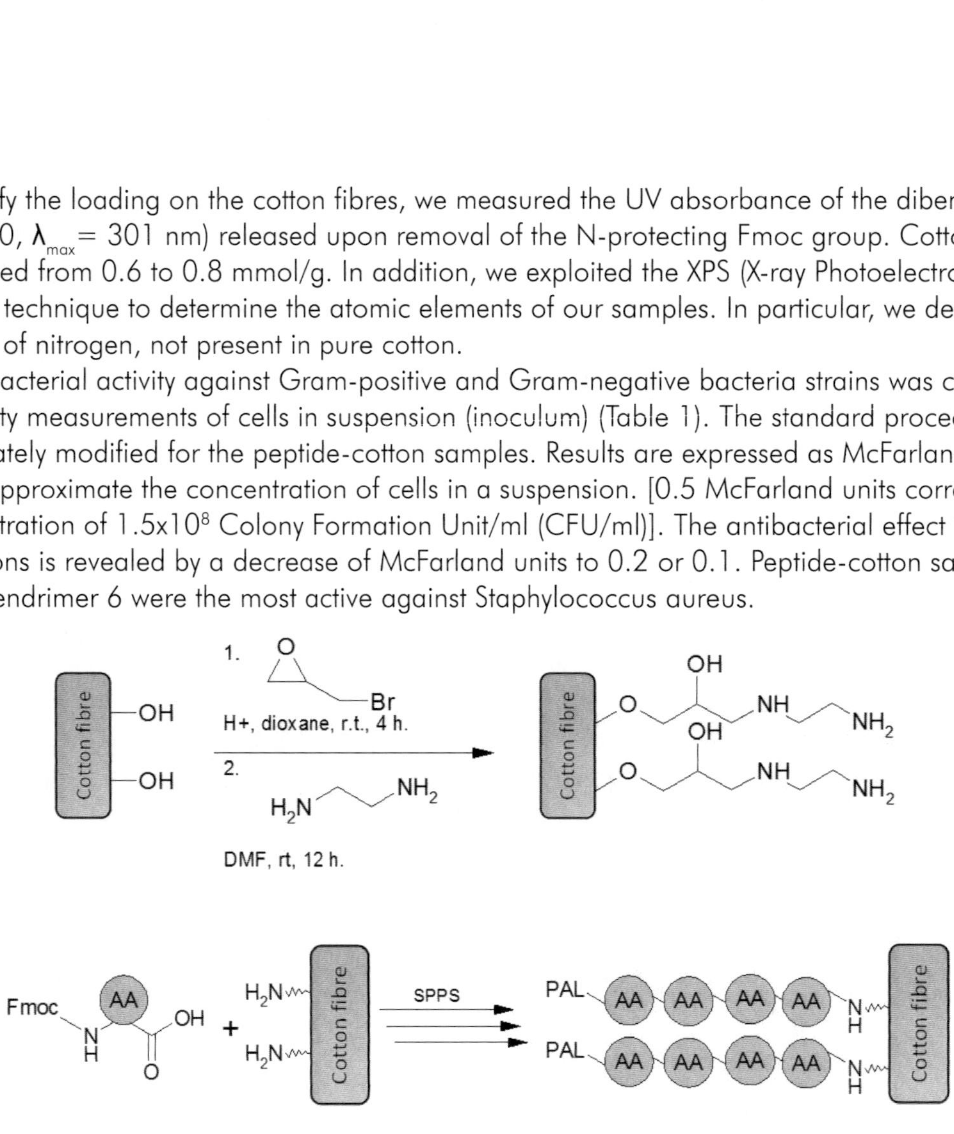

Fig. 1. Schematic view of the cellulose functionalization (top) and the SPPS-like procedure (bottom) exploited to bind our peptides to cotton fabrics [2].

As the peptides are immobilized onto cotton fabrics through a covalent bond, they remain on the fibres even after repeated washing and wearing. Indeed, we observed that the cotton-peptide conjugates can be used a number of times without losing bioactivity. In addition, they were shown to stand abrasion and sterilization and washing cycles.

In conclusion, we observed that short cationic peptides, rich in His and Arg, are particularly promising candidates as they are active against Gram-positive strains (responsible for most hospital-acquired infections) also when linked to cotton fibers.

References

01. Nakamura, M., Iwasaki, T., Tokino, S., Asaoka, A., Yamakawa, M., Ishibashi, J. Biomacromolecules 12, 1540–1545 (2011).
02. Orlandin, A., Formaggio, F., Toffoletti, A., Peggion, C. J. Pept. Sci. 20, 547-553 (2014).

A NEW FAMILY OF N-TERMINALLY TRUNCATED PEPTAIBOLS FROM THE BIOCONTROL FUNGUS TRICHODERMA HARZIANUM

Thomas Degenkolb[1], Christian René Röhrich[2], Andreas Vilcinskas[1,2], Hans von Döhren[3], Hans Brückner[4]
[1]Interdisciplinary Research Centre for BioSystems, Land Use and Nutrition (IFZ), Department of Applied Entomology, Institute of Insect Biotechnology, University of Giessen, Heinrich-Buff-Ring 26 – 32, 35392 Giessen, Germany;
[2]Fraunhofer Institute for Molecular Biology and Applied Ecology (IME), LOEWE Center for Insect Biotechnology and Bioresources (ZIB), Winchesterstrasse 2, 35394 Giessen, Germany;
[3]Biochemistry and Molecular Biology OE 2, Institute of Chemistry, Technical University of Berlin, Franklinstrasse 29, 10587 Berlin, Germany;
[4]Interdisciplinary Research Centre for BioSystems, Land Use and Nutrition (IFZ), Department of Food Sciences, Institute of Nutritional Science, University of Giessen, Heinrich-Buff-Ring 26 – 32, D-35392 Giessen, Germany

PP VI - 093

Introduction

Peptaibiotics and their subgroup of peptaibols are defined as bioactive microbial peptides containing the characteristic, non-proteinogenic α-aminoisobutyric acid (Aib). They are of permanent interest to both academics and industry owing to their manifold biological activities including unique membrane-modifying properties [1]. In a previous study [2] we analysed five commercial biocontrol agents (BCAs) formulated with recently described species of the Trichoderma harzianum complex [3]. The detection of peptaibiotics in plant-protective Trichoderma species, which are successfully used against economically important bacterial and fungal plant pathogens, corroborates their synergistic interaction with non-peptidic secondary metabolites and cell wall-degrading enzymes [2]. Here we report on the sequences of a new class of truncated peptaibol-derived peptides named brevikindins.

Results and Discussion

Using the well-established, HPLC/ESI-HRMS-based peptaibiomics approach [4] (see Figure 1), it could unequivocally be demonstrated that agar plate cultures of T. harzianum CBS 226.95 contained recurrent trichokindin-type [5] 18-residue peptaibols, i.e., peptaibiotics carrying an acetylated N-terminus the C-terminus of which is reduced to a β-amino alcohol [1] (Table 1).

No.	$[M+H]^+$	$[M+Na]^+$	t_R [min]		Residue																		Compound identical or positionally isomeric with	Name
							1	2	3	4	5	6	7	8	9	10	11	12	13	14	15	16		
1	1505.9362	1527.9203	33.2-33.3				Ala	Aib	Aib	Gln	Aib	Vxx	Aib	Ala	Aib	Aib	Pro	Lxx	Aib	Aib	Gln	Lxxol	**new**, cf. **7**	**Brevikindin I**
2	1519.9564	1541.9393	34.2-34.3				Ala	Aib	Aib	Gln	Aib	Lxx	Aib	Ala	Aib	Aib	Pro	Lxx	Aib	Aib	Gln	Lxxol	**new**, cf. **8**	**Brevikindin II**
3	1533.9743	1555.9551	35.6				Ala	Aib	Aib	Gln	Aib	Lxx	Aib	Ala	Vxx	Aib	Pro	Lxx	Aib	Aib	Gln	Lxxol	**new**, cf. **11**	**Brevikindin III**
4	1547.9362	1569.9722	36.2				Ala	Aib	Aib	Gln	Vxx	Lxx	Aib	Ala	Vxx	Aib	Pro	Lxx	Aib	Aib	Gln	Lxxol	**new**, cf. **12** and **14**	**Brevikindin IV**
5	1562.0062	1583.9862	37.9-39.4				Ala	Vxx	Aib	Gln	Vxx	Lxx	Aib	Ala	Vxx	Aib	Pro	Lxx	Aib	Aib	Gln	Lxxol	**new**, cf. **16**	**Brevikindin V**
6	1576.0178	1598.0017	40.9-41.1				Ala	Vxx	Vxx	Gln	Vxx	Lxx	Aib	Ala	Vxx	Aib	Pro	Lxx	Aib	Aib	Gln	Lxxol	**new**, cf. **18**	**Brevikindin VI**
					Residue																		Compound identical or positionally isomeric with trichokindin (TK)	Ref.
					1	2	3	4	5	6	7	8	9	10	11	12	13	14	15	16	17	18		
7	1720.0378	1742.0200	44.1-44.4	Ac	Aib	Ser	Ala	Aib	Aib	Gln	Aib	Vxx	Aib	Ala	Aib	Aib	Pro	Lxx	Aib	Aib	Gln	Lxxol	**new**	
8	1734.0568	1756.0371	45.2-45.6	Ac	Aib	Ser	Ala	Aib	Aib	Gln	Aib	Lxx	Aib	Ala	Aib	Aib	Pro	Lxx	Aib	Aib	Gln	Lxxol	**new**	
9	1748.0738	1770.0539	47.1-47.3	Ac	Aib	Ser	Ala	Aib	Aib	Gln	Vxx	Lxx	Aib	Ala	Aib	Aib	Pro	Lxx	Aib	Aib	Gln	Lxxol	Ia	
10	1748.0747	1770.0557	47.3-47.6	Ac	Aib	Ser	Ala	Aib	Vxx	Gln	Aib	Lxx	Aib	Ala	Aib	Aib	Pro	Lxx	Aib	Aib	Gln	Lxxol	Ib	
11	1748.0749	1770.0559	48.1-48.2	Ac	Aib	Ser	Ala	Aib	Aib	Gln	Aib	Lxx	Aib	Ala	Vxx	Aib	Pro	Lxx	Aib	Aib	Gln	Lxxol	IIa	
12	1762.0912	1784.0718	48.7-48.9	Ac	Aib	Ser	Ala	Aib	Aib	Gln	Vxx	Lxx	Aib	Ala	Vxx	Aib	Pro	Lxx	Aib	Aib	Gln	Lxxol	IIIa	Iida et al.,1994
13a	1762.0911	1784.0713	49.1-49.5	Ac	Aib	Ser	Ala	Aib	Vxx	Gln	Vxx	Lxx	Aib	Ala	Aib	Aib	Pro	Lxx	Aib	Aib	Gln	Lxxol	IIb, IVa	
13b	1762.0901	1784.0703	49.4-49.5	Ac	Aib	Ser	Ala	Aib	Vxx	Gln	Aib	Lxx	Aib	Ala	Vxx	Aib	Pro	Lxx	Aib	Aib	Gln	Lxxol	IIIb	
14	1762.0919	1784.0717	49.8-50.0	Ac	Aib	Ser	Ala	Aib	Aib	Gln	Vxx	Lxx	Aib	Ala	Vxx	Aib	Pro	Lxx	Aib	Aib	Gln	Lxxol	Va	
15	1762.0912	1784.0718	50.3-50.5	Ac	Aib	Ser	Ala	Aib	Vxx	Gln	Aib	Lxx	Aib	Ala	Vxx	Aib	Pro	Lxx	Aib	Aib	Gln	Lxxol	Vb	
16	1776.1085	1798.0893	51.6-51.8	Ac	Aib	Ser	Ala	Vxx	Aib	Gln	Vxx	Lxx	Aib	Ala	Vxx	Aib	Pro	Lxx	Aib	Aib	Gln	Lxxol	**new**	
17	1776.1094	1798.0890	51.9-52.6	Ac	Aib	Ser	Ala	Aib	Vxx	Gln	Vxx	Lxx	Aib	Ala	Vxx	Aib	Pro	Lxx	Aib	Aib	Gln	Lxxol	VI, VII	Iida et al., 1994
18	1790.1210	1812.1027	53.9-54.2	Ac	Aib	Ser	Ala	Vxx	Vxx	Gln	Vxx	Lxx	Aib	Ala	Vxx	Aib	Pro	Lxx	Aib	Aib	Gln	Lxxol	**new**	

Table 1. Sequences of novel 16-residue brevikindins and their 18-residue trichokindin-type precursors. Variable positions are underlined.

y, Institute of Insect Biotechnology, University of Giessen, Heinrich-Buff-Ring 26 – 32, 35392 Giessen, Germany;
d Bioresources (ZIB), Winchesterstrasse 2, 35394 Giessen, Germany;
, 10587 Berlin, Germany;
itute of Nutritional Science, University of Giessen, Heinrich-Buff-Ring 26 – 32, D-35392 Giessen, Germany

In total, twelve 18-residue trichokindin-type peptaibols and six novel truncated 16-residue peptaibiotics (see Table 1) with free N-alanyl termini, named brevikindins I – VI, were found. (Lxx, Leu or Ile; Vxx, Val or isovaline; Lxxol, leucinol or isoleucinol). Our data provide first insight into a new mechanism of formation of N-terminally truncated peptaibols in the course of biosynthesis of peptaibiotics.

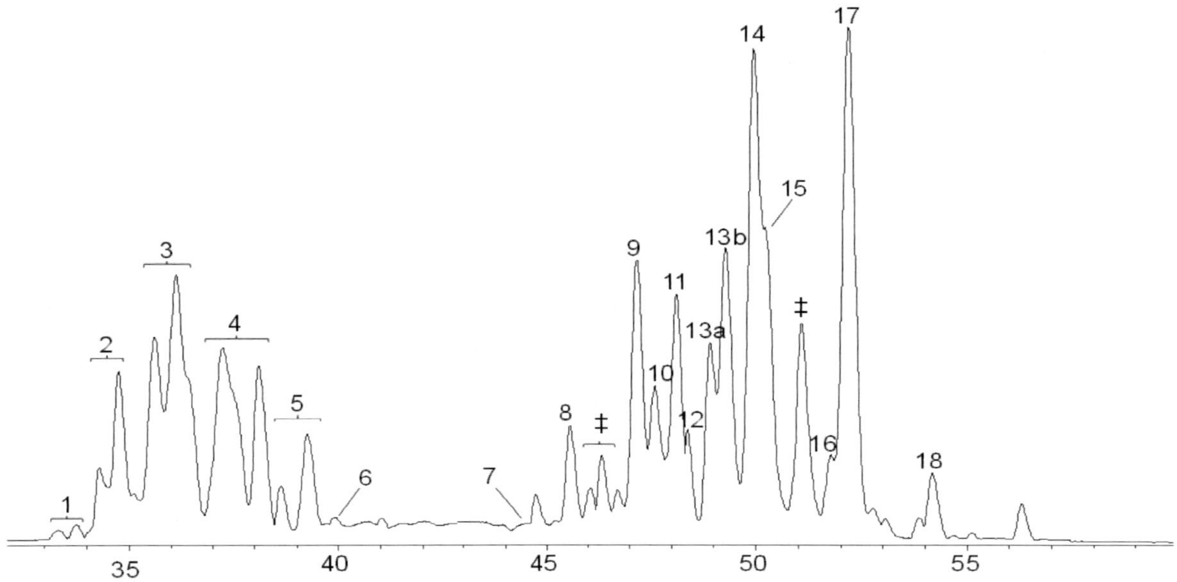

Fig. 1. EIC profile of the peptaibiotic fraction of T. harzianum CBS 226.95. Consecutive numbers of peptides correspond to those in Table 1.

Acknowledgments

We acknowledge the financial support provided by the Hessen State Ministry of Higher Education, Research and the Arts (HMWK) including a generous grant for the LOEWE research center "Insect Biotechnology and Bioresources" to A. V.

References

01. Degenkolb, T., Brückner, H. Chem. Biodivers. 5, 1817-1843 (2008).

02. Degenkolb, T., Nielsen, K.F., Dieckmann, R., Branco-Rocha, F., Chaverri, P., Samuels, G.J., Thrane, U., von Döhren, H., Vilcinskas, A., Brückner, H. Chem. Biodivers. 12, 662-684 (2015).

03. Chaverri, P., Branco-Rocha, F., Jaklitsch, W., Gazis, R., Degenkolb, T., Samuels, G.J. Mycologia 107, 558-590 (2015).

04. Röhrich, C.R., Jaklitsch, W.M., Voglmayr, H., Iversen, A., Vilcinskas, A., Nielsen, K.F., Thrane, U.,von Döhren, H., Brückner, H., Degenkolb, T. Fungal Divers. 69, 117-146 (2014).

05. Iida, A., Sanekata, M., Fujita, T., Tanaka, H., Enoki, A., Fuse, G., Kanai, M., Rudewicz, P.J., Tachikawa, E. Chem. Pharm. Bull. 42, 1070-1075 (1994).

Marek Cebrat[1], Magdalena Wierzbicka[1], Joanna Fedorowicz[2], Jarosław Saczewski[2], Zbigniew Szewczuk[1]
[1]Faculty of Chemistry, University of Wroclaw, F. Joliot-Curie 14, 50-383 Wrocław, Poland
[2]Faculty of Pharmacy with Subfaculty of Laboratory Medicine, Medical University of Gdansk, Al. Gen. J. Hallera 107, 80-416, Gdansk, Poland

PP VI – 094

Introduction

Isoxazolo[3,4-b]pyridin-3(1H)-ones and isoxazolo[3,4-b]quinolin-3(1H)-ones, which can be easily synthesized from various commercialy available 2-chloro-pyridine-3-carboxylic acids or 2-chloro-quinoline-3-carboxylic acids, react with formaldehyde and secondary amines in selective tandem Mannich-electrophilic amination reaction resulting in a new class of photostable fluorescent dyes with pyrido-triazolium or quinolino-triazolium core structures (Safirinium P and Q derivatives, respectively), which upon esterification with N-hydroxysuccinimide provide cationic, fluorescent amine-reactive probes (Figure 1) [1, 2].

Conjugation of peptides with quaternary ammonium groups increases their ionization efficiency in ESI-MS experiments and lowers the detection limit even to a low attomole range [3, 4]. Therefore, we decided to check if Safirinium analogues could be used as ionization tags for identification of peptides.

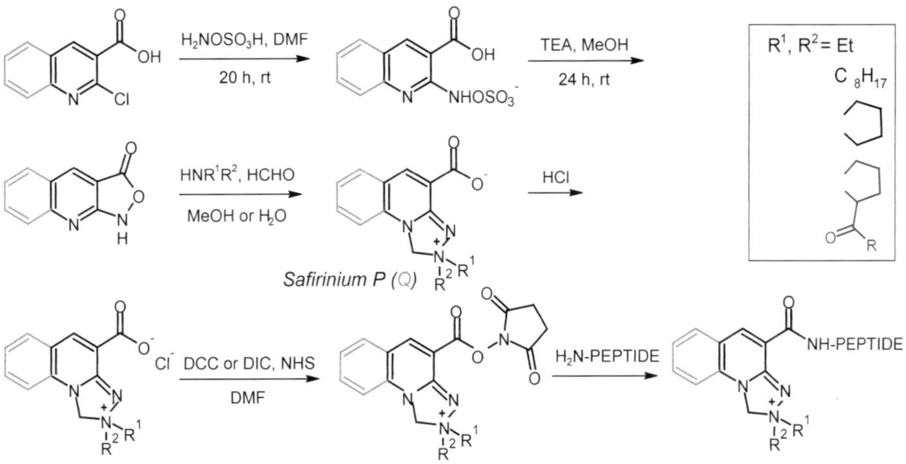

Figure 1. Synthesis scheme leading to Safirinium P and Q derivatives and their NHS esters which can be used for labeling of peptides.

Results and Discussion

N-hydroxysuccinimide esters of Safirinium P/Q efficiently react with primary amine groups of amino acids and peptides introducing fluorescent tags containing quaternary ammonium salt (QAS). This allows for an effective detection of such labeled peptides both by fluorescence and in mass spectrometric measurements where introduction of a permanent positive charge (QAS) causes enhacement of a MS signal.

Safirinium derivatives containing alkyl groups on the quaternary ammonium undergo Hofmann elimination during ESI-MS experiments which limits their use as MS tags (Figure 2). In case of spiro compounds (e.g. cyclopentyl derivative), Hofmann elimination is not possible which makes them much more resistant to CID fragmentation and more suitable as ionisation tags.

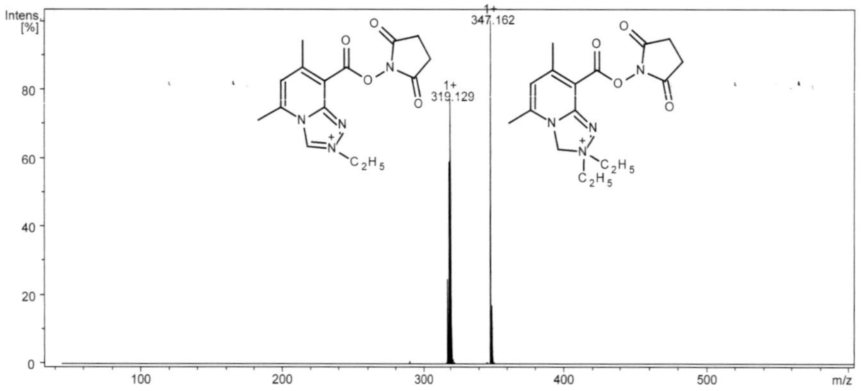

Figure 2. ESI-MS spectrum of diethyl analogue of Safirinium P N-hydroxysuccinimide ester showing Hofmann elimination of one of the ethyl groups.

LC-MS analysis of the ubiquitin hydrolizate labeled with the diethyl analogue of Safirinium P shows that the NHS ester of the tag reacts both with the N-terminal amino groups of the peptides and the side chain amino groups of lysine residues. The labeling reaction can be performed in water-containing buffer solution. The protein sequence coverage is very good in this case (ca. 80 %) and the MS signal intensity is enhanced up to 8 times for some of the labeled peptides.

If a proline residue is used as a secondary amine in a reaction of isoxazolones with formaldehyde, a new class of fluorescent peptide building blocks with rigid structure and an interesting stereochemistry is formed. The reaction can be also performed on the N-terminal proline of peptides attached to the standard solid supports. The resulting labeled peptide is stable during peptide cleavage by 95% TFA (Figure 3).

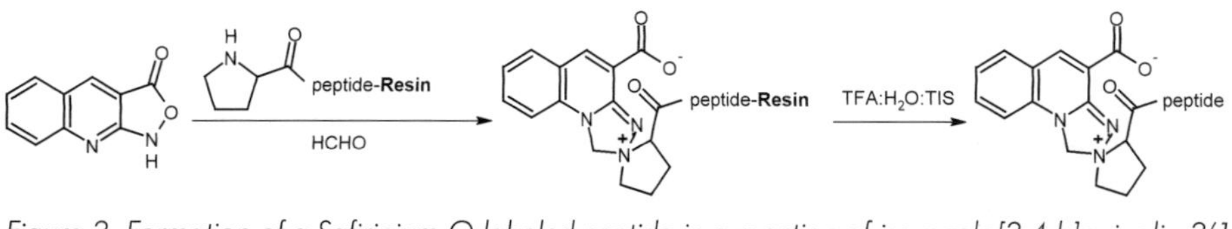

Figure 3. Formation of a Safirinium Q labeled peptide in a reaction of isoxazolo[3,4-b]quinolin-3(1H)-one with N-terminal proline residue of a resin-bound peptide.

Acknowledgements
Project supported by Wroclaw Centre of Biotechnology, programme The Leading National Research Centre (KNOW) for years 2014-2018.

References

01. Saczewski J, Hinc K, Obuchowski M, Gdaniec M. Chem. Eur. J. 2013; 19: 11531–11535.
02. Saczewski J, Fedorowicz J, Korcz M, Saczewski F, Wicher B, Gdaniec M, Konopacka A. Tetrahedron 2015; 71: 8975–8984.
03. Cydzik M, Rudowska M, Stefanowicz P, Szewczuk Z. J. Pept. Sci. 2011; 17: 445–453.
04. Bachor R, Mielczarek P, Rudowska M, Silberring J, Szewczuk Z. Int. J. Mass Spectrom. 2014; 362: 32–38.

A HIGH-THROUGHPUT RECOVERY METHOD FOR THE QUANTIFICATION OF A PEPTIDE IN PLASMA USING HPLC

Christos Kontos[1], Maria Eleni Androutsou[2], AlexiosVlamis-Gardikas[1], Theodore Tselios[1]
[1]University of Patras, Department of Chemistry, Rion Patras, 26504, Greece
[2]Eldrug S.A., Pharmaceutical Company, 26504, Platani, Greece

PP VI - 097

Introduction

The recovery of high molecular weight peptides from complex biological samples is a challenging task [1]. This work describes the development of a reliable, cost effective and rapid methodology for the quantitative determination of a myelin oligodendrocyte glycoprotein (MOG) peptide namely (Lys-Gly)$_5$ MOG$_{35-55}$ (3518 Da) in plasma. The MOG$_{35-55}$ epitope is an autoantigen for the stimulation of encephalitogenic T cells that are responsible for multiple sclerosis (MS) and chronic experimental autoimmune encephalomyelitis (EAE: animal model of MS) [2, 3]. The specific peptide conjugated to a polysaccharide (polymannose) was successful in EAE using therapeutic and vaccinated protocols making it a promising candidate for the treatment of MS [2].

Results and Discussion

Several different types of solvents and reagents were tested to precipitate plasma proteins using the protocol shown in figure 1. Quantitative precipitation of plasma proteins was achieved with an acetonitrile (AcN) / H_2O / formic acid (FA) solution (80 / 19.4 / 0.6 v/v).

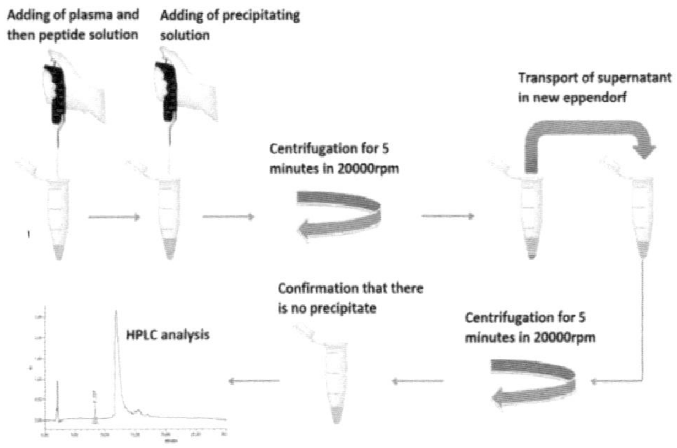

Figure 1: Experimental procedure for the precipitation of plasma proteins

The next step was the development of a method to determine the concentration of the (Lys-Gly)$_5$ MOG$_{35-55}$ peptide in plasma and calculate its recovery. Two standard curves were constructed using seven calibration points per curve at concentration ranges of 50 to 500 μg/ml. All standards were prepared in supernatants of precipitated plasma proteins (with the AcN / H_2O / FA solution). The standards were analysed by HPLC measuring A_{214} (figures 2A and 2B). Quality control (QC) samples were used to determine the accuracy and precision of the method and were independently prepared at low (150 μg/ml), medium (275 μg/ml) and high (400 μg/ml) concentrations (insets in figures 2A and 2B).

To evaluate the recovery of the peptide, plasma was spiked with peptide solution to 90.1 μg/ml after which the precipitation solution was added. Recovery was determined by comparing the peptide concentration in the precipitated plasma sample (as determined by HPLC and the constructed standard curves) to the initially added sample concentration (figures 2C and 2D). The average recovery of the peptide from plasma ranged from 84 to 91.1% (figures 2C and 2D). All experiments were performed in duplicate.

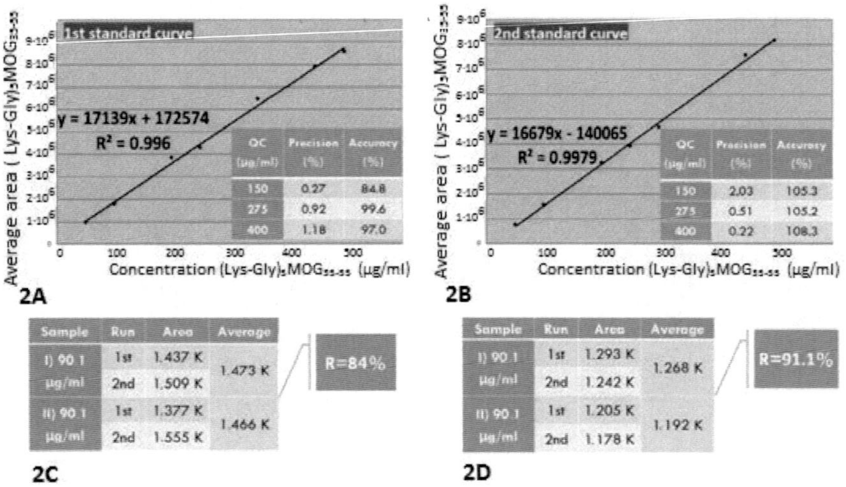

Figure 2: A) 1st standard curve and its QCs, B) 2nd standard curve and its QCs, C) samples made from the same aliquot that was used for the 1st standard curve and the calculated recovery (R), D) samples made from the same aliquot that were used for the 2nd standard curve and the calculated recovery (R).

Further experiments will investigate the pharmacokinetics of $(Lys\text{-}Gly)_5MOG_{35\text{-}55}$ in animal models while the AcN / H_2O / FA solution could be tested for other peptides with high molecular weights or/and challenging sequences.

References

01. van den Broek, I.; Sparidans, R.W.; Huitema, A.D.R.; Schellens, J.H.M.; Beijnen, J.H. J. Chromatogr. B, (2006), 837, 49-58
02. Lutterotti, A.; Yousef, S.; Sputtek, A.; Stürner, K.H.; Stellmann, J.P.; Breiden, P.; Reinhardt, S.; Schulze, C.; Bester, M.; Heesen, C.; Schippling, S.; Miller, S.D.; Sospedra, M.; Martin, R. Sci. Transl. Med. (2013), 5, 188.
03. Tseveleki, V.; Tselios, T.; Kanistras, I.; Koutsoni, O.; Karamita, M.; Vamvakas, S.S.; Apostolopoulos, V.; Dotsika, E.; Matsoukas, J.; Lassmann, H.; Probert, L. Exp. Neurol., (2015), 267, 254-267

Bartosz Setner, Alicja Kluczyk, Piotr Stefanowicz, Zbigniew Szewczuk
Faculty of Chemistry, University of Wrocław, Wrocław, Poland

PP VI - 104

De novo peptide sequencing by tandem mass spectrometry, utilizing ionization tags (such as quaternary ammonium [QAS] or phosphonium salts) is limited due to instability of the ionization tags during the MS/MS experiment.[1] Ionization tags are usually good leaving groups mostly due to Hofmann elimination and/or cyclization of a peptide derivatives via S_N2 type reaction.[2,3] We proposed a rigid scaffold: 5-azoniaspiro[4.4]nonyl, in which all bonds susceptible to cleavage are protected by cyclization, which drastically increased the stability of QAS during MS/MS fragmentation of labelled peptides.[4] The
5-azoniaspiro[4.4]nonyl moiety is the first quaternary ammonium salt which is not a good leaving group. Our goal was to examined the impact of heteroatom (sulphur or oxygen) in the 5-azoniaspiro[4.4]nonyl group (Fig. 1.) on the stability of such peptide derivatives during MS/MS experiments.

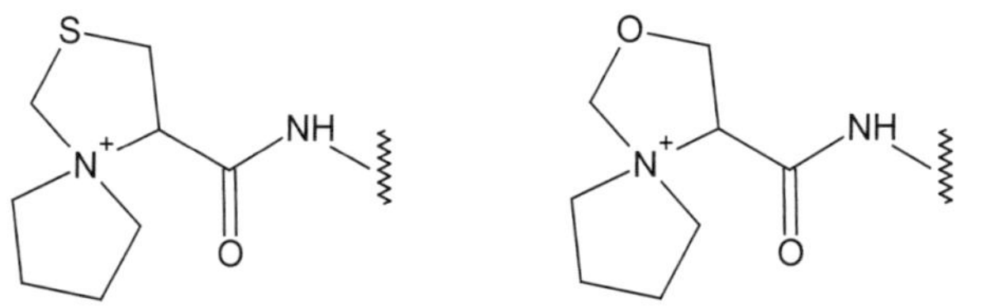

Figure 1. Structures of heterocyclic analogs of the 5-azoniaspiro[4.4]nonyl ionization tag.

A series of model tetrapeptides were synthesized using Fmoc- solid phase peptides synthesis strategy. After Fmoc- deprotection N-terminal amino acids was coupled with commercially available proline derivatives: Fmoc-thiaproline (thiazolidine-4-carboxylic acid) or Fmoc-oxaproline (oxazolidine-4-carboxylic acid) in the presence of HATU/Oxyma Pure/DIPEA. Next free N-teriminal group was reacted with 1,4-dibromobutane or α,α′-dibromo-o-xylene in the presence of DIPEA in DMF. Peptide derivatives were cleaved from the resin simultaneously with the side chain deprotection using TFA. Obtained compounds were analyzed by ESI-MS/MS.

The presence of the heteroatom in the 2-oxa/thia-benzo-5-azoniaspiro[4.4]nonyl group enables selective fragmentation of this scaffold. Intense ion at the m/z 132 is formed during fragmentation of QAS group with preservation of quaternary ammonium salt. We observed major differences between fragmentation patterns of 2-oxa and 2-thia-benzo-5-azoniaspiro[4.4]nonyl moieties during CID fragmentation of the derivatized peptides. Presence of sulphur atom in position two open competitive fragmentation pathway with characteristic neutral loss of 117 Da. On the other hand introduction of oxygen atom does not influence fragmentation pathway and provide straightforward for interpretation tandem mass spectra. Model synthetic peptide (from bovine serum albumin tryptic digest) derivatized with ionization tag based on 2-oxa-benzo-5-azoniaspiro[4.4]nonyl moiety at the ε-amino group of C-terminal lysine residue generate dominant y_i^*-ion series during low energy CID experiment which facilitate peptide sequencing (Fig. 2.). Increased collision energy is needed to observed characteristic ion at the m/z 132, which could be used in MRM (Multiple Reaction Monitoring) peptide analysis.

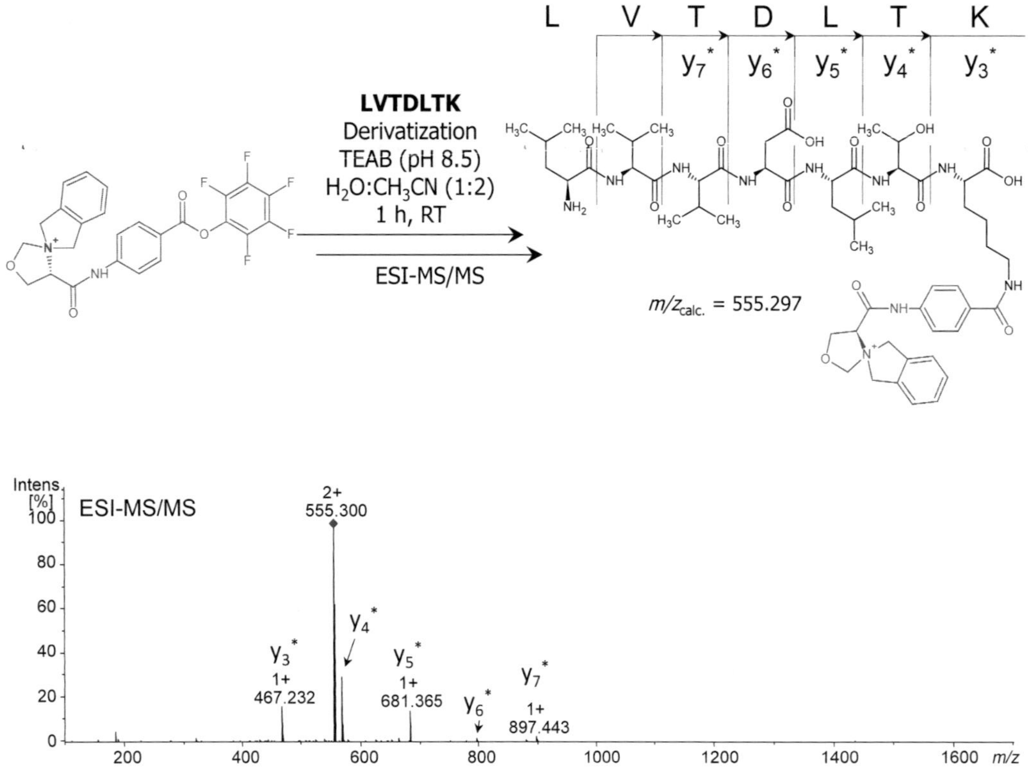

Figure 2. Derivatization and tandem mass examination of the model tryptic peptide LVTDLK.

In conclusions we developed new ionization reagent for peptide analysis by tandem mass spectrometry suitable for peptide labeling in solution due to the presence of active ester. Replacement of the methylene group in position two of the 5-azoniaspiro[4.4]nonyl moiety by sulphur or oxygen enables selective fragmentation of QAS. Characteristic reporter ion at m/z 132 is formed during

CID experiment. Presence of oxygen atom in QAS scaffold provides simple for interpretation MS/MS spectra compared to QAS with sulphur atom.

Acknowledgement

This research was supported by a grant No. 2014/15/N/ST5/00738 from the National Science Centre, Poland from the National Science Centre, Poland. The LC-ESI-MRM study were supported by Wrocław Center of Biotechnology, program "The Leading National Research Centre" (KNOW) for years 2014-2018. The authors would like to thank Andrzej Reszka (Shim-Pol, Poland) for providing the LCMS-8050 instrument.

References

01. Stefanowicz, P., Kluczyk, A., Szewczuk, Z. Amino Acids, Pept. Proteins, 2016, 40, 36-74.
02. Cydzik, M., Rudowska, M., Stefanowicz, P., Szewczuk, Z. J. Am. Soc. Mass Spectrom., 2011, 22, 2103-2107.
03. Rudowska, M., Wojewska, D., Kluczyk, A., Bachor, R., Stefanowicz, P., Szewczuk, Z. J. Am. Soc. Mass Spectrom., 2012, 23, 1024-1028.
04. Setner, B., Rudowska, M., Wojewska, D., Kluczyk, A., Stefanowicz, P., Szewczuk, Z. J. Pept. Sci. 2014, 20, S64-S65.

Mateusz Waliczek, Monika Kijewska, Magdalena Rudowska, Bartosz Setner,
Piotr Stefanowicz, Zbigniew Szewczuk
Faculty of Chemistry, University of Wroclaw, Joliot-Curie 14, 50-383 Wroclaw, Poland

PP VI – 105

Introduction

Mass spectrometry is a powerful analytical tool used in proteomic. Despite its great potential there are many problems during analysis of trace amount of peptides caused by insufficient ionization efficiency of some peptides, which results in limited sensitivity. Development and application of novel ionization enhancers is the widely used way to overcome this problem [1,2]. In this paper we present new ionization enhancer based on well-known 2,4,6-triphenylpyridinium and 2,4,6-trimethylpyridinium salts.

Application of inexpensive and commercially available pyrylium salt allows the derivatization of primary amino groups, especially these sterically unhindered, such as glycine or e-amine group of lysine.

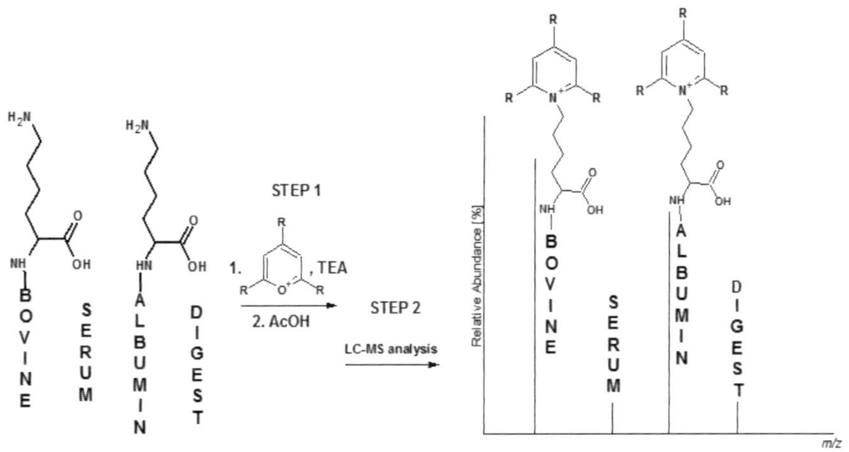

Results and Discussion

The aim of our research was to show the application 2,4,6-trisubstituded pyrylium salts as a reagent, which allows for transforming of amino groups of peptide to pyridinium moieties, which have the fixed charge. Moreover, we studied the impact of this modification on ionization efficiency in ESI mass spectrometry. In our experiments we tested 2,4,6-trimethyl and 2,4,6-triphenylpyrylium salts. The general procedure of modification consisted of dissolving of peptide (or mixture of peptides) in N,N-dimethylformamide (DMF), addition of excess of pyrylium salt and equivalent amount of triethylamine (relative to the salt) following by addition of equivalent amount of acetic acid after 20 minutes. The solvent was evaporated under stream of nitrogen gas, residue was redissolved in water and then lyophilized. The analysis of MS/MS spectrum of 2,4,6-trisubstituted pyridinium modified peptides revealed an abundant protonated 2,4,6-triphenylpyridinium cation, which is a promising reporter ion for the multiple reaction monitoring (MRM) analysis as well as for analysis in precursor ion scan mode. The fixed positive charge of the pyridinium group enhances the ionization efficiency and enables detection of peptides at attomole level. The observed increase in ionization efficiency for protein digest is in the range from 4 to 10. Fragmentation analysis performed for peptide LVNELTEFAK revealed, that pyridinium modified peptides provide full sequence coverage, which significantly facilitate analysis of such spectra due to the presence of series y ions. The high selectivity of pyrylium salt toward e-amine group of lysine makes this reagent particularly useful for derivatization of tryptic protein hydrolysate.

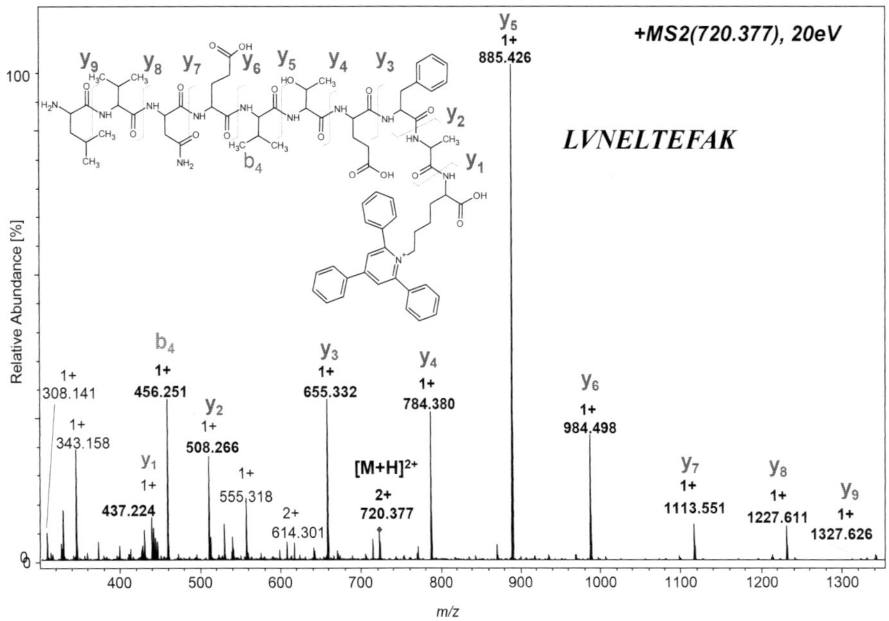

Figure 1. ESI-MS/MS spectrum obtained for peptide LVNELTEFAK.

An advantage of application of 2,4,6-trisubstituted pyrylium salts is no need to active ester formation as most of commercially common reagents. The simplicity of derivatization of peptides and the possibility of formation of the pyridinium salt both in the solid-phase as well as the solution-phase peptide synthesis are additional advantages. Moreover, several routes to obtain useful isotopologues were proposed. Thus, we presume that the application of such labeling may be useful in comparative proteomics, leading to the development of new biomarkers based on proteins of low abundance.

Acknowledgments
This work was supported by a grant No. UMO-2013/09/B/ST4/00277 from the National Science Centre, Poland.

References

01. B.Setner, M.Rudowska, E.Klem, M.Cebrat, Z.Szewczuk: J.Mass Spectrom, 49, 995 (2014).
02. R. Bachor, P. Mielczarek, M. Rudowska, J. Silberring, Z. Szewczuk: Int. J. Mass Spectrom., 362, 32 (2014)

INCREASED ANTITUMOR ACTIVITY OF A VINDOLIN DERIVATIVE CONJUGATED WITH OCTAARGININE

Z. Bánóczi[1], J. Tóvári[2], I. Szabó[3], F. L. Regenbach[3], I. Randelovic[2], A. Keglevich[4], P. Keglevich[4], Z. Lengyel[4], Gorka-Kereskényi[4], C. Szántay Jr.[5], L. Hazai[4], F. Hudecz[1,3]

[1]Department of Organic Chemistry, Eötvös Loránd University, Budapest, Hungary;
[2]National Institute of Oncology, Budapest, Hungary;
[3] MTA-ELTE Research Group of Peptide Chemistry, Eötvös Loránd University,Budapest, Hungary;
[4] Department of Organic Chemistry and Technology, University of Technology and Economics, Budapest, Hungary; 5Spectroscopic Research Division, Gedeon Richter Plc., Budapest, Hungary

PP VII – 109

Introduction

Vinca alkaloids were isolated first from the leaves of Catharanthus roseus (L.) G. Don. The major alkaloid in this plant is the vindoline (1,2). Vindoline (Vind) is the bio- and synthetic precursor of these bisindole alkaloids. Unfortunately, early studies showed lack of any biological activities of vindoline (3). In order to improve activity the structure of vindoline was modified. Interestingly, vindoline analogues (4) and their derivatives with L- or D-Trp-OMe (5) exhibited higher cytostatic activity than the free vindoline. Our aim was to study the effect of conjugation with octaarginine on the cytostatic activity of selected vindoline derivatives.

Results and Discussion

Octaarginine was built up by standard Fmoc/^{t}Bu strategy on Rink-amide resin. The purified peptide was conjugated with the vindoline derivatives in DMF using DIC and HOBt as coupling reagents. The conjugates were purified by RP-HPLC and chemically characterised by analytical RP-HPLC and mass spectrometry.

The in vitro cytostatic effect of all conjugates was studied on three human tumour cell lines: HL-60 (human leukemia), MCF-7 (human breast adenocarcinoma), and MDA-MB-231 (human triple negative breast adenocarcinoma). The in vivo effect of two conjugates (Br-Vind-(L)-Trp-Arg$_8$ and Br-Vind-(D)-Trp-Arg$_8$) were studied on two mouse tumour cells - C26 (murine colon carcinoma) and P338 (mouse leukemia) (Table 1).

Table 1 In vitro cytostatic activity of vindoline-derivatives and - conjugates on tumour cells

Compound	IC_{50} (sd) (μM)				
	HL-60[a]	MDA-MB-231[a]	MCF-7[a]	C26[b]	P388[b]
Br-Vind-(L)-Trp-OMe (1)	56.8 (9.8)[c]	n.d.	n.d.	n.d.	n.d.
Br-Vind-(D)-Trp-OMe (2)	73.8 (10.4)[c]	37.3 (13.7)	28.5 (5.7)	n.d.	n.d.
cpropyl-Vind-(L)-Trp-OMe (3)	75.3 (2.3)[c]	56.1 (4.8)	84.2 (20.2)	n.d.	n.d.
Br-Vind-Arg$_8$ (4)	11.6 (5.3)	10.3 (3.3)	12.0 (3.3)	n.d.	n.d.
Br-Vind-(L)-Trp-Arg$_8$ (5)	11.5 (4.6)	13.2 (0.4)	6.4 (3.6)	4.38	3.49
Br-Vind-(D)-Trp-Arg$_8$ (6)	15.1 (0.5)	10.0 (3.7)	5.1 (1.1)	12.31	18.82
cpropyl-Vind-(L)-Trp-Arg$_8$ (7)	14.4 (1.0)	15.1 (0.3)	8.5 (5.3)	n.d.	n.d.

[a] *The cells were incubated with the compound for 3 h and cultured in serum-containing medium for 3 days. The IC_{50} values were determined by MTT assay.* [b]*The cells were incubated with the compound for 72 h and the IC_{50} values were determined by MTT assay.* [c] *From the literature (5); n.d. means not determined*

; 5Spectroscopic Research Division, Gedeon Richter Plc., Budapest, Hungary

The conjugates 5, 6 and 7 had higher in vitro cytostatic activity on HL-60, MCF-7 and MDA-MB-231 cells than those of the free derivatives 1, 2 and 3. This effect of the peptide conjugates can be explained by a) the attached cell-penetrating peptide increased the internalisation or b) in addition enhanced the binding ability of vindoline derivatives to the tubulin (These investigations are in progress.). In two cell lines (HL-60 and MDA-MB-231) the conjugates 4, 5, 6 and 7 had similar activity, while on MCF-7 cells only conjugates with Trp (5, 6, and 7) showed increased effect. It seems that the lack of Trp moiety influences negatively the cytostatic activity on this cell line. As the examination of the in vivo activity of conjugates 5 and 6 was studied in mouse tumor models their in vitro toxicity was determined on the relevant cells too. The L- and D-Trp conjugates exhibited the similar cytostatic activity on both cell cultures. It is interesting to mention that conjugate 5 was more active. The effect of the configuration of Trp was noticed only on these cell cultures.
The in vivo effect of conjugates 5 and 6 was studied on mice bearing C26 and P338 tumor xenografts. P388 and C26 cells were injected intraperitoneally or subcutaneously into BDF1 or BALB/c mice, respectively. The animals were treated at Day1 and Day6 after tumour cell inoculation with the tested compound at doses of 10, 20, or 40 mg/kg (Figure 1).

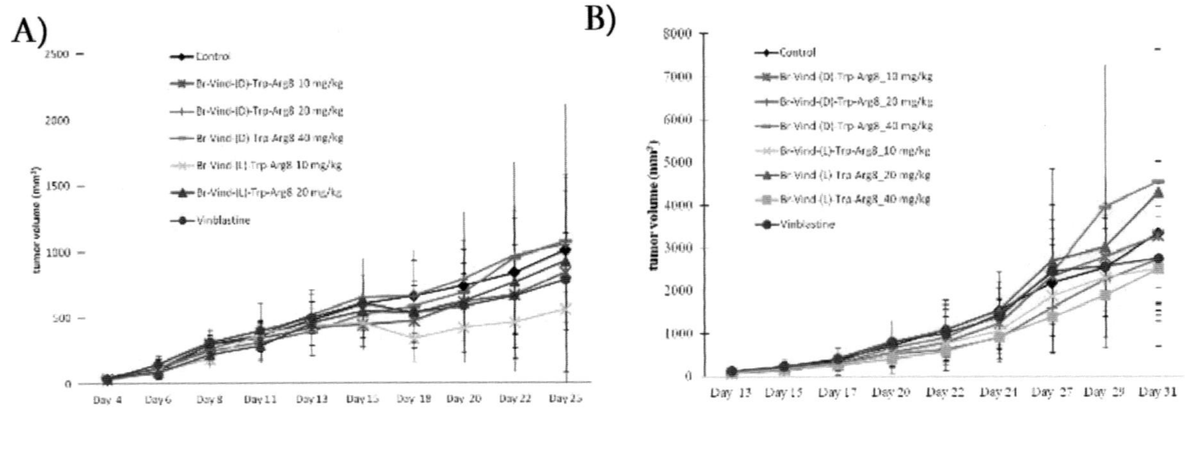

Figure 1. In vivo effect of conjugates on the growth of A) mouse leukaemia (P388) and B) murine colon carcinoma (C26) cells

The data showed that only the conjugate 5 could inhibit the growth of P388 xenograft at the lowest concentration, but it was ineffective on the C26 xenograft. It was interesting that the vinblastine used as control could not inhibit either of the growth of these xenografts. The effect of conjugates 5 and 6 as well as vinblastine on the survival of mice inoculated intraperitoneally by mouse leukaemia cells was also studied. Only the vinblastine showed increased the life span, none of the two conjugates was effective.
In conclusion, we found that the conjugation with octaarginine increased significantly the in vitro cytostatic activity of two vindoline derivatives. Based on our knowledge these are the first in vitro active derivatives of vindoline. Although the conjugation with octaarginine could not inhibit the tumour growth or the life span in vivo, these conjugates could be promising candidates for further development. The study of the mechanism of action may reveal the possible direction of structural modification.

Acknowledgements
This study was supported by a grant: OTKA K104385. Z. Bánóczi acknowledges the support of Foundation for the Hungarian Peptide and Protein Research, Budapest, Hungary.

References
01. Gorman, M. et al. J. Am. Chem. Soc. 1962, 84, 1058
02. Moncrief, J. W. et al. J. Am. Chem. Soc. 1965, 84, 4963
03. Prakash, V. et al. Biochemistry 1991, 30, 873
04. Gorka-Kereskényi, Á. et al. Heterocycles, 2007, 71, 1553
05. Keglevich, Péter et al. Heterocycles, 2013, 87, 2299

Levente E. Dókus[a], Ildikó Szabó [a], Szilvia Bősze[a], Ferenc Hudecz[a,b], Zoltán Bánóczi[b]
[a]MTA-ELTE Research Group of Peptide Chemistry, H-1117, Budapest, Pázmány P. sétány 1/A; [b]ELTE Department of Organic Chemistry, H-1117, Budapest, Pázmány P. sétány 1/A.

Introduction

Cell-penetrating peptides (CPP's) may be a useful tool to deliver different bioactive molecules - for example intracellular enzyme inhibitors, antitumor drugs - into cells and thus improve their penetration, alter their activity [1]. Octaarginine, a well-studied CPP, can penetrate very effectively and can deliver a wide range of cargos into cells, but the shorter oligoarginines have very poor cell-penetrating ability [2]. The position and st eric structure of the interacting residues may be important. This hypothesis is supported with data of the increased cellular-uptake of RGRRGRRGRR [3]. Furthermore, some proline-rich cell-penetrating peptides were described in which the presence of proline influenced the internalisation ((VXLPPP)n , where X=His, Arg or Lys and n=1–3) [4]. Our earlier results showed that the presence of 4-((4-(dimethylamino)phenyl)azo)benzoyl (Dabcyl) group increased the internalization of octaarginine conjugate [5]. In this study our aim was to design and synthesise new oligopeptides with four Arg residues which contain glycine or proline in the sequence.

Results and Discussion

Seven compounds were prepared by solid phase peptide synthesis with Fmoc/^{t}Bu strategy using side chain protected Nα-Fmoc-amino acid derivatives, diisopropyl-carbodiimide (DIC) and 1-hydroxybenztriazole (HOBt) in dimethyl-formamide (DMF) (Figure, Table.). The fluorescent dye was attached to the ε-amino group of lysine residue built into at the C-terminal. For studying the effect of the position of fluorescent dye, RPRPRPRK and RGRGRGRK peptides were labelled at the Nε- amino group of C-terminal Lys residue (compounds 1.,2. and 5.,8). Peptides were purified by semi-preparative RP-HPLC and were characterised by analytical RP-HPLC and ESI-MS (Table).
The cellular-uptake of peptides was studied on HL-60 (human promyelocytic leukaemia) cells. Cells were grown in RPMI-1640 supplemented with 10% FCS, L-glutamine (2 mM) and gentamicin (160 μg/ml). 10^5 cells per well were plated on 24-well plates. After 24 h incubation at 37 °C, cells were incubated with the compounds dissolved in the corresponding serum-free media at 1, 5 and 10 μM concentrations for 90 min. Cells treated with serum-free media was used as control. After the incubation the solution was removed and cells were treated with 100 μl trypsin (C= 0.5 g/l) for 10 min. The fluorescence intensity of cells were measured by flow cytometry and was given as the percentage of the fluorescence intensity of cells were treated by tetraarginine derivative (Table 1).
The internalisation of the four arginine containing peptides were higher than the tetraarginine with C-terminal Lys (compound 1), but lower than that of the hexaarginine derivative (compound 2). Tetraarginine with two proline residues between the arginine moieties (compound 3) internalised more efficiently than derivative with only one proline residue between two arginines (compound 4).

But the effect of insertion Gly between the arginines was opposite (compounds 6 and 7). The position of the fluorescent dye (Cf) (e.g. compounds 7 and 8) influenced on the cellular-uptake (Table).
Table: The chemical characterisation and cellular uptake of the synthesised derivatives
(Relative fluoresce= (Fluorescence of tested compound / Fluorescence intensity of compound 1.) x100).

Compounds 1 µM		Relative fluorescence intensity Frel.= (Fx / $FAcR_4K(Cf)$) x100			Rt_{min} [a] M calc.	ESI-MS[b]	
		5 µM	10 µM			M meas.	
1.	Ac-RRRRK(Cf)	100	100	100	12.8	1170.4	1170.5
2.	Ac-RRRRRRK(Cf)	207	315	295	11.9	1482.2	1482.7
3.	Cf-RPPRPPRPPR-NH_2	144	264	259	13.3	1581.4	1581.8
4.	Cf-RPRPRPR-NH_2	112	117	130	12.2	1290.1	1290.1
5.	Ac-RPRPRPRK(Cf)-NH_2	105	108	100	12.3	1462.2	1461.5
6.	Cf-RGGRGGRGGR-NH_2	102	155	151	12.2	1341.1	1341.7
7.	Cf-RGRGRGR-NH_2	123	228	227	12.2	1169.8	1170.7
8.	Ac-RGRGRGRK(Cf)-NH_2	103	120	109	12.1	1341.5	1341.9
9.	Dabcyl-RRRRK(Cf)-NH_2	147	519	830	15.6	1379.2	1379.7
10.	Dabcyl-RGRGRGRK(Cf)-NH_2	316	474	8276	14.9	1550.1	1550.3

[a]Column: Agilent Zorbax SB-C18 4.6mm x150mm, 100Å; Gradient: 0 min-0% B, 2 min-0% B, 22 min-90% B; Eluents: 0.1 V/V% TFA/H_2O (A), 0.1 V/V% TFA/80V/V% acetonitrile/20 V/V% H_2O (B); Flow rate:1ml/min; λ_{det}=220 nm. [b] ESI-MS: Bruker Esquire 3000 plus (Germany). The sample was dissolved in acetonitrile-water (50:50, v/v), 0.1% acetic acid

Derivatives bearing the Cf-group at the N-terminal of the oligopeptide (compounds 4 and 7) have a slightly higher cell-penetrating ability than compounds 5 and 8 containing Cfgroup on the side-

chain of lysine at the C-terminal. The presence of the Dabcyl-group resulted in increased internalisation of tetraarginine with C-terminal Lys (compound 9).
The effect of Dabcyl group was more pronounced in case of compound 10 (Table). At high concentration (10 μM) the internalisation of this compound was the highest among peptides studied.
In conclusion, we observed, that the insertion of glycine or proline did not increased significantly the internalisation of oligopeptides with four arginines. We found that the commonly used chromophore, Dabcyl group can enhance remarkably the internalisation of oligopeptides with four arginines but its effect was more significant in case of glycine modified derivative, as compared with those having glycine residues.

Acknowledgements
This study was supported by grant from: OTKA K104385 and Japanese-Hungarian Intergovernmental program (TÉT_12_JP-1- 2014-0023). L.E. Dókus acknowledges the support of Centenary Foundation of Gedeon Richter Plc., Budapest, Hungary, Foundation for the Hungarian Peptide and Protein Research, Budapest, Hungary and of Pázmány-Eötvös Foundation, Budapest, Hungary.

References
01. P. A. Wender, et. al. PNAS (2000) 97, , 13003-13008.
02. S. Futaki, et.al. J. BIOL. CHEM. (2001) 276, 5836-5840.
03. J. B. Rothard, et.al. J. MED. CHEM. (2002) 45, 3612-3618.
04. S. Pujals, et.al. ADV. DRUG DELIV. REW. (2008) 60, 473-484.
05. Z. Bánóczi et.al. From cell-penetrating peptides to Nanoparticles for Cellular delivery, July 1-3, 2015, Paris, France. Abstract book: p2.

MULTIVALENT PEPTIDE-POLYMER CONJUGATES AS INHIBITORS FOR PROTEIN-PROTEIN INTERACTIONS

C. Fischer[1], K. Koschek[2], O. Krylova[2], M. Wieczorek[3], S. Gupta[3], L. Henning[3], M. Bertazzon[3], V. Durmaz[4], C. Freund[2,3], M. Weber[4], J. Rademann[1,2]
[1]Institut für Pharmazie Freie Universität Berlin, Königin-Luise-Str. 2+4, 14195 Berlin
[2]Leibniz-Institut für molekulare Pharmakologie, Robert-Rössle-Str. 10, 13125 Berlin
[3]Institut für Chemie und Biochemie der Freien Universität Berlin, Thielallee 63, 14195 Berlin
[4]Konrad-Zuse-Zentrum für Informationstechnik Berlin, Takustr. 7, 14195 Berlin

PP VII - 113

For rigid arrangements of multivalent receptors and ligands an exponential increase of the binding affinity can be expected. In nature, however, many multivalent receptors are characterized by flexible arrangements of binding sides. This flexibility seems to be important for the function of the proteins. It can be introduced by regions of inherent structural flexibility (e.g. between receptor domains of multi-receptor proteins). Furthermore, the insertion of binding sites into membranes can procure the relative mobility.[1] The key challenge in the design of multivalent ligands for flexible receptors is to match the flexibility of the ligand and the receptor. In other words, the enthalpic gain has to be balanced with the entropic loss of the system. In this study various peptide-polymer conjugates were investigated as flexible, multivalent ligands for a bivalent protein receptor, the tandem-WW domain of the pre-mRNA splicing factor FBP21 (Fig. 2).[2]

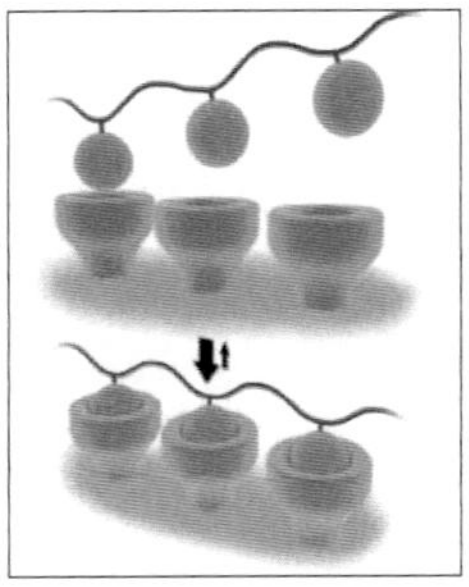

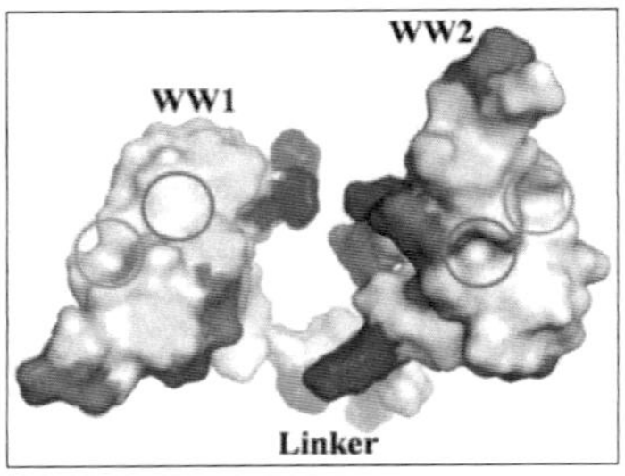

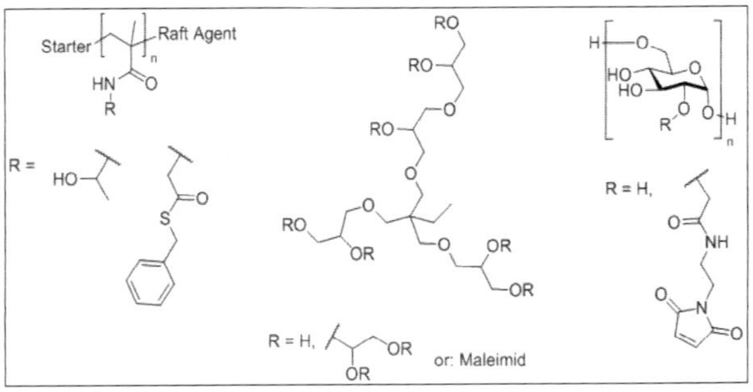

Figure 1: Multivalent binding of ligand-receptor systems; Figure 2: tandem-WW domain of FBP21; Figure :3 Polymeric backbone structures tested for use in peptide-polymer conjugates

The employment of peptide-polymer conjugates (PPC) as ligands for multivalent protein targets leads to the inhibition of intracellular protein-protein interactions. It has been demonstrated that the multivalent presentation of the ligands enhances binding affinities and therefore increases the inhibitory effect of the conjugates.[3] Previous work on the use of different polymeric supports (pHPMA, hyperbranched polyglycerol and dextrans, Fig. 3) showed dextrans to be the most promising polymeric backbones for these peptide-polymer conjugated ligands.[1]
The selective modification of the dextran conjugates could be achieved in good yields via synthesis of maleimido dextran and a subsequent maleimide-thiol coupling. The mono- and bivalent functional peptide ligands W and W_2 had been identified as suitable binders of the target tandem-WW domain of FBP21 beforehand using phage display. For further biological studies, the use of cell-penetrating peptides TAT and Nona arginine (R_9) as well as a Rhodamin B-based fluorophore was investigated.

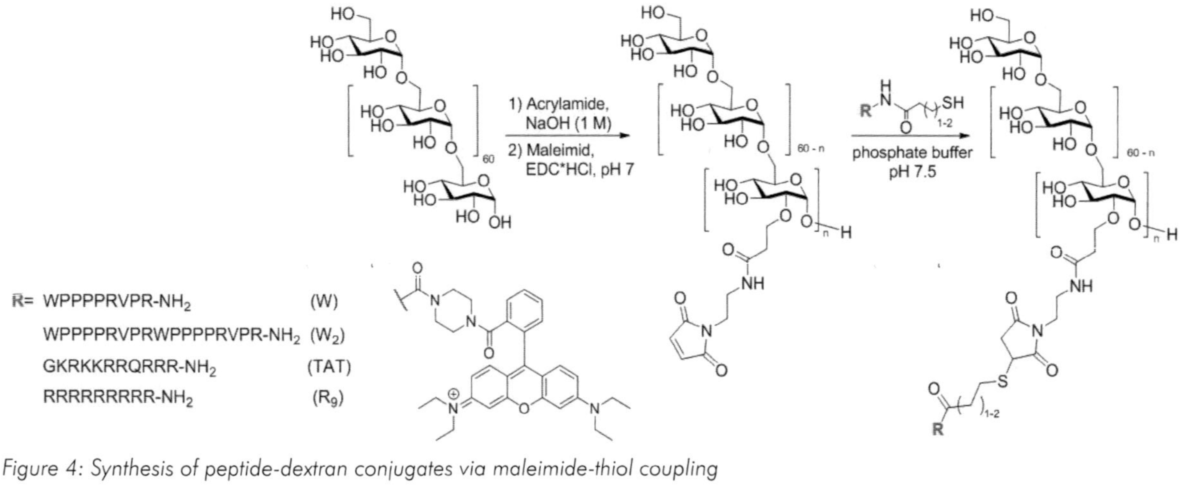

Figure 4: Synthesis of peptide-dextran conjugates via maleimide-thiol coupling

The synthesized peptide-dextran conjugates were measured using isothermal titration calorimetry yielding their respective K_D values (Fig. 5, Tab. 1)

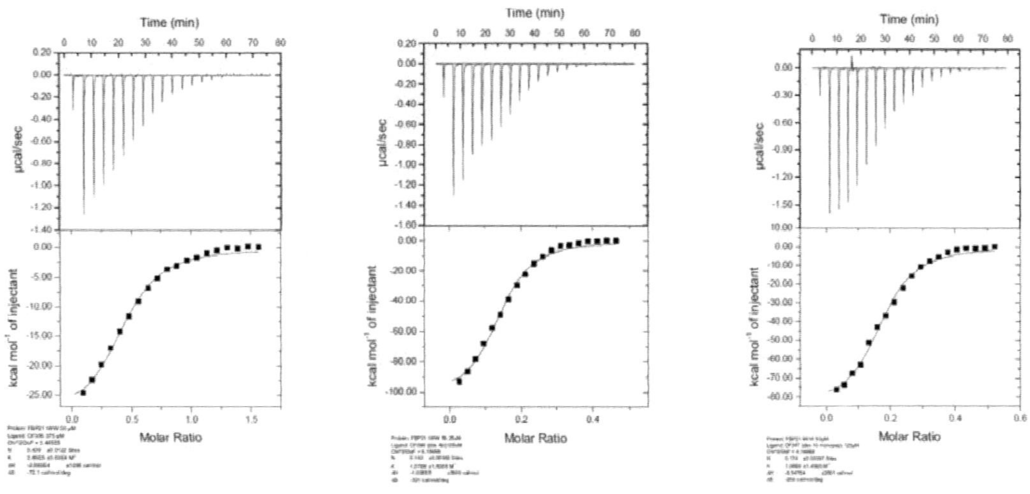

Figure 5: selected data from ITC measurements showing effects of multivalency in KD values: 4 W-ligands (KD = 56.6 µM) 10 W-ligands (KD = 9.90 µM), 4 W2-ligands (KD = 3.36 µM)

OP
PP I
PP II
PP VI
PP VII
PP VIII
PP IX
PP X
PP XI
PP XII
PP XIII
PP XIV

The monovalent, free peptide WPPPPRVPR (W) showed a moderate binding affinity (K_D = 85.2 μM) which could be enhanced by the coupling of multiple copies of the peptide onto the dextran backbone. Furthermore, through the use of the bivalent peptide W_2 the binding affinities of the peptide-dextran conjugates to the target protein could be enhanced even more.

	# peptides	Loading density [%]	$K_{D,rel}$ [μM]	$K_{D, PPC}$ [μM]
$W_{monovalent}$	1	-	85.2	85.2
	6	9.7	9.43	56.6
	8	12.9	2.88	23.0
	10	16.1	0.99	9.90
$W_{2\ bivalent}$	1	-	3.44	3.44
	4	6.5	0.84	3.36

Table 1: binding affinities of peptide-polymer conjugates as measured by ITC

Currently, the reported findings are employed for the development of cell-permeable multivalent protein ligands and inhibitors. For this purpose, conjugates bearing fluorophores and cell-penetrating peptides in addition to the functional peptides were synthesized. The cellular uptake and distribution is now being tested using confocal microscopy. It could be seen that the use of nona arginine (R_9) as a cell-penetrating peptide facilitated the uptake of the conjugates via endosomes into HEK293 cells, but could not release the cargo upon delivery (Fig. 6). The use of the cell-penetrating peptide TAT enables the release of the conjugates into the cytosol after uptake into the cells, showing a diffuse distribution of the rhodamine-labeled conjugate (Fig. 7). For a further uptake into the cell-nucleus (where the protein target is localized) a new uptake strategy of a combined cell-penetrating peptide and nuclear localization sequence is investigated at the moment.

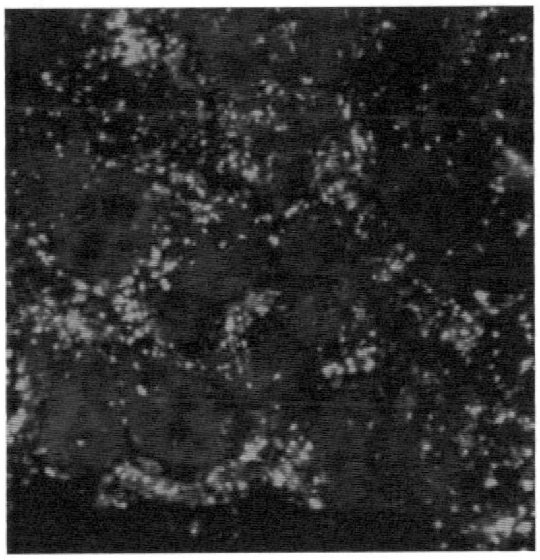

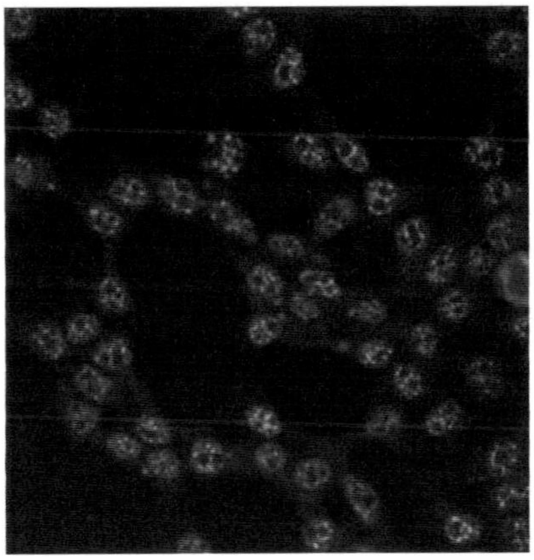

Figure 6: FAM-labeled conjugates (R_9) taken up into endosomes (green: FAM, blue: DAPI); Figure 7: rhodamine-labeled conjugates (TAT) taken up into the cell cytosol (red: rhodamine, green: Anti-SC35 (nuclear speckles), blue: DAPI)

References

01. K. Koschek, Beilstein J. Org. Chem. 2015, 11, 837.
02. X. Huang, Biol. Chem. 2009, 284, 25375-25387
03. M. Richter, Chem. Eur. J. 2012, 18, 16708-16715.
04. M. Barth, Angew. Chem. Int. Ed. 2005, 44, 1560-1563.
05. K. Koschek, M. Dathe, J. Rademann, ChemBioChem 2013, 14, 1982-1990.
06. A. Chakrabarti, Biochem. Biophys. Ac. Biomembranes 2014, 1838, 3097-3106.

MODIFICATION OF MAGNETIC FERRITE NANOPARTICLES WITH ANTIANGIOGENIC AND ANTITUMOR PEPTIDE A7R

Anna Niescioruk[1], Dorota Niecięcka[1], Anna Puszko[1], Gerrard Y. Perret[2], Pawel Krysinski[1], Aleksandra Misicka[1]
[1] Faculty of Chemistry, University of Warsaw, Warsaw, Poland
[2] Université Paris 13, Sorbonne Paris Cité, INSERM U1125, Bobigny, France

OP
PP I
PP II
PP VI
PP VII
PP VIII
PP IX
PP X
PP XI
PP XII
PP XIII
PP XIV

Introduction

Angiogenesis plays a key role in various physiological and pathological conditions. It was first suggested by Folkman that angiogenesis is an essential component of tumour progression [1]. The fundamental proangiogenic signaling molecule is vascular endothelial growth factor ($VEGF_{165}$). Recently, many reports have suggested that neuropilin (NRP-1) may serve in tumour cells as a separate receptor for VEGF. It has been shown by Starzec et al. [2] that heptapeptide ATWLPPR (A7R) selectively inhibits $VEGF_{165}$ binding to NRP-1 and in vivo decreases breast cancer angiogenesis and growth [3]. Nanotechnology is a constantly growing field of science that offers promising applications for cancer detection, diagnosis and treatment. In recent years, magnetic ferrite nanoparticles with different coatings have been recently widely investigated due to their desirable magnetic properties in biomedicine and bioengineering fields [4]. One of the potential applications is their use in targeted drug delivery. We report the synthesis, physicochemical characterization studies and preliminary in vitro studies of magnetic ferrite nanoparticles modified with heptapeptide A7R. Such a multi-purpose conjugate can be effectively guided to and maintained within the area of tumor with help of an external magnetic field, whereas the A7R ligand inhibits angiogenesis by interaction with specific receptors (NRP-1) located in a large amount on the surface of some cancer cells.

Results and Discussion

The synthesis of A7R peptide was carried out manually on the Wang resin, by the Fmoc solid-phase method, with the use of TBTU/6-Cl-HOBt as the coupling reagents, and controlling presence of a free amino group by the Kaiser or chloranil tests. The final peptide was cleaved from the resin by TFA and purified by preparative RP-HPLC using C_{12} column. Its structure was confirmed by ESI-MS. Magnetic nanoparticles, with the general formula $Ni_{0.5}Zn_{0.5}Fe_2O_4$, were synthesized by a co-precipitation method [5,6] and modified with sebacic acid as a linker. Carboxylic groups of the linker allowed conjugation of A7R peptide via amide bond using EDC as a coupling reagent (Figure 1). Due to the fact that COOH group of the C-terminal arginine and four C-terminal residues (LPPR) play a crucial role in inhibitory effect of A7R [2] we conjugated peptide to magnetic nanoparticles by N-terminal amine group of alanine.

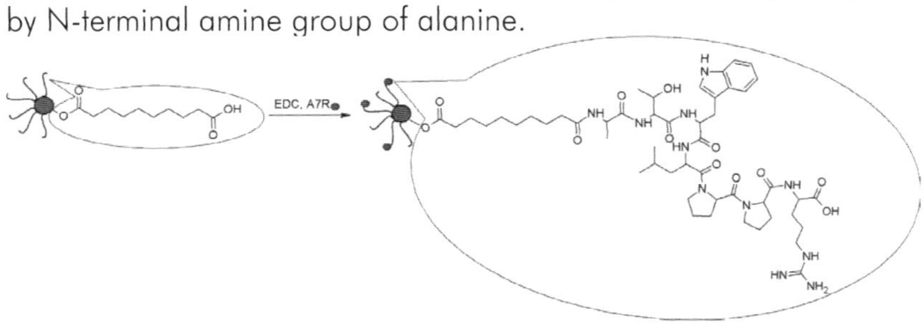

Figure 1. Reaction scheme of A7R peptide coupling to nanoparticles modified with sebacic acid

Successful conjugation of A7R peptide to iron oxide-based nanoparticles was confirmed by complementary physicochemical analysis techniques (FTIR, DLS, TEM, and TGA). Magnetic nanoparticles modified with sebacic acid and magnetic nanoparticles functionalized with A7R (conjugate) were also tested in vitro (MTS assay) for the potential cytotoxic effect against two cell lines, cancer (MDA-MB-231) and healthy (HUVEC) with NRP-1 expression. Cytotoxicity studies showed that magnetic nanoparticles modified with sebacic acid do not display significant cytotoxic activity against cancer and healthy cells. However, after the conjugation of A7R peptide with magnetic nanoparticles cell viability decreased, especially on HUVECs cells. One of possible explanation for this decrease in cell viability may be that these nanoparticles are taken up by the cells more intensively as a result of A7R binding to the NRP-1 receptors on the surface of cells.
We plan to perform more in vitro assay (e.g. angiogenic and migration) to prove antiangiogenic and antitumor activity of obtained conjugate.

Acknowledgements
This work was supported by National Science Centre (NCN) grant N204 350940 and co-financed by the EU from the European Regional Development Fund under the Operational Programme Innovative Economy, 2007-2013, and with the use of CePT infrastructure financed by the same EU program and a grant from the University of Warsaw for young researchers, no. 120000-501/86-DSM-110200.

References
01. J. Folkman, The New England Journal of Medicine, 1971, 285, 1182-1186
02. A. Starzec, P. Ladam, R. Vassy, S. Badache, N. Bouchemal, A. Navaza, C. Hervé du Penhoat, G. Y. Perret, Peptides, 2007, 28, 2397-2402
03. A. Starzec, R. Vassy , A. Martin, M. Lecouvey, M. Di Benedetto, M.Crépin , G. Perret, Life Sciences, 2006, 79, 2370–2381
04. A. K. Gupta, M. Gupta, Biomaterials, 2005, 26, 3995–4021
05. M. Brzozowska, P. Krysinski, Electrochimica Acta, 2009, 54, 5065-5070
06. P. Majewski, P. Krysinski, Chem. Eur. J., 2008, 14, 7961-7968

SYNTHESIS, CYTOTOXICITY AND CELLULAR UPTAKE OF NEW, BRANCHED POLYMER CONJUGATES CONTAINING HYDROPHOBIC AMINO ACIDS OR ARGININE AND METHOTREXATE

Rita Szabó[1], Mónika Sebestyén[1], György Kóczán[1], Ferenc Hudecz[1,2]
[1] MTA-ELTE Research Group of Peptide Chemistry, Pázmány P. st. 1/A, H-1117, Budapest, Hungary
[2] Department of Organic Chemistry, Eötvös L. University, Pázmány P. st. 1/A, H-1117, Budapest, Hungary

PP VII - 130

Selective killing of intracellular parasites causing the severe tropical disease, leishmaniasis, can be achieved via inhibition of their folate metabolism, which is different from humans [1]. Methotrexate (MTX, L-4-amino-N[10]-methylpteroyl-glutamic acid) is a specific inhibitor of dihydrofolate reductase enzyme of the parasites. In our research group the anti-leishmanial effect of methotrexate coupled to poly[L-lysine] based branched polypeptide carriers were studied earlier. Several conjugates containing amfoteric or cationic polypeptide carrier – (poly[Lys(X_i-DL-Ala_m)], XAK or poly[Lys(DL-Ala_m-X_i)], AXK) – were tested, and it was established that treatment with MTX conjugated to ALK (poly[Lys (MTX_j-DL-Ala_m-Leu_i)]) polypeptide containing a hydrophobic leucine in the side chains resulted in a reduced parasite load in L. donovani infected peritoneal macrophages in vitro as well as in the liver of Balb/c mice infected with L. donovani amastigotes in vivo [2].

Based on these findings, we synthesized a set of new polypeptide-MTX conjugates containing branched chain polypeptide carrier with hydrophobic amino acid (valine, leucine, isoleucine or norleucine) or arginine in the side chains (Figure 1.). The polymeric backbone and the oligo-Ala side chains were synthesized by polymerization of the corresponding N-carboxyanhydrides [2]. The terminal amino acids were attached as the carboxybenzyl (Z) protected active esters. The cleavage of Z protecting groups was performed by HBr/acetic acid. MTX was coupled by BOP/HOBt method, and fluorescence labeling was done by the NHS-ester of 5(6)-carboxyfluorescein. The polymers were purified by dialysis, the composition was determined by amino acid analysis.

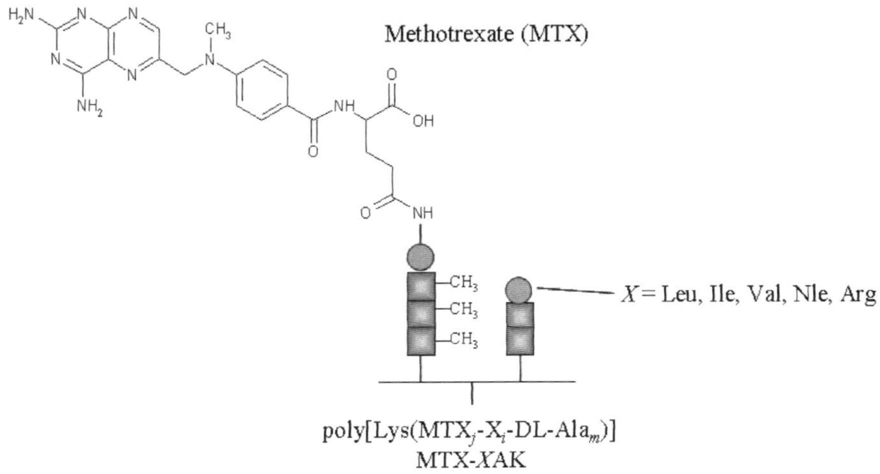

Figure 1. Chemical structure of the new MTX-polypeptide conjugates

In vitro cytotoxicity of the polypeptides and conjugates was examined by MTT assay; cellular uptake of the fluorescently labeled polypeptides and conjugates was characterized by flow cytometry and fluorescent microscopy on bone marrow derived murine macrophages (BMDM) of Balb/c origin.

Results indicate that none of the MTX conjugates were toxic to BMDM cells after 1 hour incubation, but two polypeptides, the cationic RAK (poly[Lys(Arg_i-DL-Ala_m)]) and the hydrophobic L_iK (poly[Lys(Leu_i)]) polypeptides proved to be toxic to the cells. Among the MTX conjugates, MTX-ALK (poly[Lys(MTX_i-DL-Ala_m-Leu_i)]) was slightly toxic, whereas MTX-RAK (poly[Lys(MTX_i-Arg_i-DL-Ala_m)]) and MTX-L_iK (poly[Lys(MTX_i-Leu_i)]) – similarly to the polypeptide carriers – elicited a more pronounced cytotoxicity to the macrophages after 24 hours. The cellular uptake of the polypeptides as well as the conjugates proved to be dependent on the concentration, although a decreased internalization could be observed after coupling MTX to the polypeptide carriers. The polypeptides and conjugates were localized in the cytoplasm of the macrophages following to the uptake (Figure 2).

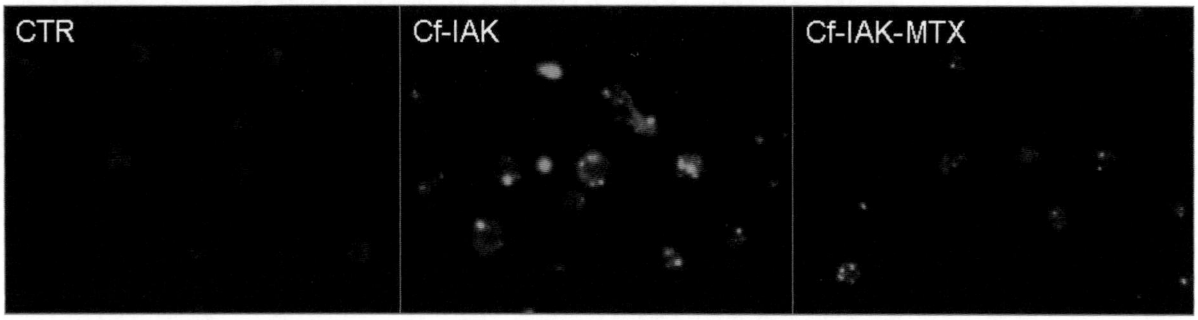

Figure 2. Intracellular localization of Cf-IAK polypeptide and Cf-IAK MTX conjugate

Cellular uptake was influenced by the side chain composition of the polypeptide carriers and on the distance of amino acid X (and MTX) from the polylysine backbone.

Acknowledgments
This work was supported by the grant from the Hungarian National Research Fund (OTKA K104385); Rita Szabó was supported by the Postdoctoral Fellowship of Hungarian Academy of Sciences (HAS).

References
01. Schüttlekopf, A.W., Hardy, L.W., Beverley, S.M., Hunter, W.N. J. Mol. Biol.,2005, 352: 105–116
02. Kóczán, Gy., Ghose, A.C., Mookerjee, A., Hudecz, F. Bioconjug. Chem., 2002, 13, 518-524.

INDOLOAZEPINONE-CONTAINING OLIGOMERS AS CELL-PENETRATING (NON)PEPTIDES: SYNTHESIS, STRUCTURATION AND IN VITRO INTERNALIZATION

Olivier Van der Poorten,[1] Baptiste Legrand,[2] Lubomir L. Vezenkov,[2] Nadir Bettache,[2] Jean Martinez,[2] Marcel Garcia,[2] Dirk Tourwé,[1] Muriel Amblard,[2] and Steven Ballet[1*]
[1]Research Group of Organic Chemistry, Vrije Universiteit Brussel, Pleinlaan 2, B-1050, Brussels, Belgium
[2]Institut des Biomolécules Max Mousseron, UMR5247 CNRS, Universités Montpellier, ENSCM, 15 avenue Charles Flahault, 34000 Montpellier, France

Introduction

The hydrophobic nature of cellular membranes often prevents polar bioactive molecules such as peptides, proteins, liposomes and oligonucleotides from efficient cell entry. One strategy to traverse the phospholipid bilayer and to deliver therapeutic agents into cells involves the use of peptide vectors.[1] Such vectors, referred to as cell-penetrating peptides (CPPs), have shown the intrinsic ability to efficiently cross the cell membrane while carrying a variety of (non-)covalently linked bioactive cargoes.[2,3] A recent discovery showed that noncationic dihydro-1,5-benzothiazepin-4(5H)-one (DBT) oligomers (Figure 1) penetrate cells more efficiently than the well-established polyarginine peptide vectors.[4] However the synthesis of the DBT scaffold does not allow an easy introduction of amino acid-based appendages to influence, for instance, the polarity and/or the solubility of the resulting oligomers. Hence, we synthesized oligomers encompassing azepinone-constrained amino acids which allow easier structural and functional fine-tuning for improved transport of bioactive cargoes.

In our study, various amino-indoloazepinone (Aia) dipeptidic scaffolds of type Aia-Xxx were prepared in solution starting from commercially available L-Trp following a previously reported procedure.[5] Overall, the aim was to examine the influence of the second amino acid in the Aia-Xxx dipeptides on the folding properties of $[\text{Aia-Xxx}]_n$ oligomers. Varying the type of amino acid, their chirality and their substitution pattern (e.g. α-AA ↔ β^3-h-AA) enables us to gain insight into the structural behavior of oligomers consisting of constrained amino acids and the influence thereof on their cell-penetrating properties.

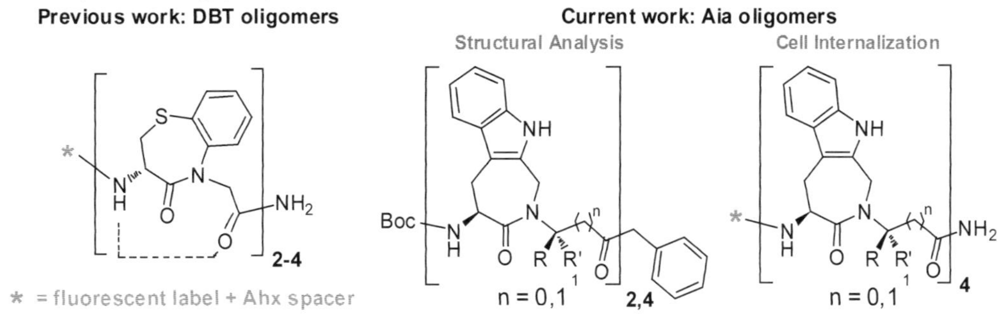

Vezenkov, L. *et al.* *Bioconjugate Chem.* **2010**, *21*, 1850.

Figure 1. Dihydro-1,5-benzothiazepin-4(5H)-one (DBT) oligomers (left) versus amino-indoloazepinone (Aia) oligomers (right).

Results and Discussion

The dipeptidic Aia-Xxx building blocks were coupled on $SOCl_2$-preactivated 2-Cl trityl and on Rink amide resins followed by Fmoc-based solid phase peptide synthesis (SPPS) using DIC/HOBt activation. After overnight TFE/CH_2Cl_2 cleavage from the 2-Cl trityl resin, C-terminal carboxylic acids were successfully benzylated using $BnBr/Cs_2CO_3$ in DMF at room temperature. In foldamer crystallogenesis, most of the crystal structures have N-terminal Boc and C-terminal benzyl ester capping groups.[6]

Conformational studies were performed on the various Boc-[Aia-Xxx]$_n$-OBn oligomers (Figures 2-3). The solution structures of the Aia-Gly, homochiral Aia-L-Ala and heterochiral Aia-D-Ala dimers and tetramers were investigated by CD and NMR spectroscopy. As expected for peptide mimetics, the CD signatures of the various oligomers in MeOH at 20 °C were atypical with globally positive bands around 220 (strong) and 250 nm, and negative ones about 235 and 272 nm (very large) (Figure 2).

Well-resolved 1D and 2D NMR spectra (COSY, TOCSY, ROESY, ^{15}N and ^{13}C-HSQC) were obtained in CD_3OH at 298 K. In contrast to the Aia-Gly and Aia-D-Ala dimers, the homochiral Aia-L-Ala dimer showed numerous NOE correlations in particular between the two Aia residues. Structural calculations using NMR-derived distance restraints converged toward a turn conformation mainly stabilized by the aromatic π-stacking between the indole rings of the Aia residues (Figure 3).[7]

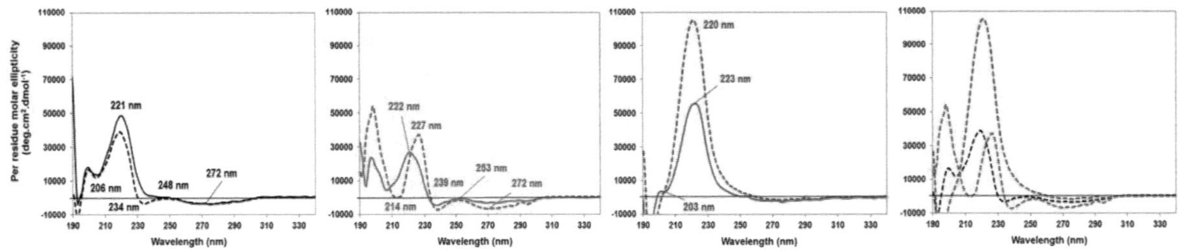

Figure 2. CD spectroscopy of Aia-Gly (black), Aia-L-Ala (red) and Aia-D-Ala (violet) dimers (plain) and tetramers (dashed) in MeOH at 20°C.

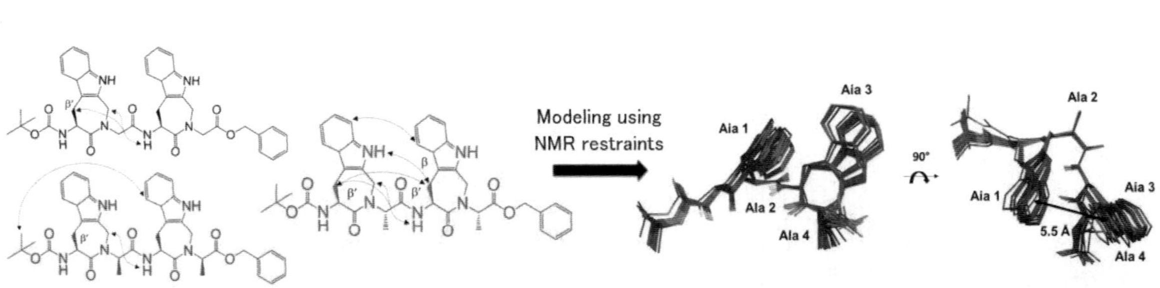

Figure 3. Comparison of the inter-residue NOE correlations and the superimposition of the 20 best NMR structures.

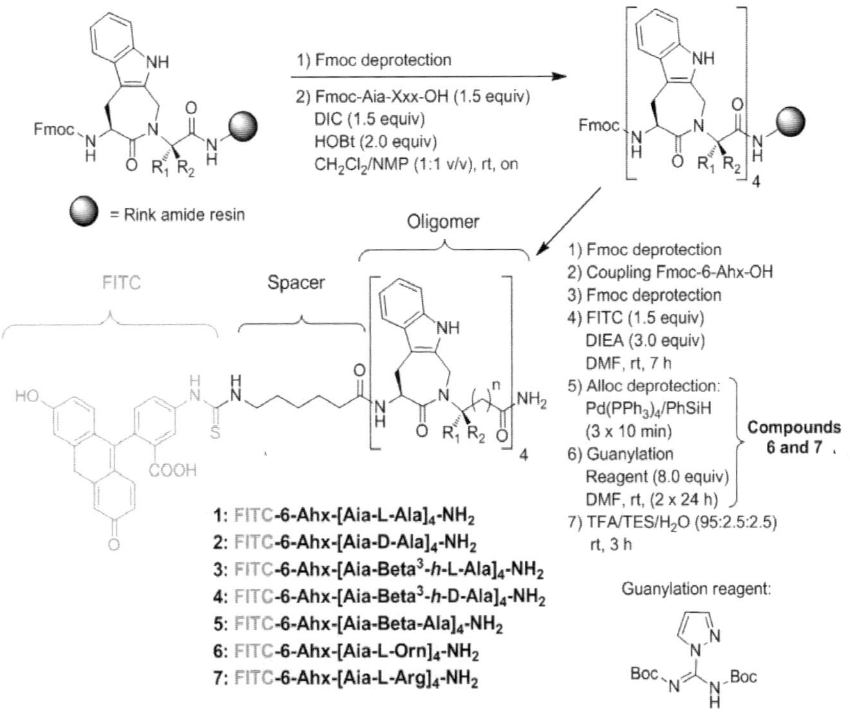

Scheme 1. SPPS approach to FITC-labeled [Aia-Xxx]$_4$ oligomers.

In order to gain insights into the cell-penetrating properties of the designed oligomers, Aia dipeptidic scaffolds were repeatedly coupled to present their corresponding tetramers on Rink amide resin, and labeled at the N-termini with fluorescein isothiocyanate. When coupling the Fmoc-Aia-Xxx-OH dipeptides, mild base-free DIC/HOBt activation was used since the use of HATU/DIEA resulted in ± 15 % racemization per coupling step. The hydrophobic amino-caproyl spacer was used to separate both the fluorophore and the vector and N,N′-bis-Boc-1H-pyrazole-1-carboxamidine was the reagent of choice for the conversion of Orn residues into Arg(N^{ω},$N^{\omega'}$-bis-Boc) on solid support (Scheme 1).

Preliminary in vitro internalization results using human breast cancer MDA-MB-232 cells indicated significantly improved cell permeation in case of the [Aia-Xxx]$_4$ oligomers in comparison with Penetratin and [DBT]$_4$ as positive references (not shown). Confocal microscopy images confirmed the high internalization of the Aia-oligomers.

Acknowledgments

OVDP and SB are grateful to the Flanders Innovation & Entrepreneurship (VLAIO) for the financial support.

References

01. Koren E.; Torchilin, V. P. Trends. Mol. Med. 2012, 18, 386-393.
02. Bechara, C.; Sagan, S. FEBS Lett. 2013, 587, 1693-1702.
03. Milletti, F. Drug. Discov. Today 2012, 17, 850-860.
04. Vezenkov, L. L.; Maynadier, M.; Hernandez, J.-F.; Averlant-Petit, M.-C.; Fabre, O.; Benedetti, E.; Garcia, M.; Martinez, J.; Amblard, M. Bioconjugate Chem. 2010, 21, 1850-1854.
05. Pulka, K.; Feytens, D.; Van den Eynde, I.; De Wachter, R.; Kosson, P.; Misicka, A.; Lipkowski, A.; Chung, N. N.; Schiller, P. W.; Tourwé, D. Tetrahedron 2007, 63, 1459-1466.
06. Legrand, B.; André, C.; Moulat, L.; Wenger, E.; Didierjean, C.; Aubert, E.; Averlant-Petit, M. C.; Martinez, J.; Calmes, M.; Amblard, M. Angew. Chem. Int. Ed. 2014, 53, 13131-13135.
07. Knuhtsen, A.; Legrand, B.; Van der Poorten, O.; Amblard, M.; Martinez, J.; Ballet, S.; Kristensen, J. L.; Pedersen, D. S. Chem. Eur. J. 2016, 22, 1-8.

THE HELICAL SCREW SENSE OF BLUE-COLORED, BIS-NITRONYL NITROXIDE PEPTIDES AS REVEALED BY A VIBRATIONAL CIRCULAR DICHROISM ANALYSIS

Marta De Zotti[1]*, Karen Wright[2], Edouard d'Aboville[2], Antonio Toffoletti[1], Claudio Toniolo[1], Giovanna Longhi[3], Giuseppe Mazzeo[3], Sergio Abbate[3], Fernando Formaggio[1]
[1] Department of Chemistry, University of Padova, 35131 Padova, Italy, e-mail: marta.dezotti@unipd.it
[2] Institute Lavoisier de Versailles, UMR 8180, University of Versailles, 78035 Versailles, France
[3] Department of Molecular and Translational Medicine, University of Brescia, 25123 Brescia, Italy

Introduction

The Ullman imidazolinyl nitronyl nitroxide (NN) monoradicals [1] have been extensively investigated, in particular as spin probes and organic magnetic materials. We previously reported the synthesis, configurational and conformational assignments, and physico-chemical properties of a tripeptide with a central, chiral, blue-colored (R)-Aic(NN) residue (Figure 1), where Aic(NN) is the helicogenic 2-amino-5-nitronylnitroxideindan-2-carboxylic acid [2]. Here, we discuss the synthesis and characterization of two peptides, each with two pendant, chiral, nitronyl nitroxide free radical units at positions i and i+3 (Figure 1). In particular, to unambiguously identify the screw sense adopted by our helical peptides, we exploited vibrational CD (VCD), as the conformational analysis on the basis of electronic CD (ECD) in the far-UV region is not helpful (the ECD spectrum is heavily biased by the nitronyl nitroxide absorptions). Instead, in the IR absorption region the strong vibrational transition of the nitronyl nitroxide chromophore (near 1360 cm^{-1}) does not overlap the amide

Boc-(*R*)-Aic(NN)-(L-Ala)$_2$-(*R*)-Aic(NN)-L-Ala-OMe **(PENTA)**

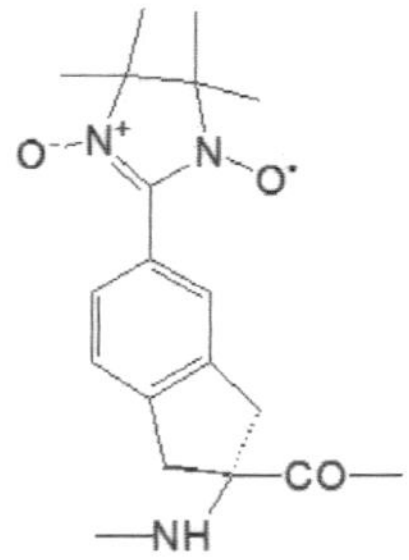

Boc-[L-Ala-(*R*)-Aic(NN)-L-Ala]$_2$-OMe **(HEXA)**

(A, I and II) absorptions.

Fig. 1. Amino acid sequences of the Aic(NN) penta- and hexapeptides discussed in this work and the chemical structure of the blue-colored (due to the broad and weak NN n→π absorption band near 590 nm) (R)-Aic(NN) residue.*

Results and Discussion

The FT-IR absorption spectra of the two bis-Aic(NN) peptides in diluted (1mM) $CDCl_3$ solution revealed high ratios for the areas under the peaks at 3330 cm^{-1} (H-bonded NHs) and 3415 cm^{-1} (free NHs), indicative of the onset of highly folded, intramolecularly H-bonded molecules. The 600 MHz 2D-NMR analysis of the simpler to study, synthetic precursor of the pentamer [where each of the two side-chain imidazolinyl nitronyl nitroxide moieties is replaced by a nitrile (CN) function] in CD_3OH allowed us to conclude that our conformationally restricted, short peptides adopt, as expected, 3_{10}-helical structures. This result is based on the observation of all of the i→i+1 NH-NH and i→i+2 αCH-NH correlations diagnostic of this conformation. We attribute the helical propensities to the presence of two Cα-tetrasubstituted α-amino acids in each peptide sequence [3].

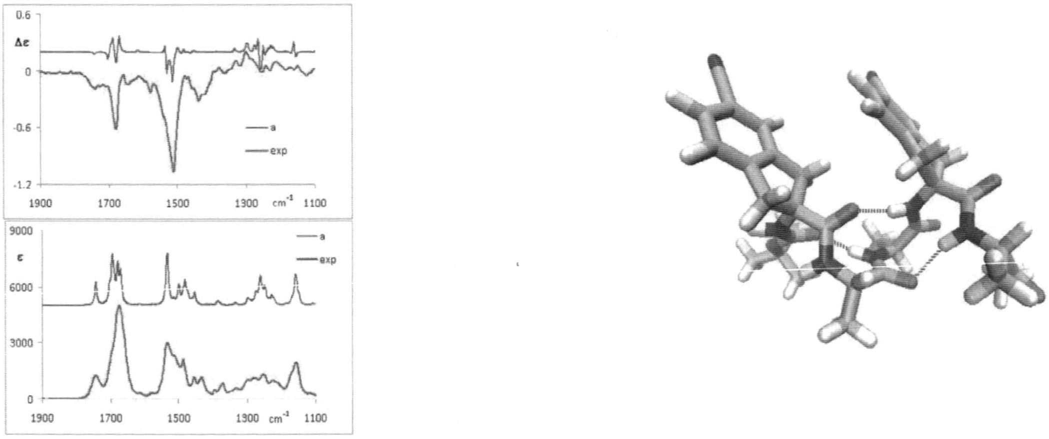

Fig. 2. Left: experimental (red) and calculated (blue) IR absorption (bottom) and VCD (top) spectra of the bis-Aic(CN) pentapeptide. Computations of the VCD spectra were performed on the basis of the lowest-energy conformer shown on the right.

The VCD analysis was carried out on the bis-Aic(CN) pentapeptide (Figure 2) and the Aic(N-N)-hexapeptide. Our experimental data, combined with DFT calculations, strongly support the view that our peptides are folded in right-handed 3_{10}-helices. This conclusion is primarily supported by the occurrence of the strong and negative amide II band (near 1520 cm^{-1}) [4,5].

Finally, from the experimental and simulated cw-EPR spectra of the bis-Aic(NN) hexapeptide in toluene solution (where nine hyperfine lines were observed) and in frozen toluene (120 K), and using the point dipole approximation, we obtained a value of 19.4 Gauss for the interaction parameter D. This result allowed us to estimate a value 11.0 Å for the distance between the middle of the two nitroxide N-O bonds, which is compatible with a 3_{10}-helical peptide with the two Aic(NN) residues located on the same face of the helix.

References

01. Osiecki, J.H.; Ullman, E.F. J.Am.Chem.Soc. 1968, 90, 1078-1079.
02. Wright, K.; d'Aboville, E.; Scola, J.; Margola, T.; Toffoletti, A.; De Zotti, M.; Crisma, M.; Formaggio, F.; Toniolo, C. Eur. J. Org. Chem. 2014, 1741-1752.
03. Toniolo, C.; Crisma, M.; Formaggio, F.; Peggion, C. Biopolymers (Pept. Sci.) 2001, 60, 396-419.
04. Yasui, S.C.; Keiderling, T.A.; Bonora, G.M.; Toniolo C. Biopolymers 1986, 25, 79-89.
05. Yasui, S.C.; Keiderling, T.A.; Formaggio, F.; Bonora, G.M.; Toniolo C. J.Am.Chem.Soc. 1986, 108, 4988-4993.

CHEMICAL SYNTHESIS AND EPR INVESTIGATIONS ON SPIN-LABELED CHALCIPORIN A ANALOGS

Barbara Biondi[1], Chiara Pignaffo[1], Cristina Peggion[1], Marta De Zotti[1], Fernando Formaggio[1*], Annalisa Dalzini[1], Marco Bortolus[1], Victoria N. Syryamina[2], Yuri D. Tsvetkov[2], Sergei A. Dzuba[2], Claudio Toniolo[1]
[1] Department of Chemistry, University of Padova, 35131 Padova, Italy, e-mail: fernando.formaggio@unipd.it;
[2] Institute of Chemical Kinetics and Combustion, 630090 Novosibirsk, Russian Federation

Introduction

Chalciporin A is a 14-amino acid peptaibol isolated from a strain of Sepedonium chalcipori [1]. Peptaibols (also termed peptaibiotics) [2] are relatively short peptides, each characterized by the presence of several, strongly helicogenic Aib residues [3], and blocked at the Nterminus by an acyl moiety and at the C-terminus by an 1,2-amino alcohol. Thanks to their helical structures, peptaibols display a relevant ability to interact with phospholipid bilayers. Consequently, not surprisingly, they have been found to exhibit antimicrobial, antifungal or anticancer activity, but they can also target healthy human cells. Therefore, the study of their mechanism of action may lead to the design of new drugs.

In this Communication, we describe the chemical syntheses and EPR properties of chalciporin A and two analogs thereof containing the nitroxide free radical, conformational constrained, helicogenic α-amino acid TOAC [4]. The amino acid sequences of the three peptides examined are given in Figure 1.

1 Ac-Trp^1-Val-Aib-Val-Ala^5-Gln-Ala-Aib-Ser-Leu^{10}-Ala-Leu-Aib-Gln^{14}-Lol
2 Ac-------Ala^2--TOAC^{13}---------
3 Ac-------Ala^2-TOAC^3--TOAC^{13}---------

Fig. 1. Amino acid sequences of peptides synthesized and studied in this work (Ac, acetyl; Aib, α-aminoisobutyric acid; Lol is the 1,2-amino alcohol leucinol; TOAC, 2,2,6,6-tetramethylpiperidine-1-oxyl-4-amino-4-carboxyilic acid).

Results and Discussion

A synthetic protocol on solid-phase was set up for chalciporin A (1) and its two TOAC-containing analogs (2 and 3). In the [TOAC^{3,13}] analog (3), the native Val^2 residue was replaced by the less sterically demanding Ala, because the originally attempted coupling reaction of the extremely substituted Val^2-TOAC^3 peptide bond failed. Therefore, for a more stringent comparison, the Ala^2 replacement was inserted also in the [TOAC^{13}] analog (2). In this specific sequence 1, a Val→Ala replacement is not expected to alter significantly the overall peptide preferred conformation.

Chalciporin A and its two analogs exhibit CD spectra typical of a helical structure, dictated by the Cα-tetrasubstituted α-amino acids Aib and TOAC, in all environments (MeOH, TFE, POPC vesicles) investigated. The experimental R values [5] are indicative of a largely prevailing α-helical conformation for all these peptides. This conclusion was further supported by a 600 MHz 2D-NMR study of chalciporin A (1) in MeOH, d_3 solution looking specifically at the $NH_i \rightarrow NH_{i+1}$ and the $C\alpha H_i \rightarrow C\alpha H_{i+n}$ (where n is 1-3) cross-peaks.

Membrane permeability properties for all three peptides, tested in small unilamellar vesicles mimicking mammalian cytoplasmatic and Gram-negative bacterial membranes were found to be significant.

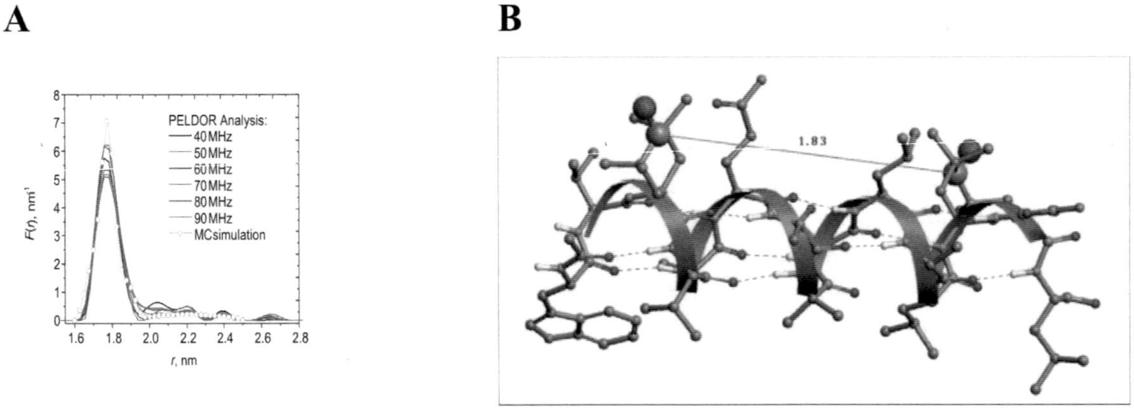

Fig.2. (A) Distance distribution functions between the two nitroxide free radicals of [Ala[2], TOAC[3,13]] chalciporin A (3) from the PELDOR analysis. (B) Model of the α-helical peptide 3 highlighting the radical···radical distance.

We performed a cw EPR study on the TOAC-labeled peptides (2 and 3) in POPC vesicles: both compounds showed a slow-motion component (peptide interacting with the membrane) and a rapid motion component (free peptide in solution). A comparison between the experimental spectrum of peptide 2 and its simulation suggested that 96% of the peptide molecules was bound to the phospholipid bilayer.

PELDOR (DEER) experiments performed on the doubly-labeled peptide 3 indicated a prevailing (90% of the molecules) α-helical conformation (interspin distance: 1.76 ±0.01 nm), with the other 10% adopting a more extended conformation (Figure 2A). A radical···radical distance (1.83 nm) (Figure 2B) fits nicely with the experimental result.

References

01. Neuhof, T.; Berg, A.; Besl, H.; Schwecke, T.; Dieckmann, R.; von Döhren, H. Chem. Biodivers. 2007, 4, 1103-1115.
02. Peptaibiotics II, Topical Issue, Brückner, H.; Toniolo, C., Eds. Chem. Biodivers. 2013, 10, 731-961.
03. Toniolo, C.; Crisma, M.; Formaggio, F.; Peggion, C. Biopolymers (Pept. Sci.) 2001, 60, 396-419.
04. Toniolo, C.; Crisma, M.; Formaggio, F. Biopolymers (Pept. Sci.) 1998, 47, 153-158.
05. Toniolo, C.; Polese, A.; Formaggio, F.; Crisma, M.; Kamphuis, J. J. Am. Chem. Soc. 1996, 118, 2744-2745.

FRAGMENTS OF HUMAN PROTEIN NEDD4L AS A NEW EPITOPES FOR THE DIAGNOSIS OF RHEUMATOID ARTHRITIS

Beata Kolesinska[1], Inga Relich[1], Justyna Fraczyk[1], Iwona Konieczna[2], Wieslaw Kaca[2], Aleksandra Kaczmarek[3], Dariusz Timler[3], Zbigniew J. Kaminski[1]
[1] Institute of Organic Chemistry, Lodz University of Technology, Zeromskiego 116, 90-924 Lodz, Poland,
[2] Department of Microbiology, Institute of Biology, Jan Kochanowski University, Swietokrzyska 15, 25-406 Kielce, Poland,
[3] Copernicus Regional Specialist Hospital, Pabianicka 62, 93-51, Lodz, Poland

Introduction

Rheumatoid arthritis (RA) is a common autoimmune disease associated with progressive disability, systemic complications and early death [1]. The origin of RA is still unknown which means that both the diagnosis and the prognosis is hampered. The main pathogenic triggers associated with the development of rheumatoid arthritis are bacteria. The chronic survival of H. pylori in humans is possible because of an overall downregulation of the immune's system due to molecular mimicry.

Results and discussion

Taking into account that H. pylori urease has been suggested as dominant antigen detected in infected patients and phenomena of molecular mimicry [2] we were looking for human proteins with motifs similar to H. pylori urease fragment 327-334 recognized by antibodies. We focused our attention on human protein NEDD4L (neural precursor cell expressed developmentally downregulated gene 4-like), an ubiquitin ligase containing 62-67 fragment similar to H. Pylori urease 327-334 fragment. NEDD4L modulates gene transcription of metalloproteinase 1 and 13. MMP-1 and MMP-13 which degrade type II collagen in cartilage, thereby contributing to the development of rheumatoid arthritis [3]. To select immunologically active fragments the epitope mapping of NEDD4L protein has been performed using polyclonal antibodies against Jack bean urease. In the first step, a 125-elements library of non-overlapping decapeptides covering NEDD4L was synthesized on cellulose matrix by using triazine coupling reagent [4] and isocyanuric linker for anchoring peptides to the cellulose. It has been selected 18 fragments forming immune complexes with antibodies against Jack bean urease. In the next stage, it has been synthetized 162-elements of overlapping decapeptides with reading frame shifted by 4 amino acid residues from both N- and C-terminus. Based on the dot blot test it was found that the epitopes recognized by the anti-Jack bean antibodies are fragments: 11-20, 91-100, 261-270, 291-300, 341-350, 441-460, 551-570, 621-630, 671-680, 711-720, 771-780, 981-990, 1081-1100, 1141-1150, 1171-1180. From selected immunologically active fragments of NEDD4L protein, 24 decapeptides: 10-19 (A1), 95-14 (A2), 262-271 (A3), 292-301 (A4), 293-302 (A5), 342-351 (A6), 441-450 (A7), 450-459 (A8), 451-460 (A9), 452-461 (A10), 555-564 (A11), 562-571 (A12), 624-633 (B1), 671-680 (B2), 672-681 (B3), 712-721 (B4), 772-781 (B5), 981-990 (B6), 984-993 (B7), 1081-1090 (B8), 1082-1091 (B9), 1087-1096 (B10), 1145-1154 (B11), 1167-1178 (B12) were chosen to studies with antibodies of rheumatoid arthritis patients sera and sera of healthy blood donors. The intensity of the formation of immune complexes between immobilized on a cellulose epitopes and RA patients sera was higher in comparison to healthy volunteers sera (Fig. 1a). Only in th e case of peptides: A3, A11, A12 and B2 observed differences were small. It has been found (Fig. 1b) that 5 fragments indicated by ANOVA test have potential diagnostic value: ^{292}SYQTSHQFII^{301}, ^{293}YQTSHQFIIS^{302}, ^{624}YSNQAFPSPP^{633}, ^{981}YFKFIGRVAG^{990} and ^{1145}YSANHQVIQW^{1154}.

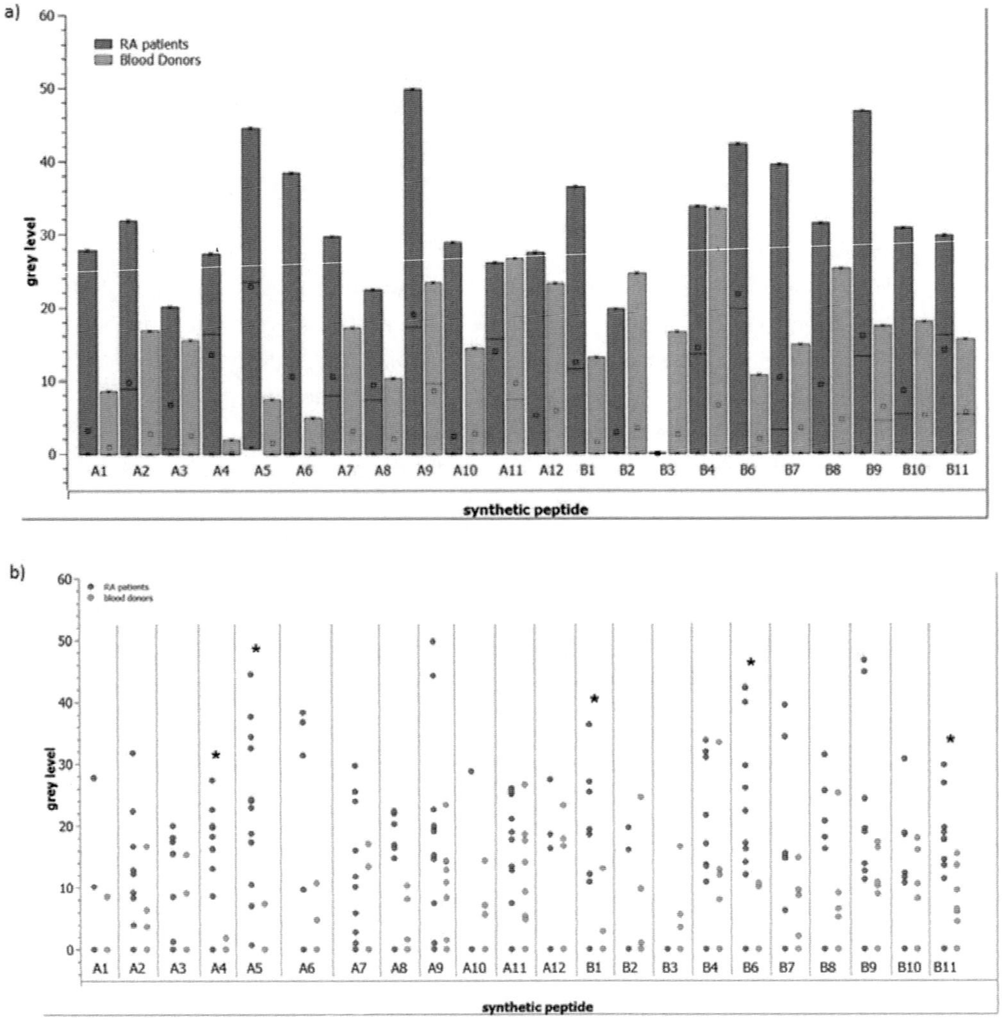

Figure 1. a) The intensity of interactions of selected peptides with RA patients sera and healthy volunteers sera; b) results of Anova analysis selected NEDD4L fragments.

The obtained results confirmed that it is possible to design a diagnostic test for rheumatoid arthritis consisting of human protein epitopes, for which can be determined the difference between the strengths of reaction with RA sera patients and healthy donors. This can be a starting point in search for diagnostic tools useful in both prediction of RA and for monitoring of the treatment.

Acknowledgements
This work was supported by grant NCN UMO-2012/05/N/ST5/01460

References

01. McInnes, I.B., Schett, G., N. Engl. J. Med. (2011), 365, 2205-2219.
02. Arabski, M., Konieczna, I., Sołowiej, D., Rogon, A., Kolesinska, B., Kaminski, Z.J., Kaca, W., Clin. Biochem., (2010), 43, 115-123.
03. Vincenti, M.P., Brinckerhoff, C.E., Arthritis Res., (2002), 4, 157-164.
04. Kolesinska, B., Rozniakowski, K.K., Fraczyk, J., Relich, I., Papini, A.M., Kaminski, Z.J., Eur. J. Org. Chem., (2015), 401–408.

NMR AND MOLECULAR DYNAMICS CONFORMATIONAL ANALYSIS OF PROTEOLIPID PROTEIN (PLP) PEPTIDE ANALOGUES

Golfo Kordopati[1], Haralambos Tzoupis[1], Anastasios N. Trogkanis[2], Gerasimos M. Tsivgoulis[1], Simona Golic Grdadolnik[3], Theodore Tselios[1],
[1] University of Patras, Department of Chemistry, Rion Patras, 26504, Greece
[2] Department of Biological Applications and Technology, University of Ioannina, 45110, Ioannina, Greece
[3] Laboratory of Biomolecular Structure, National Institute of Chemistry, 1001, Ljubljana, Slovenia

PP VIII - 166

Introduction

Proteolipid Protein (PLP) is one of the main proteins of myelin sheath that are destroyed during the progress of Multiple Sclerosis (MS) [1]. PLP has been shown to induce chronic Experimental Autoimmune Encephalomyelitis (EAE, the best well known animal model of MS). The immunodominant $PLP_{139-151}$ epitope ($H^{139}SLGKWLGHPDKF^{151}$) elicits immune response in SJL/J mice inducing chronic EAE [2,3]. The amino acids at positions 144 and 147 are recognized by T cell receptor (TCR) during the formation of trimolecular complex between TCR, peptide-antigen and Major Histocompability Complex (MHC) which is responsible for the EAE induction [4]. NMR and Molecular Dynamics (MD) simulations studies of linear $PLP_{139-151}$ and cyclic(139-151)(L^{144}, R^{147}) $PLP_{139-151}$ analogues in aqueous solution were carried out to explore their conformational and stereochemical characteristics aiming to the rational design of altered peptide ligands (APLs) or non-peptide mimetics EAE inhibition.

Results and Discussion

NMR conformational analysis: The NMR spectra were recorded on a Bruker AVANCE 400 MHz and Varian DirectDrive 800 MHz spectrometers at 298 K, using H_2O/D_2O (9:1 v/v) as solvent. In our studies, TOCSY and NOESY experiments were performed for resonance assignment of the protons, the identification of the amino acid sequencing and the establishment of the NOE connectivities. Strong NOE connectivities corresponding to intra-residue $H_{\alpha(i)}$-$H_{N(i)}$ and short range $H_{\alpha(i)}$-$HN_{(i+1)}$ NOE cross-peaks were observed in NOESY spectrum of both peptides.

NMR characterization of $PLP_{139-151}$: A medium $H_{N(i)}$-$H_{N(i+1)}$ cross-peak was observed between the NH protons of Leu^{145} and Gly^{146}. In addition, $H_{\beta(i)}$-$H_{N(i+1)}$ cross-peaks, medium between His^{139}-Ser^{140} and Ser^{140}-Leu^{141} and weak between Leu^{141}-Gly^{142}, Lys^{143}-Trp^{144} and Leu^{145}-Gly^{146} were found. The observation of four cross-peaks of $H_{\alpha(i)}$-$H_{N(i+2)}$ (medium), $H_{\beta2(i)}$-$H_{N(i+2)}$ (medium), $H_{\beta1(i)}$-$H_{N(i+2)}$ (weak) and $H_{R(i)}$-$H_{N(i+2)}$ (weak), between the Leu^{145} and His^{147} indicate the presence of a bend between the 145 and 147 residues.

NMR characterization of cyclic(139-151)[L^{144},R^{147}]$PLP_{139-151}$: Medium $H_{N(i)}$-$H_{N(i+1)}$ cross-peaks were observed between the NH protons corresponding to Gly^{146}-Arg^{147}, Lys^{150}-Phe^{151} and Phe^{151}-His^{139}. NOE cross-peaks, weak between $H_{\beta(i)}$-$H_{N(i+1)}$ His^{139}-Ser^{140}, Lys^{143}-Leu^{144}, Pro^{148}-Asp^{149}, Lys^{150}-Phe^{151} and Phe^{151}-His^{139} and medium between Ser^{140}-Leu^{141} and Asp^{149}-Lys^{150} were observed. Finally, between $H_{\gamma(i)}$- $H_{N(i+1)}$ of Leu^{141}–Gly^{142} and Leu^{145}-Gly^{146} weak NOE connectivities were found.

Molecular Dynamics Simulation: The small length of the peptide -only 13 residues- allows greater mobility in the MD simulation of the linear $PLP_{139-151}$ in aqua solution.

Clustering analysis: The clustering analysis of the linear $PLP_{139-151}$ showed the presence of three distinct clusters (Fig. 1, blue, magenta and red). The sigmoidal form of the peptide is present for 65% of the simulation time, while the cyclic conformation (Fig. 1, red) is present for approximately

25%. The cyclic(139-151)[L[144], R[147]]$PLP_{139-151}$ does not allow extensive changes in its conformation. The most important observation of the clustering analysis is that the linear form of the peptide may adopt a cyclic conformation in aqueous solution that closely resembles that of the cyclic analogue (Fig. 1, red and black). Also, the substitution of His^{147} with Arg^{147} in the cyclic analogue does not alter the orientation of these residues in the representative conformation of linear and cyclic peptides (Fig. 1, red and black).

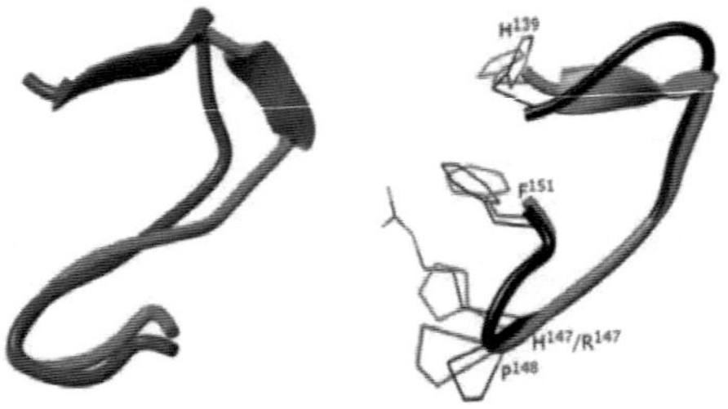

Figure 1: Representative conformations of the linear $PLP_{139-151}$ (blue, magenta and red) and cyclic (139-151) [L^{144}, R^{147}] $PLP_{139-151}$. (black) analogues based on the clustering analysis performed on the MD simulations.

Conclusion

The analysis of the MD simulations for the linear $PLP_{139-151}$ peptide, revealed the presence of different conformations in aqueous environment (Fig. 1). As expected, the linear form of the peptide is not retained during the simulation time and instead the peptide adopts a sigmoidal (semi-extended) conformation (Fig. 1, blue). The most important characteristic observed in our MD simulation is the presence of a cyclic form adopted by the linear $PLP_{139-151}$ in aqueous environment (Fig. 1, red). The analysis of the distances between the backbone hydrogen atoms during the MD simulation time, were found to be in good agreement with the NOE data.

Acknowledgment

This work is financially supported by the "Cooperation" Program 09SYN21-609 and by "Cooperation Greece-Israel Program" ISR-3148 O.P. Competitiveness & Entrepreneurship (EPAN II), ROP Macedonia- Thrace, ROP Crete and Aegean Islands, ROP Thessaly- Mainland Greece- Epirus, ROP Attica and Slovenian Research Agency (grand no. P1-0010).

References

01. L. Steinman, Cell, 1996, 85, 299-302.
02. J.M. Greer, V.K. Kuchroo, R.A. Sobel, M.B. Lees, J. Immunol., 1992, 149, 783-788.
03. J.M. Greer, R.A. Sobel, A. Sette, S. Southwood, M.B. Lees, V.K. Kuchroo, J. Immunol., 1996, 156, 371-379.
04. V.K. Kuchroo, J.M. Greer, D. Kaul, G. Ishioka, A. Franco, A. Sette, R.A. Sobel, M.B. Lees, J. Immunol., 1994, 153, 3326-3336.

Johannes Lach, Ralf Kling, Tim Clark and Jutta Eichler
Department of Chemistry and Pharmacy, University of Erlangen-Nürnberg, Schuhstraße 19, 91052 Erlangen, Germany

PP VIII – 169

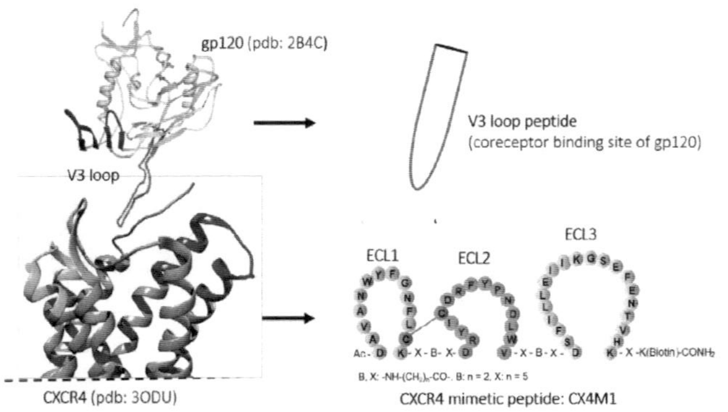

Figure 1: Structure based design of peptides mimicking the gp120 V3 loop and CXCR4

We have previously designed a soluble synthetic peptide (CX4-M1) that functionally mimics the HIV-1 coreceptor CXCR4, a chemokine receptor that also serves as a coreceptor for HIV-1 entry. This CXCR4 mimetic peptide presents the three extracellular loops (ECLs) of the receptor in a single peptide. In binding assays involving recombinant proteins, as well as in cellular HIV-1 infection assays, CX4-M1 was found to selectively recognize gp120 from HIV-1 strains that use CXCR4 for cellular entry (X4 tropic HIV-1). Recently, we could show that the selectivity of CX4-M1 pertains not only to gp120 from X4 tropic HIV-1, but also to synthetic peptides presenting the V3 loops of these gp120 proteins [2], which is thought to be an essential part of the coreceptor binding site of gp120 for the coreceptor.

Aiming at exploring the structural basis for the CX4-M1 - V3 loop interaction, we have now generated, using molecular dynamics (MD) simulation, a structure model of the complex of both peptides. Both peptide sequences were pre-oriented based on a previously published MD structure of CXCR4 –V3 loop complex [3], which served as a template for subsequent simulations. Unlike the individual peptides, which remain unstable over the entire simulation period, the CX4M-1 – V3 loop complex stabilizes, after forming serveral meta-stable conformations, within 1400 ns.

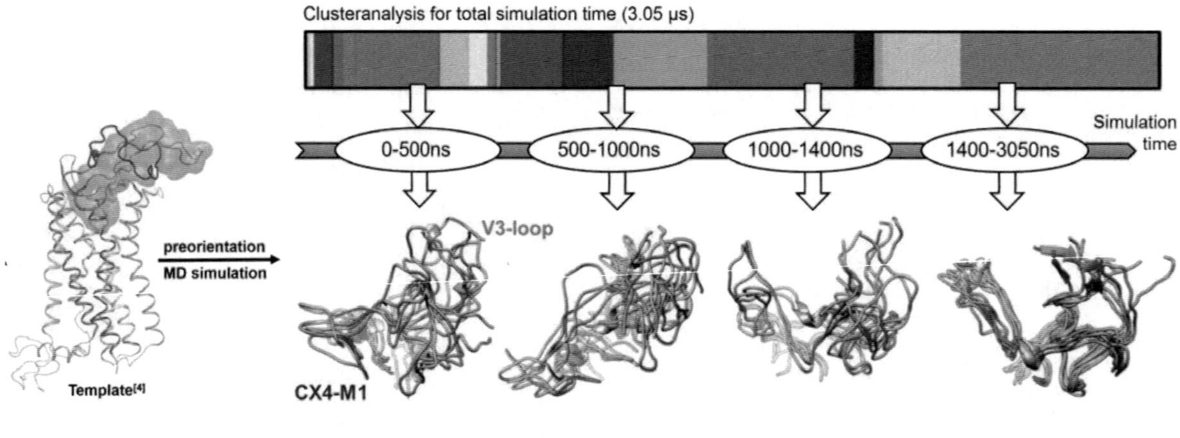

Figure 2: Time evolution and cluster analysis for MD simulation ofthe CX4M1 : V3-loop peptide:peptide complex

Between 1400 and 3050 ns, a very stable complex with a calculated free binding energy of about -150 kJ/mol (MM-PBSA) is formed.

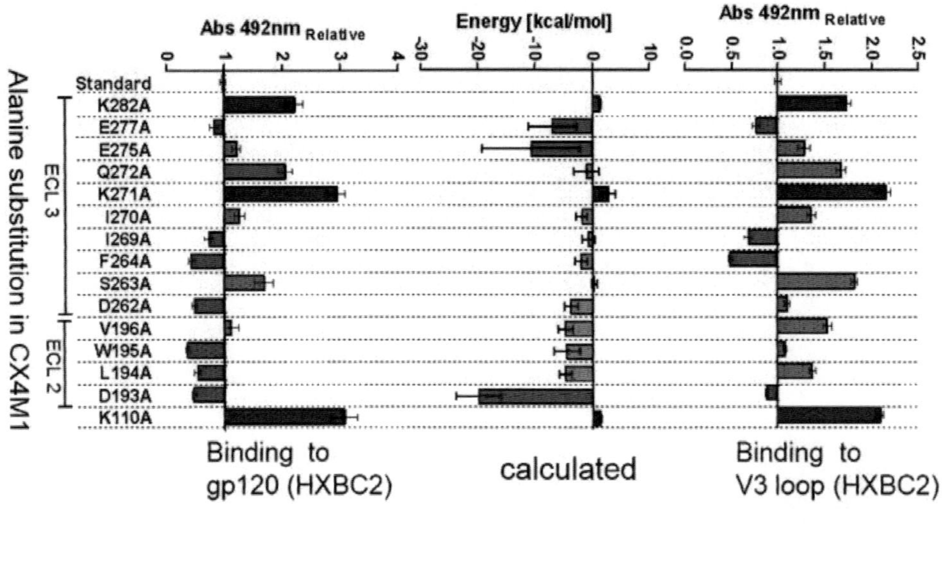

Figure 3: Calculated and experimentally determined influence of Ala-substitutions in CX4-M1

Furthermore, the calculated energy contributions of selected individual amino acids are in good agreement with experimental SAR data obtained using alanine substitution variants of both peptides, validating the structure model. The established role of charged, acidic residues (D,E) within the three ECLs of CXCR4, as well as positively charged amino acids (R,K) in the V3 loop of gp120, could be verified in this complex structure, and the experimentally determined binding data for both cases binding of CX4M1 to recombinant gp120 protein and synthetic V3-loop peptide. Furthermore, separate MD simulation of the two peptides after removal from the stable complex between 1400 and 3050ns revealed only small changes for the V3-loop peptide, whereas the CXCR4 mimicking peptide CX4M1 undergoes strong conformational changes when removed from the V3-loop peptide. These results are a strong indication of a "mutually induced fit" binding mechanism of two intrinsically unfolded peptides, which, once in contact with each other, fold into a stable complex. Ongoing studies are aimed at experimentally validating this notion.

References:

01. K. Möbius, R. Dürr, C. Haußner, U. Dietrich, J. Eichler. Chem. Eur. J. 18, 8292 (2012).
02. A. Groß, K. Möbius, C. Haußner, N. Donhauser, B. Schmidt, J. Eichler. Front. Immunol. 4, 257 (2013).
03. P. Tamamis, C.A. Floudas, Biophys J , 105, 1502-1514 (2013).

TOWARDS β-ARRESTIN BIASED β2AR LIGANDS: RATIONAL DESIGN OF NANOBODY LOOP MIMETICS

Martin, C.[1]; Betti, C.[1]; Fabris, C.[1]; Danielsen, M.[2]; Mosolff Mathiesen, J.[2]; Pardon, E.[3]; Peyressatre, M.[4]; Moors, S.L.C.[5]; De Proft, F. [5];Morris, M.C.[4]; Devoogdt, N.[6,7]; Caveliers, V[7]; Steyaert, J.[3]; and Ballet, S.[1]

[1] Research Group of Organic Chemistry, Vrije Universiteit Brussel, Brussels, Belgium

[2] Zealand Pharma, Department of Molecular Pharmacology, Glostrup, Denmark

[3] Structural Biology Brussels, Vrije Universiteit Brussel, Vlaams Instituut voor Biotechnologie, Brussels, Belgium

[4] Institut des Biomolécules Max Mousseron-IBMM-CNRS-UMR 5247, Faculté de Pharmacie, Montpellier, France

[5] Eenheid Algemene Chemie (ALGC), Vrije Universiteit Brussel, Brussels, Belgium

[6] In Vivo Cellular and Molecular Imaging lab, Vrije Universiteit Brussel, Brussels, Belgium

[7] Cellular and Molecular Immunology Lab, Brussels, Belgium

Introduction

G protein-coupled receptors (GPCRs) represent one of the most important classes of drug targets. [1] Most of these drugs interact with the extracellular part of the protein, and either increase or reduce G protein-mediated signaling, ultimately leading to a therapeutic response. The conventional concept - that GPCRs mediate signals only through G proteins - has changed since it was demonstrated that alternative mechanisms exist which are mediated through arrestin binding, resulting in arrestin signaling and the activation of other cellular processes.[2] The so-called biased ligands can selectively activate the G protein- or arrestin-mediated pathways.

In recent years, X-ray structures of the Nanobody(Nb)-stabilized β_2-adrenergic receptor (β_2AR) in its active state have been determined.[3] Remarkably, Nanobody (Nb80) was shown to bind to the intracellular side of the receptor. Additionally, the complementary determining region (CDR3) domain of the Nanobody adopts a β-hairpin conformation that penetrates the G protein binding cavity, providing a structural surrogate for the G protein Gs.

In this study, we aim to synthesize structural and functional peptidomimetics of Nanobody CDR3 domains of the β_2AR. We hypothesize that such small compounds, that bind the intracellular G protein binding cavity of the receptor, can work cooperatively with agonists that bind to the extracellular orthosteric site. If such compounds compete with G protein binding and trigger arrestin recruitment, they could act as β-arrestin biased ligands.

Results and Discussion

Based on the crystal structure of Nb80 with the β_2AR receptor, where the CDR3 of interest adopts a β-hairpin conformation, a series of cyclic peptides of different macrocycle size was designed and contains a DPro-Pro motif to stabilize the desired conformation (Table 1).[4] The incorporation of such template into loop sequences favors macrocyclization efficiency due to its strong β-hairpin inducing properties and appears to improve antigen-peptide binding. The peptide analogues were synthesized using classic SPPS using the 2-chloro-trityl chloride resin as solid support and the cyclization was performed in solution by use of a combination of fluorinated alcohol (trifluoroethanol)-dichloromethane solvent system and a DIC/additive coupling mixture.

Table 1: Cyclic peptide mimics of the CDR3 domain of Nb80

N°	Peptide analogues
CFPP1	c[Leu-Tyr-DPro-Pro-Ala-Val]
CFPP2	c[Leu-Tyr-Glu-DPro-Pro-Gly-Ala-Val]
CFPP3	c[Leu-Tyr-Glu-Tyr-DPro-Pro-Tyr-Gly-Ala-Val]
CFPP6	$_c$[Leu-Tyr-Glu-Tyr-DPro-Pro-Asp-Tyr-Gly-Ala-Val]
CM87	$_c$[Leu-Tyr-Glu-Tyr-Asp-DPro-Pro-Asp-Tyr-Gly-Ala-Val]
CM86	$_c$[Leu-Tyr-Glu-Tyr-Asp-Tyr-DPro-Pro-Lys-Asp-Tyr-Gly-Ala-Val]
BC484	$_c$[Leu-Tyr-Glu-Tyr-Asp-Tyr-DPro-Pro-Val-Lys-Asp-Tyr-Ala-Val]
CM85	$_c$[Leu-Tyr-Glu-Tyr-Asp-Tyr-DPro-Pro-Val-Lys-Asp-Tyr-Gly-Ala-Val]
CM92	$_c$[Leu-Tyr-Glu-Tyr-c[Asp-Tyr-DPro-Pro-Val-Lys]-Asp-Tyr-Gly-Ala-Val]
CM114	$_c$[Leu-Tyr-Glu-Tyr-Asp-Tyr-Trp-Gly-DPro-Pro-Cys-Asn-Val-Lys-Asp-Tyr-Gly-Ala-Val]

To gain structural insight into the peptidomimetics, molecular dynamics calculations were carried out. This study showed a high overlap between the backbone of the Nb loop and the backbone of the peptidomimetic analogue CM85 (Conformational analysis in Figure 1), whereas the automated docking of the latter confirmed the desired pose of the CDR3 mimetic, i.e. at the binding site of the Nb CDR3 domains.

Having a clear indication that the peptidomimetics should adopt the same conformation as the native CDR3 domain of the Nb, a functional assay was performed. The peptidomimetics were tested for their ability to block isoproterenol (ISO) induced cyclic adenosine 3′,5′-monophosphate (cAMP) signaling. cAMP is an important second messenger whose production is regulated by the activation of GPCRs. This assay is based on the quantification of cAMP as a measure of β_2AR activation by

ISO in membrane fragments by FRET microscopy.[5] The relationship between the cAMP concentration and the FRET ratio enables to determine inhibition of cAMP production by the peptide ligand. Within the current study, only the largest cyclic peptides could inhibit the cAMP accumulation at higher concentration (100 μM), as compared to Nb80 (10 μM) (cAMP accumulation assay in Figure 1).

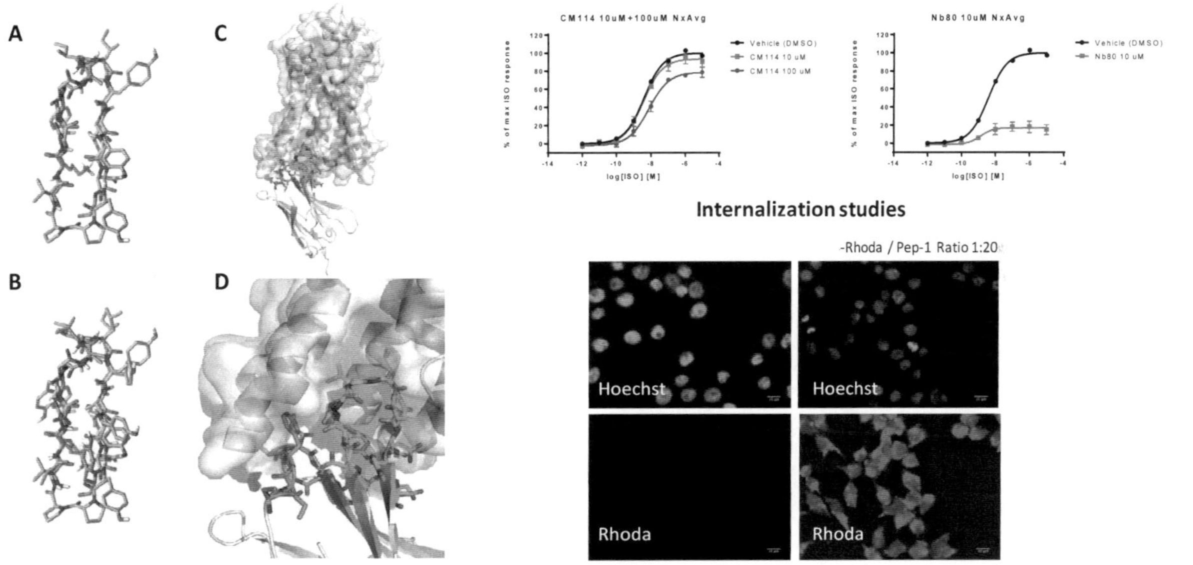

Figure 1: Conformational analysis: Superimposition between a cyclic peptide mimetic (cyan) and the backbone of the CDR3 loop region in Nb80 (orange) (Panel A), the peptidomimetic (CM85) and the Nb80 CDR3 domain with side chains (Panel B), binding pose of Nb80 into the G-protein cavity, with the CDR3 colored green (Panels C and D). cAMP accumulation assay: Concentration-response curves of peptidomimetic CM114 and Nb in HEK293 cell membranes overexpressing the β_2AR, using Isoproterenol (ISO) as orthosteric agonist ligand. Experiments were performed using the Cisbio Bioassay's cAMP kit. Internalization studies: Fluorescence microscopy pictures at (10 ms) of 2 μM rhodamine-labelled peptide (CM114-Rhoda) overlaid onto HeLa cells in DMEM supplemented with 10% glucose.

In view of a binding at the cytosolic side of the receptor, cell internalization experiments were performed on the best cyclic analogue (CM114). The cyclic peptide was labelled with the rhodamine maleimide via the cysteine residue. When overlaid onto cells alone, the cyclic analogue was unable to cross the cell membrane (I nternalization studies in Figure 1). However, when complexed with the amphipathic peptide carrier Pep-1 at a molar ratio of 1:20 (CM114-Rhoda : Pep1), the peptidomimetic analogue was efficiently delivered into living cells.[6] Pep-1 technology presents the advantage that no chemical modification or denaturation of the peptidomimetic is required.
In conclusion, we have designed a series of peptidomimetics of different sizes, and obtained a proof-of-principle that cyclic peptide analogues can structurally mimic the CDR3 domain of a Nanobody, while blocking agonist-induced cAMP formation at high peptidomimetic concentration.

Acknowledgements
We thank the Research Foundation Flanders (FWO Vlaanderen) and the Strategic Research Program – Growth funding of the VUB for the financial support.

References

01. Violin, J. D. et al., Trends. Pharmacol. Sci. 2014, 7, 308
02. Whalen, E. J. et al., Trends Mol. Med. 2011, 17, 126
03. Rasmussen, S. G. S. et al., Nature 2011, 477, 549; Rasmussen, S. G. S. et al., Nature 2011, 469, 175
04. Robinson J. A., Synlett 1999, 4, 429
05. http://www.cisbio.com
06. Morris, M. C. et al., Nat. Biotech. 2001, 19, 1173

Rina Nakamura[1], Motomi Konishi[1], Ryuichiro Tanaka[1], Masanari Taniguchi[1], Aya Kojima[2], Tadashi Yamamoto[2,3] Toshifumi Akizawa[1]
[1] Setsunan University, Pharmaceutical Sciences, Hirakata, Japan
[2] Ritsumeikan University, Pharmaceutical Sciences, Kusatsu, Japan
[3] Okinawa Institute of Science and Technology Graduate University, Cell Signal Unit, Onna-son, Japan

PP VIII – 177

Introduction

We found the novel peptide which is consisted only 9 amino acids termed to JAL-TA9 (YKGSGFRMI) derived from Tob1 protein BoxA domain possessing auto-digested by its self and cleaved Amyloid-β protein (Aβ42). Furthermore these activities are inhibited by AEBSF of the serine protease inhibitor [1, 2]. On the basis of these results, we proposed that the general name of enzyme peptide like JAL-TA9 is termed Catalytide (Catalytic Peptide). The N-terminal region of Tob/BTG family proteins is conserved and included three homologous regions, BoxA, BoxB and BoxC. In the aspect of function, the Tob/BTG family proteins are involved in cell cycles and regulation in a variety of cells such as T lymphocytes, fibroblasts, epithelial cell, and germ cells [3, 4]. But there is no report about enzyme activity of Tob1.

It is generally accepted that specific conformation is required to form catalytic domain, in consequence it should be large molecules. In this study, we examined whether JAL-TA9 can form a conformation to act as serine protease by using NMR analysis, computer modeling and circular dichroism (CD) spectra.

Result and Discussion

JAL-TA9 which was a derived from Tob1 BoxA domain was synthesized by Applied Biosystems Japan 430A peptide synthesizer by F-moc methods, purified by high performance liquid chromatography (HPLC) with reversed-phase column. Major peak was collected and identified as JAL-TA9 by MS analysis. To confirm the amino acids sequence, 10 mM JAL-TA9 peptide dissolved in aqueous solution containing 20 % D_2O was analyzed by two-dimensional proton nuclear magnetic resonance spectroscopy (2D NMR) by JNM-ECA 600 spectrometer. All signals including amide protons were assigned on the basis of ^1H-^1H correlated spectroscopy (COSY) and rotating frame Overhauser effect spectroscopy (ROESY) data.

We examined whether JAL-TA9 forms a special conformation in the solution. The ROESY is useful to determine the proton signals being close to each other in space even if they are not bonded. To confirm the secondary structures, ROESY spectrum of JAL-TA9 were analyzed. Judging from ROESY spectra, two NOEs were identified between ε-H of ^{2}Lys and γ-H of ^{8}Met, and between β-H of ^{4}Ser and δ-H of ^{7}Arg (Figure 1a). We observed the NOE between β-H ^{4}Ser and δ-H ^{7}Arg is stronger than that of ε-H of ^{2}Lys and γ-H of ^{8}Met. This correlation indicates that those ^{4}Ser and ^{7}Arg are very closed each other, and conformation of JAL-TA9 is very compact. Put otherwise the important amino acids to start nucleophilic attack is very closed each other like serine proteases. We constructed molecular structure model of JAL-TA9 by HGS biochemistry model being applied by those NOEs data. The structure shows JAL-TA9 forms compact structure.

We next conducted computer modeling of JAL-TA9 with the Software CSC Chem3D Ultra™ ver.9.0. We did three steps before calculation by MM2 parameter. At first all peptide bond angles and dihedral angles fix to 180°. Second, the six atoms organizing a peptide bond were arranged in one plane, and then settled the bond length. Based on NOEs we settled the length 5 Å between ε-H of ^{2}Lys and γ-H of ^{8}Met and 4 Å between β-H ^{4}Ser and δ-H ^{7}Arg. After settled these, we carried out calculation by the structural optimization and energy minimization according to MM2 parameters (bond length, bond angles, torsion, dipole-moment, and van der Waals values). The result suggested that JAL-TA9 forms catalytic triad and oxyanion hole which required to act serine protease (Figure. 1b). Structure obtained by MM2 calculation is similar to the structure constructed by HGS model. We also calculated by MM2 same methods written above but without NOEs. Interestingly JAL-TA9 forms almost same structure absence or presence of NOEs date (Figure. 1c). Moreover the steric energy absence of NOE date and presence of NOE is 9.6 kcal/mol and -12.7 kcal/mol respectively. This data indicated that JAL-TA9 structure defined by primary sequence. These findings as useful for drug design in silico.

CD analysis was done using a J-805 (JASCO). CD spectra in a wavelength range of 190–260 nm with 0.2 cm length of cell were measured in 10 mM Tris-HCl (pH 7.5) and 10 mM JAL-TA9. This spectrum showed a positive at around 190 nm and two negative peaks around 222 and 203 nm (Figure 2). Compared with spectrum of unfold conformation, the negative minimum was red–shifted to 203 nm and the level of the negative ellipticity at 222 nm, which is characterized of the α-helix conformation, was increased. This suggested the presence of the α-helical structure of the peptide.

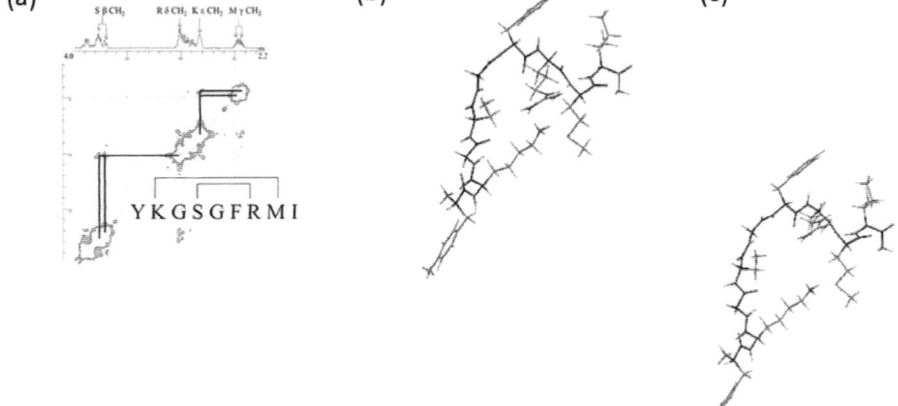

Figure 1. Computer modeling of JAL-TA9

ROESY spectrum, (b) Presence of NOEs, (c) Absence of NOE

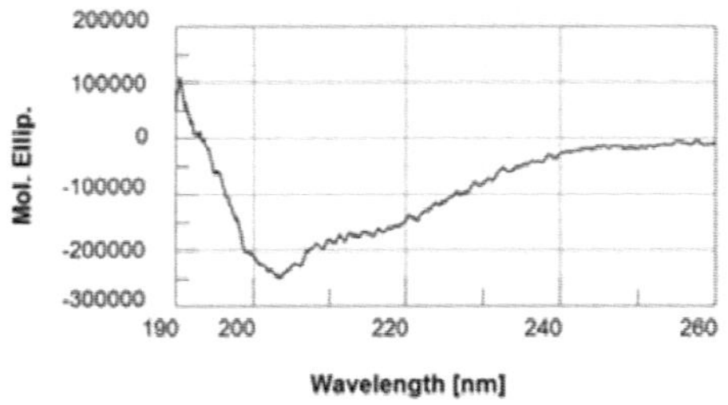

Figure 2. CD spectrum

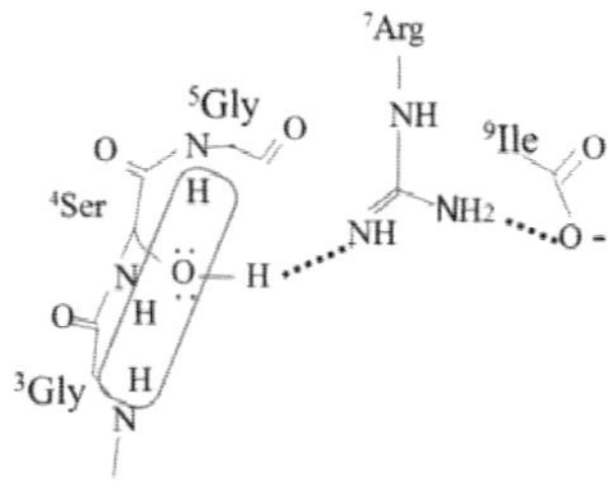

Figure 3.catalytic triad

The classical serine proteases uses Ser/His/Asp catalytic triad mechanism, where serine is nucleophile, histidine is general base and acid, and the aspartate helps to orient the histidine residue and neutralized the charge that develops on the histidine during the transition states [5]. In the structure study suggested that the three-dimensional spatial distance of [4]Ser and [7]Arg are close so we thought [7]Arg can acts as a base to abstract a proton from [4]Ser, C-terminal carboxyl group helps to orient the histidine groups, and start nucleophilic attack to substrate same as serine protease (Figure 3). Taken together, even small molecules peptides can form the special structure for showing enzymatic activity. We think that our study could be useful to design novel Catalytide.

References

12. Yamamoto, T., Akizawa, T., 2016 NOVEL PEPTIDE Patent Pending No US62/275,599
13. Yamamoto, T., Akizawa, T., 2016 NOVEL HYDROLASE-LIKE PEPTIDE AND ITS USE, No JP2016-0668496
14. Koga, N., Tatsumi-Koga, R., Liu, et al., Nature, 2012, 491(7423): 222.
15. Winkler, G.S., J. Cell. Physiol. 2010, 222, 66.
16. Ekici OD., Paetze M., Dalbey RE., Protein Sci. 2008 Dec; 17(12): 2023–2037.

Olga V. Ovdiichuk,[1,2] Olga V. Hordiyenko,[1] Marie-Christine Averlant-Petit,[2] Axelle Arrault[2]
[1] Department of Chemistry, Kyiv National Taras Shevchenko University, 64/13, Volodymyrska str., 01601, Kyiv, Ukraine;
[2] Laboratoire de Chimie Physique Macromoléculaire, Université de Lorraine, 1 rue Grandville, BP 20451, 54001 Nancy, France.

PP VIII – 180

The combination of aromatic or heterocyclic rings with a peptide motif represents a strategy towards pseudopeptides with conformational restrictions which leads to a more stable and bioavailable product, hence favoring recognition and pharmacological properties. In this study, the effect of the introduction of pyridine and pyrazine rings on the conformational behaviour of peptides was investigated. Recently, we have developed a convenient synthesis of new nicotinic acid and pyrazine-based pseudopeptides bearing amidoxime function as a replacement of amidine one.[1] The amidines are important pharmacophores since they can mimic arginine residue in biological structures.[2] This study, therefore, describes the results on further chemical functionalization of these scaffolds.

The used approach includes transformation of 2(3)-cyano heteroaromatic acid (1) into the corresponding amidoximes (4) and N-acylamidrazones (5) in two steps. The first condensation led to a mixture of substituted amino acids 2 and cyclic intermediates 3. Derivatives 2 and 3 or their mixture gave amidoximes 4. The pyrrolidine ring opening with amino acid hydrazide provides a new convenient method for the preparation of N-acylamidrazones 5. Amino acid-derived 3,5-disubstituted-1,2,4-triazoles 6a′,b′ were synthesized by cyclodehydration under MW from 5a′ and 5b′. We were not able to obtain pyridine-linked 1,2,4-triazoles due to high thermo and solvent sensibility of 5a,b. We suggested hydrogen bonding between NH-amide and nitrogen of the pyrazine ring can stabilize the pyrazine derivatives 5a′ and 5b′ (Scheme 1).

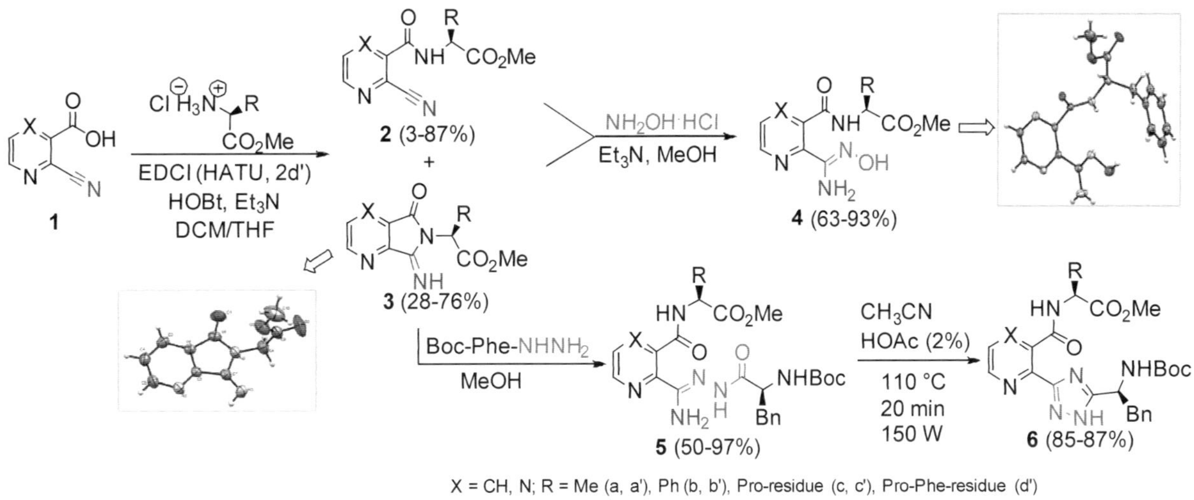

Scheme 1. Synthesis of amidoximes 4 and 1,2,4-triazoles 6

Crystal structures of some compounds were analyzed by X-Ray diffraction study (Scheme 1). Pyridine-based product 3 adopts the E configuration at C=N bond and pyrazine amidoxime 5 - Z configuration at C=N bond.
Microwave assisted synthetic route for preparation of a new variety of chiral α-amino acid derived 1,2,4-oxadiazoles (8) from corresponding amidoxime esters (7) was performed. In addition, we have accomplished the condensation of amidoxime 4 with Boc-protected hydrazide of amino acid affording hydrazide modified turn mimics 9 (Scheme 2).

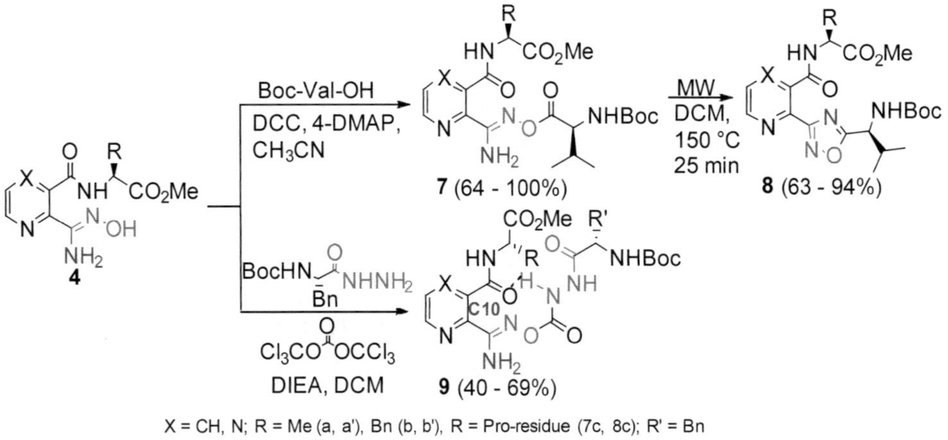

X = CH, N; R = Me (a, a'), Bn (b, b'), R = Pro-residue (7c, 8c); R' = Bn

Scheme 2. Synthesis of 1,2,4-oxadiazoles 8 and hydrazide modified turn mimics 9

The solution structure of new Pro-pseudotripeptide 4d' was determined by a combination of FTIR, NMR spectroscopic studies with molecular dynamics simulations and showed two hydrogen bonds: between carbonyl oxygen atom of the C-terminal ester group and the hydrogen atom of the amidoxime OH, and between the oxygen atom of the amide carbonyl next to the pyrazine ring and the hydrogen of the amide group of phenylalanine. All hydrazide modified peptidomimetics revealed similar NH-bonded stretching vibrations in the FT-IR spectra and low solvent sensibility of the NH protons therefore adopt a turn structure in $CDCl_3$ solution stabilized with the hydrogen bond forming C_{10}-pseudocycle (Scheme 2).

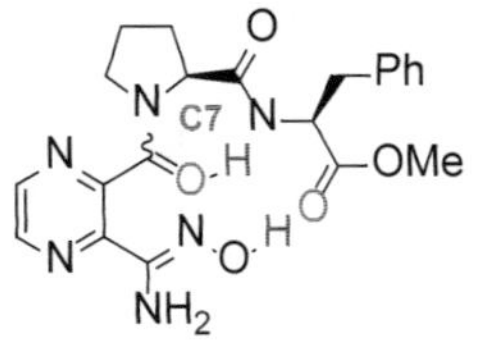

4d'

In summary, we have developed the syntheses of various novel 2,3-substituted pyrazine and pyridine-based non-peptidic turn structures possessing amidoxime, hydrazide modified or esterified with amino acid amidoxime, chiral 1,2,4-oxadiazole and 1,2,4-triazole residues. The 1,2,4-oxadiazole and 1,2,4-triazole ring closure was performed under optimized microwave-assisted conditions Conformational studies confirmed that these heterocyclic moieties can be used to increase rigidity by adopting of a seven- or ten-membered γ-turn conformations, and the pyrazine core could stronger effect on conformation stabilization.

References

01. a) Ovdiichuk O. V.; Hordiyenko O. V.; Medviediev V. V.; Shishkin O. V.; Arrault A. Synthesis 2015, 47, 2285. b) Ovdiichuk O. V.; Hordiyenko O. V.; Arrault A. Tetrahedron 2016, 72(24), 3427-3435.
02. Peterlin-Mašič L.; Cesar J.; Zega A. Curr. Pharm. Des. 2006, 12, 73.

SPLICED ANALOGUES OF TRYPSIN INHIBITOR SFTI-1 AND THEIR APPLICATION FOR ELUCIDATION OF THE PEPTIDE SPLICING, TRACING PROTEOLYSIS AND DELIVERY OF CARGOS TO THE CELLS

Natalia Ptaszynska[1], Magdalena Filipowicz[1], Dawid Debowski[1], Agata Gitlin-Domagalska1, Anna Łegowska1, Timo Burster[2], Krzysztof Rolka[1]
[1] Department of Molecular Biochemistry, Faculty of Chemistry, University of Gdansk, Poland; Correspondence: natalia.ptaszynska@phdstud.ug.edu.pl
[2] Department of Neurosurgery, University Medical Center, Ulm, Germany

Introduction

The low-molecular mass sunflower trypsin inhibitor (SFTI-1) is an attractive scaffold for the design of novel protease inhibitors and/or cargo peptides with therapeutic potential due to its relative stability, and demonstrated ability to penetrate cells [1]. In our previous studied we have shown that double-sequence SFTI-1 analogues undergo serine proteinase catalyzed peptide splicing, in which the middle fragment is released and the monocyclic SFTI-1 is formed [2]. Moreover, the SFTI-1 sequence was also shown to be an extremely valuable scaffold for the molecular-grafting concept applied in drug design [3,4].

In this work we describe, possibilities of applying rationally designed SFTI-1 analogues, which can undergo peptide splicing, and be used as vehicles for the introduction of peptide sequences with potential therapeutic or diagnostic relevance into cells.

Results and Discussion

We have designed and synthesized series of monocyclic SFTI-1 analogues with grafted RGD sequence. Selected examples are shown in Fig. 1. In peptide I, a hexapeptide GRGDNP (bolded) was flanked by a trypsin-sensitive Lys-Ser sequence whereas in peptide II trypsin-resistant D-Lys-Ser was introduced. In order to shorten the structure, the analogue of peptide II, deprived of Ser6 (assigned as [desSer^6]II), was synthesized. The last analogue (III) contained a pair of fluorescent groups: donor (fluorescein derivative, F) and acceptor of fluorescence (rhodamine derivative, R) separated by a distance that allows fluorescence resonance energy transfer (FRET). The acceptor of fluorescence is located within the fragment that can be released from the SFTI-1 structure by a trypsin-specific proteinase to produce a fluorescence emission. This gives the possibility to trace the proteolysis of peptide III in biological systems.

Our intention was to design peptides, which upon penetration into the cells underwent proteolysis in their interior with the release of the active peptide (RGD sequence) or middle fragment with rhodamine to produce a fluorescence emission and the simultaneous creation of a monocyclic SFTI-1 - trypsin inhibitor.

&^{1}Gly-Arg-Cys(&^{2})-Thr-Lys-Ser-Ile-Pro-Pro-Ile-Cys(&^{2})-Phe-Pro-Asp&^{1} **wild type SFTI-1**

Gly-D-Arg-Cys&-Thr-Lys^5-Ser-**Gly-Arg-Gly-Asp-Asn-Pro**-Lys^{13}-Ser-Ile-Pro-Pro-Ile-Cys&-Phe-Pro-Asp **I**

Gly-D-Arg-Cys&-Thr-D-Lys^5-Ser-**Gly-Arg-Gly-Asp-Asn-Pro**-D-Lys^{13}-Ser-Ile-Pro-Pro-Ile-Cys&-Phe-Pro-Asp **II**

Gly-D-Arg-Cys&-Thr-D-Lys^5-**Gly-Arg-Gly-Asp-Asn-Pro**-D-Lys^{13}-Ser-Ile-Pro-Pro-Ile-Cys&-Phe-Pro-Asp **[desSer^6]II**

βAla-D-Arg-Cys&-Thr-Lys^5-Ser-Ile-Lys-Thr-Lys^{10}-Ser-Ile-Pro-Pro-Ile-Cys&-Phe-Pro-Asp **III**
F R

Fig.1. Primary structure of SFTI-1 analogues.

The studies of proteolytic susceptibility have shown, that peptides I and III undergo peptide splicing and the middle fragments were released.The peptides containing the GRGDNP motif displayed cytotoxic effects (towards the human glioblastoma cell line, U87-MG) that were stronger than the reference linear hexapeptide. Interestingly, one of the studied peptides, [desSer^{6}]II, appeared to be exceptionally highly cytotoxic under experimental conditions, significantly higher than the reference peptide GRGDNP described in the literature [5]. Our results are in line with published data on biologically active peptides (angiogenic [4] and tumor-targeting peptides [3]) grafted into the SFTI-1 structure. It has been shown that the incorporation of these peptides into the SFTI-1 molecule either increased or prolonged their biological activity. By applying fluorescence microscopy (Fig. 2) and flow cytometry we were able to show that the peptide was internalized into the cells, where it was degraded and as a result the high fluorescence of donor was observed. The analysis (RP-HPLC with fluorescence detection and mass spectrometry) of the cell lysates after the incubation with analogue III have shown that two products were formed: N-terminal fragment with fluorescein F-β-ARCTK and the middle fragment containing rhodamine (SIK(R)TK). The presence of both products in the lysate confirmed that peptide entered the cells, where the disulfide bond was reduced and the peptide was proteolyzed. We have shown that FRET displaying peptides are potentially useful tools for tracing the proteolysis taking place inside the cells. We also proved that SFTI-1 provides an excellent base structure allowing the introduction of peptides having the desired biological activity into the cell while improving its stability as well as affecting the growth of their activity.

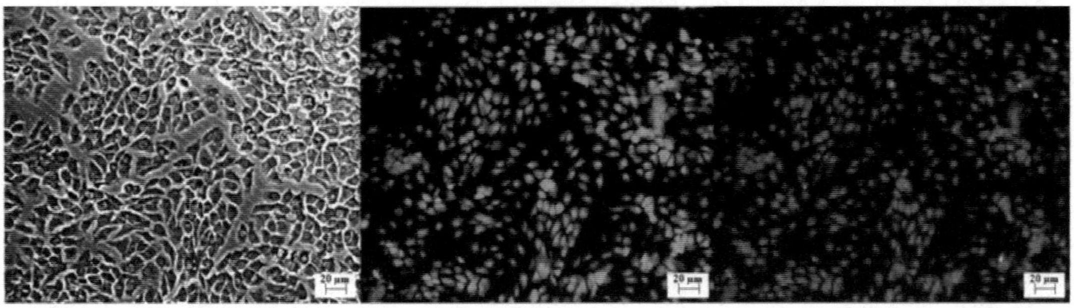

Fig.2. Transmitted light and fluorescent images of HaCaT cells incubated with FRET peptide (peptide III, 30min, 20mM).

References:

01. Cascales L., Henriques S.T., Kerr M.C., Huang Y-H., Sweet M.J., Daly N.L., Craik D.J., J BiolChem, 286, 36932-43, 2011
02. Karna N., Łegowska A., Malicki S., Debowski D., Golik P., Gitlin A., Grudnik P., Władyka B., Brzozowski K., Dubin G., Rolka K., ChemBioChem, 16, 2036-2045, 2015.
03. Zoller F., Markert A., Barthe P., Zhao W., Weichert W., Askoxylakis V., Altman A., Mier W., Haberkorn U., Angew. Chem. Int. Ed. Engl., 51, 13136-13139, 2012
04. Chan L.Y., Gunasekera S., Henriques S.T., Worth N.F., Le S.J., Clark R.J., Campbell J.H., Craik D.J., Daly N.L., Blood, 118, 6709-6717, 2011.
05. Buckley C.D., Pilling D., Henriquez N.V., Parsonage G., Threlfall K., Scheel-Toellner D., Simmons D.L., Akbar A.N., Lord J.M., Salmon M., Nature, 397,534-539, 1999.

A. Schrimpf, A. Geyer
Philipps-Universität Marburg, Hans-Meerweinstraße 4, 35032 Marburg, Germany, geyer@staff.uni-marburg.de

PP VIII - 190

Rheumatoid arthritis (RA) is a widespread inflammatory autoimmune disease, the demand for specific and sensitive serological markers is high. Citrullinated peptides can be utilised for the early diagnosis of RA.[1,2] Our β-hairpin peptides are expected to differentiate between suspected subpopulations of a patient's autoantibodies. They are systematically modified by variation of stereochemistry, macrocyclic ring size and charge patterns. Shape and dynamics are analysed via modern NMR experiments and CD spectroscopy (Figure 1A). Our double D design[3] is inspired by the cyclic, antibiotic lipopeptide Daptomycin which shows a pairwise correlated double D motif (Figure 1B). The naturally inspired design principle is merged with a conformationally robust framework and a biologically active component to create new peptide folds.

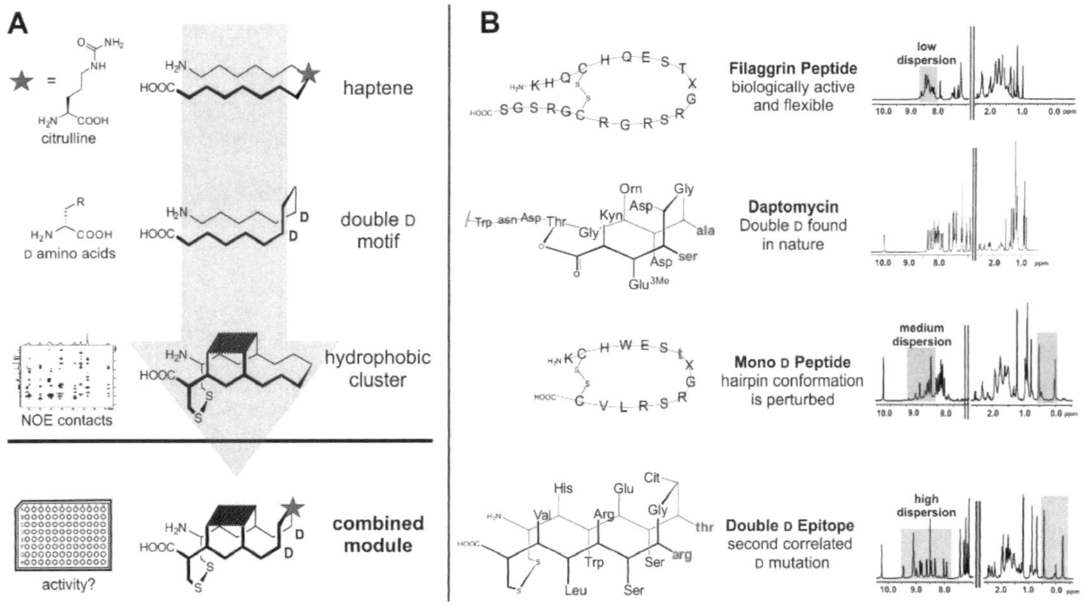

Figure 1. A (left): Epitope modules. The peptidic epitope is assembled from independent modules which are: cyclisation position, hydrophobic cluster, stabilising interactions of pairwise opposed amino acids on the β-strands, and others. B (right): Underlying sequences. Daptomycin contains the double D motif. While a single D mutation disrupts this secondary structure, we show that the correlated double D mutation of two opposing amino acids compensates this destabilizing effect.[3]

A constricted peptide epitope is characterised by a well-defined ^{1}H-NMR which shows a high dispersion of chemical shifts, similar to a folded protein. A flexible peptide structure with fast conformational averaging shows a much lower chemical shift dispersion. Other NMR parameters like ^{3}J coupling constants and NOEs corroborate this observation. In a first approximation, rigid peptides show a lock-and-key receptor binding while flexible peptides are better characterised by a conformational selection binding accompanied by a larger receptor promiscuity (figure 2).

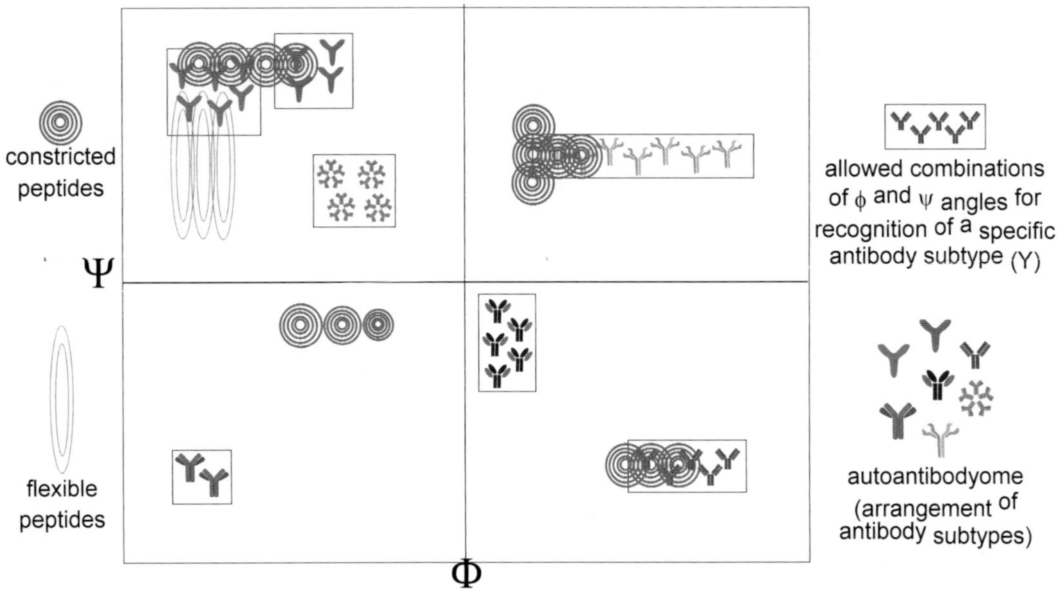

Figure 2. This hypothetic Ramachandran plot simplifies the multidimensional mobility of a hairpin peptide to a two-dimensional representation. The contour lines lie closer for rigid peptides than for flexible ones. Antibody binding is only possible when these contour lines overlap with the binding pocket of the antibody which is represented by a rectangular field. What appears like "playing battleship" is the search for a readily available set of peptides which is able to differentiate subtypes of the autoantibodyome of a RA patient.

In a first approach we were able to differentiate antibodies obtained from rabbit vaccination from a human autoantibody.[2] Currently our peptides are able to differentiate between two different human RA autoantibodies.[3] We intend to identify characteristic autoantibody patterns of individuals by these means, leading to a as complete as possible mapping of the autoantibodyome (the entity of all autoantibodies) of any RA patient.

Acknowledgement: We thank the Fonds der Chemischen Industrie im Verband der Chemischen Industrie for financial support.

References

01. G. Schellekens, H. Visser, B. de Jong, F. H. van den Hoogen, J. M. Hazes, F. C. Breedveld, W. J. van Venrooij, Arthritis Rheum. 2000, 43, 155–63.
02. S. Fischer, A. Geyer, Angew. Chem. Int. Ed. 2014, 53, 3849–3853.
03. A. Schrimpf, A. Geyer, ChemBioChem 2016, accepted. doi: 10.1002/cbic.201600479.

THE CHEMICAL SYNTHESIS AND ANALYTICAL INVESTIGATION OF MAXADILAN PEPTIDES

Szolomájer J.[1], Hegyi O.[2], Kele Z.[1], Tóth G.[1]
[1] University of Szeged, Department of Medical Chemistry, Szeged, Hungary
[2] University of Pannonia Georgikon Faculty, Department of Animal Sciences and Animal Husbandry, Keszthely, Hungary

Introduction

The identification of Maxadilan

Blood-feeding arthropods produce vasoactive compounds in their salivary glands, they serve to counteract the hemostatic processes of the host, when the arthropod obtains a blood meal. The bite of the New World sand fly Lutzomyia Longipalpis, a vector of the protozoan disease leishmaniasis, results in the rapid development of a small area of long-lasting redness and erythema at the site. In 1991 Lerner et al. analyzing the salivary glands of the sand fly Lutzomyia Longipalis, isolated the vasoactive principle responsible for this effect. The erythema-inducing fraction shown to contain a 61-amino acid containing vasodilatory peptide which is proved to be 500 times more potent vasolidator than the calcitonin gene-related peptide (CGRP) and named, because of its high potency, Maxadilan. CGRP and the N-terminus of Maxadilan also shared slight sequence similarity. While the receptor for Maxadilan was found to be a G-protein coupled receptor (GPCR), it was not that for CGRP, amylin or adrenomedullin. The receptor for Maxadilan was found, in 1996, to specifically be the PAC1 receptor. At that time, it was referred to as the PACAP type I receptor. [1]

2. The structure of Maxadilan

The Maxadilan is a 61 amino-acid containing polypeptide. The Maxadilan contains 4 cysteine residues in 1, 5, 14 and 51 positions. Disulfide bonds are present between the cysteines at positions 1–5 and 14–51. Removal of the first ring or substituting the cysteines in this ring with alanines did not affect activity. Substitution with alanine of the cysteine residues in the second ring, individually or together, resulted in the loss of activity. This observation demonstrated the importance of this second ring in maintaining the structural integrity of the peptide in a certain conformation to present necessary residues to the PAC1 receptor. Deletion of the amino acids between positions 25 and 41 in the larger disulfide loop generated M65, a potent and specific PAC1 antagonist. Threonine residues appear to be cooperatively involved in the activation of the PAC1 receptor with threonine-33 being the most important. Taken together, a model was proposed in which C-terminal lysine residues initiate interaction with the PAC1 receptor, while threonine residues maybe responsible for receptor activation. [2]

3. Aims

Based on our previous work on PACAP 1-38, 1-27 and PACAP 6-38, 6-27 nonselective antagonist peptides, our aims were the chemical synthesis and MS characterization of Maxadilan (Maxa61) and Maxadilan65 (Maxa65) peptides respectively, using different synthesis methods and the investigation of their activity on PAC1 receptor.

4. Synthetic work

The Maxadilan peptide containing 61 amino acids was synthesized on solid phase applying Fmoc/tBu strategy, and the synthesis was carried out using a CEM® microwave assisted fully automated peptide synthesizer. After reaching the desired peptide sequence, the peptide was cleaved from the resin using a standard peptide cleaving protocol (TFA 90%, water 5%, TIS 2.5%, DTT 2.5%) and the crude peptide was purified using RP-HPLC. The sequence of the synthesized Maxa61 peptide:
CDATCQFRKAIDDCQKQAHHSNVLQTSVQTTATFTSMDTSQLPGNSVFKECMKQKKKEFKA
To obtain the desired disulfide bridges the purified Maxa61 peptide was oxidized. The oxidation step was carried out by using iodine, and the reaction was monitored by using RP-HPLC. The MS spectrometry analysis of the oxidized Maxa61 successfully proved the supposed disulfide connectivity, thus the presence of the S-S bonds between cysteine residues 1-5 and 14-51.

5. Work in progress

The low yield of the Maxa61 product obtained by using the stepwise synthesis prompted us to use native chemical ligation (NCL). NCL introduced by Kent and co-workers is a revolutionary method for the synthesis of relatively long peptides and proteins. Native chemical ligation a peptide -thioester (N-terminal fragment) and a cys-peptide (C-terminal fragment) is based on a thiol catalyzed thiol-thioester exchange and subsequent S-N acyl transfer (amide bond formation). [3]
The synthesis of the Maxa65 and Maxa61 peptide fragments compatible with native chemical ligation were carried out using manually solid phase peptide synthesis and Boc chemistry. The C-terminal fragment was prepared by using a standard DCC/HOBT coupling protocol. The crude peptides were purified by semipreparative HPLC, and the integrity of the products was verified by mass spectrometry. The ligation of peptide fragments was performed in 0,1M ammonium acetate buffer (pH 7,5) containing 3% of thiophenol at 5-6 mg/ml peptide concentration at 40°C.The reaction was monitored by analytical RP-HPLC. Depending on the sequence of the peptide the ligation reaction reached completion in 5-6h in all cases. The Acm protecting group removal was carried out in 2%anisole/TFA solution in the presence of silvertrifluoromethanesulfonate (50eq.) and reached completion in 2-3h. The oxidation of the linear Maxa65 peptide was carried out using different oxidation conditions but unfortunately the formation of the desired folded Maxa65 was not observed. The synthesis of the Maxa61 peptide using native chemical ligation is in progress.

6. Summary

The synthesis of the two disulfide bridges containing Maxa61 peptide by using a fully automated microwave assisted peptide synthesizer was carried out successfully. The MS spectrometry analysis of the oxidized Maxa61 proved the supposed disulfide connectivity, thus the presence of the S-S bonds between cysteine residues 1-5 and 14-51. We have successfully synthesized the Maxa65 PAC1 antagonist peptide by using native chemical ligation procedure, but unfortunately the oxidation and the identification of the formed disulfide bonds were failed.

References

01. Ethan A. Lerner , Aurel O. Iuga, Vemuri B. Reddy; peptides 28 (2007) 1651 – 1654
02. Vemuri B. Reddy, Yhong Li, Ethan A. Lerner; J Mol Neurosci (2008) 36:241–244 DOI 10.1007/s12031-008-9079-1
03. Györgyi Váradi, Gábor K. Tóth, Zoltán Kele, László Galgóczy, Ádám Fizil, Gyula Batta; Chem. Eur. J. 2013, 19, 12684 – 12692

PLASMIN SPECIFIC INHIBITORS: OPTIMIZATION OF THE P2 AND P1′ RESIDUES

Naomichi Iwasa[1] Takashi Masuda[1] Mototsugu Kuno[1] Koushi Hidaka[1] Keiko Hojo[1] Keigo Gohda[2] Naoki Teno[3] Keiko Wanaka[1] Yuko Tsuda[1]
[1] Kobe Gakuin University, Pharmaceutical Sciences, Kobe, Japan
[2] Computer-Aided Molecular Modeling Research Center, Kansai, , Nishinomiya, Japan
[3] Hiroshima International University, Faculty of Clinical Nutrition, Kure, Japan

PP VIII – 198

Plasmin (Plm) plays a dominant role in the fibrinolysis pathway and the Plm inhibitors have been used in the treatment of bleeding. Additionally, the localized activation of plasminogen (Plg) and action of Plm on the specific regions of the extracellular milieu to digest many proteins on the localized cell surface results in inducing cell inversion and metastasis and alternating the expression of cytokines. In fact, we disclosed that Plm inhibitor, YO-2 [IC_{50} = 0.53 and 5.3 mM for Plm and urokinase Plg activator (uPA), respectively] [1], attenuated both aGVHD- and colitis-associated lethality in mice [2, 3]. Development of selective Plm inhibitors has been required to study the precise role of Plm and its relationship in treating several diseases. The design of Plm inhibitors that could be applicable to treat inflammation or inhibit cancerous growth or development is a challenge for the future.

Docking experiments with the uPA- or Plm-YO-2 complexes revealed the structural differences in the binding pockets between uPA and Plm [4]. The binding pocket of uPA, an insertion loop exits between the S2 and S3 areas, and the absence of the S2/S3 extra pocket. In contrast, the binding pocket of Plm, an insertion loop is absent; instead, an extra open area is found. A larger P2 residue would be tolerated in the Plm binding pocket, but it would not be accommodated by uPA. Furthermore, Plm has a wider S1′ area than that of uPA. Recently, BMS reports imidazole (S)-1, which consists of a basic tranexamic acid (Tra) (P1), Phe (P1′) and aminoindazole (P2′), slightly inhibited Plm (K_i = 8.4 mM) [5]. Incorporation of the imidazole scaffold into the P1′ moiety of YO-2 may produce new interactions with Plm to affect its binding affinity. In this report, we describe the modification of YO-2 at the Tyr(OPic) and octylamine residues.

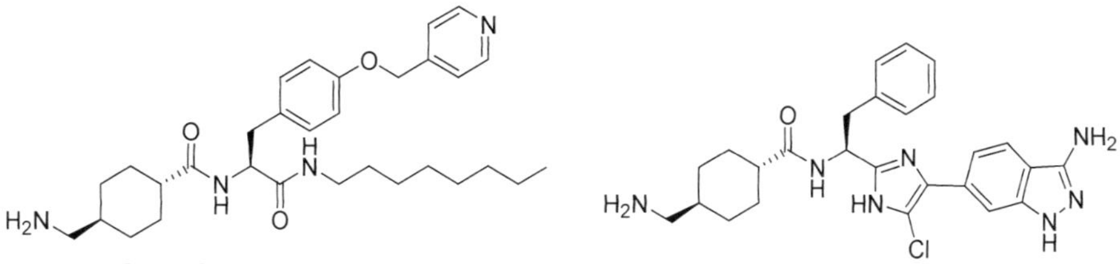

YO-2 (IC_{50} = 0.53 μM) **Imidazole(*S*)-1** (K_i = 8.4 μM)

Fig. 1. The structure of YO-2 and Imidazole(S)-1 as known Plm inhibitors.

The imidazole core is created from a coupling of N-protected a-amino acid and a -bromoketone. Cyclization of the resulting ketoester in the presence of ammonium acetate in refluxing xylene yielded imidazole derivatives [6]. Protected peptides were synthesized by a solution method using Boc-chemistry. Final products were identified by analytical HPLC and MALDI (ESI)-TOF mass spectrometry.

First, the (4-pyridinyl)methyl (Pic) residue on OH-Tyr (P2 residue) was modified using aromatic moieties. To evaluate the effect of the acidic function, 4-carboxybenzyl group (1) was incorporated and followed by ethylesterification (2) and ethylamidation (3). Similarly, subsequent introduction of a 3-carboxybenzyl group (4) was converted into an ethylester (5) or an ethylamide (6). Furthermore, 4 was condensed with ethanolamine, 3-aminopyridine, 3-aminomethylpyridine or 4-(aminomethyl) phenol to give compounds 7-10. While, compound 11 had an aliphatic extension, a (benzylamincarboxy)methyl group. All compounds inhibited Plm with a mM range (IC_{50} = 0.20-2.7 mM). Compounds 1, 4, 8 and 9 showed relatively weak Plm inhibition (IC_{50} = 2.7, 0.53, 0.70 and 0.44 mM, respectively), suggesting that acidic and basic functions did not interact with the S2 site. The extension, which might result in additional hydrogen bonds (13 and 16), did not affect the Plm inhibition (IC_{50} = 0.80 and 0.41 mM, respectively). Even the H-bonds do not appear to be a key factor for binding at the S2 site. Among 1-11, 6 and 11 made a slight improvement in Plm inhibition (IC_{50} = 0.20 and 0.25 mM, respectively). The S2 binding pocket of Plm has a preference for the hydrophobic moiety regardless of aromatic or aliphatic elements.

Concerning the imidazole derivatives (12-17), compound 15 exhibited the highest Plm inhibition (IC_{50} = 2.19 mM), which was 4-fold stronger than imidazole (S)-1 itself, but still less active compared to YO-2. Compound 15 has a basic tranexamic acid (Tra) (P1), Tyr(OPic) (P2) and aminoindazole attached with a benzene moiety (P1′). Replacement of Phe with Tyr(OPic) led to a 350-fold increase in the Plm inhibition (12 to 15). In the series of containing Phe (12-14), the inhibition was relatively weak (IC_{50} =772 to 75.4 mM) and depended on the P1′ residue, seen with benzene, 3-metoxybenzene and 4-metoxybenzene. In contrast, in the series of Tyr(OPic) analogues (15-17), Plm was inhibited with similar IC_{50} values (2.19-3.26 mM) in the absence of co-relationship. These observations on the new inhibitors indicated that the imidazole moiety could interact with Plm by coordinating to the P2 residue.

In conclusion, the modification of P2 and P1′ residues has been explored, and their Plm inhibition has been evaluated. Higher binding activities were obtained with a non-polar extension on OH-Tyr. The Plm binding pocket could accept non-polar expansions at the P2 residue. Additionally, the substitution of octylamide with imidazole scaffold retained the Plm inhibitory activity. Heterocyclic scaffolds can provide Plm inhibitors with new physical properties.

This research was supported in part by KAKEN(16K08333) and MXET-Supported Program for the Strategic Research Foundation at Private Universities, 2012-2016.

References
01. Okada, Y., Tsuda, Y., Tada, M., Wanaka, K., Hijikata-Okunomiya, A., Okamoto, S. Chem. Pharm. Bull., 48, 1964-1972 (2000).
02. Sato, A., Nishida, C., Sato-Kusubata, K., Ishihara, M., Tashiro, Y., Gritli, I., Shimazu, H., Munakata, S., Yagita, h., Okumura, K., Tsuda, Y., Okada, Y., Tojo, A., Nakauchi, S., Takahash, S., Heissattori, K. Leukemia, 29, 145-156 (2015).
03. Munakata, S., Tashiro, Y., Nishida, C., Sato, A., Komiyama, H., Shimazu, H., Dhahri, D., Salama, Y., Eiamboonsert, S., Takeda, K., Yagita, H., Tsuda, Y., Okada, Y., Nakauchi, H., Sakamoto, K., Heissig, B., Hattori, K. Gastroenterology, 148, 564-578 (2015).
04. Gohda, K., Teno, N., Wanaka, K., Tsuda, Y. J. Enzyme Inhib. Med. Chem., 27, 571-577 (2012).
05. Corte, J. R., Fang, T., Hangeland, J. J., Friends, T. J., Rendina, A. R., Luettgen, J. M., Bozarth, J. M., Barebera, Frank. A., Rossi, K. A., Wei, A., Ramamurthy, V., Morin, P. E., Seiffert, D. A., Wexler, R. R., Quan, M. L. Bioorg. Med. Chem. Lett., 25, 925-930 (2015).
06. Contour-Galcera, M. –O., Poitout, L., Moinet, C., Morgan, B., Gordon, T., Roubert, P., Thurieau, C. Bioorg. Med. Chem. Lett., 11, 741-745 (2001).

APPLICATION OF DIFFERENT PROTECTING STRATEGIES FOR THE SYNTHESIS OF THE ANTIFUNGAL PROTEIN AFP OF ASPERGILLUS GIGANTEUS

Györgyi Váradi[1], Gyula Batta[2], László Galgóczy[3], Dorottya Hajdú[2], Ádám Fizil[2], Máté Virágh[3], Zoltán Kele[1], Gábor K. Tóth[1]
[1] Department of Medical Chemistry, University of Szeged, Dóm tér 8, 6720 Szeged, Hungary,
[2]Department of Organic Chemistry, University of Debrecen, Egyetem tér 1, 4010 Debrecen, Hungary and
[3]Department of Microbiology, University of Szeged, Közép fasor 52, 6726 Szeged, Hungary
E-mail: varadi.gyorgyi@med.u-szeged.hu

Introduction

The basic, 51 amino acid containing antifungal protein (AFP) isolated from the mold Aspergillus giganteus inhibits the growth of fungi without having effect on mammalian cells[1,2]. The structure of AFP is stabilized by 4 disulfide bridges. Although the solution structure of the protein has been determined by NMR, there is no direct evidence for interlocking disulfide bond pattern characteristic for the vast majority of multiple disulfide bond containing antimicrobial peptides and proteins[3] (Fig. 1). Preparation of analogues of AFP for further structural and functional investigations requires an efficient chemical method.

Fig. 1. Sequence and supposed disulfide bond pattern of AFP.

The purpose of the work was the development of chemical synthesis of AFP. The strategy allows us to prepare analogues of AFP and examine structure-activity relationships as well as the importance of disulfide bond pattern on the antifungal effect of the protein.

Results and Discussion

The synthesis of AFP was performed by native chemical ligation[4]. Two fragments of the protein were prepared by solid-phase method applying ᵗBoc chemistry for the N-terminal part and Fmoc chemistry for the C-terminal fragment. Using the acid-labile ᵗBoc protecting group, the thioester prepared on the previously described Cys-SH resin, remained intact during the course of the synthesis[5].
Two different protecting strategies were applied for the thiol groups of the cysteine residues: uniform and selective protection. At the first attempt, the protecting groups of all cysteines were removed during the cleavage of the peptides from the resins. In this case, native chemical ligation produced a peptide having eight free sulfhydryl groups. Air oxidation led to the formation of correctly folded AFP as major product, as well as two misfolded minor products. The disulfide isomers were isolated and examined by NMR and antifungal susceptibility tests. At the second attempt, Acm (acetamidomethyl) group was used for half of cysteine residues allowing us the selective formation of the first two and the second two disulfide bridges. Oxidation was carried out with iodine in an acidic medium. Having four free SH groups, it is mathematically possible to form two disulfide bridges in three different ways. In the first oxidation step, all of the three analogues could be isolated. Further Acm cleavage and oxidation of them led to the formation of two major products in all cases. The proteins were isolated and subjected to NMR and microbiological investigations. Some of the misfolded analogues of AFP treated with glutathione redox system and could be refolded successfully.

Synthetic AFP was characterized by analytical HPLC, MS, NMR and in vitro broth microdilution antifungal susceptibility test. To confirm the disulfide-bridge pattern of AFP variants, they were subjected to enzymatic digestion. While misfolded proteins could be cleaved to fragments, the well-folded synthetic and native AFP remained intact against a trypsin-chymotrypsin enzyme cocktail. Different NMR techniques (^{1}H-NMR, ^{13}C-HSQC and 2D-NOESY) were used for investigating the 3D structure of AFP. These measurements prove that the structures of synthetic and native AFP are identical. Moreover, two minor components that cannot be seen on the HPLC profile could be detected in all AFP samples, and these are also identical in the synthetic and native AFP. Supposedly, the minor components originate from a yet uncharacterized thermal equilibrium. Antifungal susceptibility tests revealed that the in vitro antifungal activity of synthetic AFP against Aspergillus niger SZMC 601 is the same as exerted by native AFP[6]. Two misfolded variants of AFP tested under the same conditions did not show significant growth inhibition. Determination of the fungal growth of A. niger SZMC 601 in the presence of AFP variants after 48 hours of incubation at 25 °C on YPG media proved identical antifungal activity of synthetic and native AFP. Colony diameters in the presence of misfolded AFP variants found to be the same as that of the untreated control.

Acknowledgements

^{15}N-AFP was kindly provided by Dr. Florentine Marx (Innsbruck Medical University).

References

01. Nakaya, K., Omata, K., Okahashi, I., Nakamura, Y., Kolkenbrock, K., Ulbrich, N., Eur. J. Biochem. 1990, 193, 31-38.
02. Lacadena, J., Martínez del Pozo, A., Gasset, M., Patino, B., Campos-Olivas, R., Vázquez, C., Martínez-Ruiz, A., Mancheno, J. M., Onaderra, M., Gavilanes, J. G., Arch. Biochem. Biophys. 1995, 324, 273-281.
03. Campos-Olivas, R., Bruix, M., Santoro, J., Lacadena, J., Martínez del Pozo, A., Gavilanes, J. G., Rico, M., Biochemistry 1995, 34, 3009-3021.
04. Dawson, P. E., Muir, T. W., Clark-Lewis, I., Kent, S. B. H. Science 1994, 266, 776-779.
05. Váradi, G., Tóth, K. G., Kele, Z., Galgóczy, L., Fizil, Á., Batta, G., Chem. Eur. J. 2013, 19, 12684-12692.
06. Kaiserer, L., Oberparleiter, C., Weiler-Gorz, R., Burgstaller, W., Leiter, E., Marx, F. Arch. Microbiol. 2003, 180, 204-210.

PROGRAMMED BACTERIAL CELL DEATH IS A SOURCE OF PHYSIOLOGICALLY ACTIVE PEPTIDES IN MACROORGANISM.

Voronina O.L.[1], Aksenova E.I.[1], Kunda M.S.[1], Semenov A.N.[1], Ryzhova N.N.[1], Zamyatnin A.A.[2,3], Gintsburg A.L.[1]
[1] N.F. Gamaleya Federal Research Center of Epidemiology and Microbiology, Ministry of Health, Russia
[2] Federal Research Centre Fundamentals of Biotechnology, RAS, Russia
[3] Universidad Técnica Federico Santa Maria, 1680 av. España, Valparaiso, 110-V, Chile

Introduction

Phenoptosis is now described not only in eukaryotes, but in prokaryotes too and is discussed in the context of regulation of a multicellular bacterial community [1]. However the products of bacterial cell degradation can participate in macroorgamism's process regulation in case of bacterial infection. Peptides are big part of this pool of regulators. Our previous investigation demonstrated that analogs of the endogenous oligopeptides can be the natural fragments of grape proteins [2] or proteins of food [3]. How many physiologically active peptides can be formed by proteolysis during the phenoptosis of bacterial cells? We can predict it by bioinformatics analysis of the cell proteome, using EROP-Moscow database as a source of endogenous peptides structures [4].

Results and Discussion

Mycobacterium tuberculosis complex infection is an actual public and animal health problem. M. bovis, the member of this complex, was the object of our study. Sequenced genome M. bovis BCG Russia 368 strain (GenBank Accession Number CP009243) was annotated by a set of bioinformatics tools, than the proteome was analyzed. We performed computer comparison of all proteins sequences with all known functionally characterized oligopeptides included in EROP-Moscow database (13915 entries, 01-Sep-2016 [5]). 2886 from 4287 annotated proteins contained 7761 natural fragments corresponding to 549 physiologically active peptides. Ten peptide sequences repeated more than 100 times in the proteome, 146 sequences were found 11-100 times, 235were revealed 2-10 times.

PE_PGRS family proteins, which contained polymorphic GC-rich repetitive sequence, had the most natural fragments with different functions. Table 2 demonstrated the example of PE-PGRS54, which can generate 41 sequences analogous to endogenous oligopeptides. As expected, the most frequent was the neuropeptide GGGG, which was found in 131 proteins too. Less frequent were antimicrobial peptides and enzymes inhibitors sequences in PE-PGRS54. In whole proteome we revealed 128 antimicrobial peptide sequences. Most of them were active against Gram-positive bacteria or Fungi.

VLIAP and AGSS peptides had very important function - inhibiting angiotensin I-converting enzyme (ACE). VLIAP peptide was presented only in tree proteins, while AGSS sequence was found in 66 M. bovis proteins. We revealed 1329 natural fragments of ACE inhibitors. It was 20.8% from all natural fragments found in the proteome. They differed in functional activity and had IC50 for ACE from 3.8x10-7 (E04543) to 9.4x10-4 (E07457) mol/L [5]. If maximal mycobacterium cell size is 3.5 μm3 [6], concentration of 1329 found fragments may be 1,6x10-4mol/L, that is near the higher border of activity.

ACTIVE PEPTIDES IN MACROORGANISM.

Physiologically active oligopeptides predicted in PE-PGRS54 of Mycobacterium bovis

Peptide functional group	Unknown function							Neuropeptides				Enzymes inhibitors		Anti-micro-bial
EROP-Moscow Accession Number	E00563	E00563	E00563	E00563	E00563	E01667	E01668	E04741	E04739	E04740	E09280	E07059w	E09362	E05756
Peptide sequence	DGGAG	NAGAG	DAGAG	DAGVG	SAGAG	WGDV	VLGGV	GGGG	GGGFGG	GGGFG	NAGA	VLIAP	AGSS	ACGAG
Peptides found in BCGR_3852 (PE-PGRS54)	6	5	3	3	1	1	1	11	2	2	4	1	1	2
CP009243:peptides found in proteome	164	27	14	8	11	9	9	326	9	12	53	3	70	2
CP009243: proteins with found peptides	53	18	12	6	10	9	9	131	7	10	43	3	66	1

These data are additive step in understanding the molecular mechanisms involved in mycobacterial hematogenous spread. If M. tuberculosis bacteria have RD1 region in genome, and so stimulate angiogenic factor VEGF secretion in human macrophages, inducing the formation of new blood vessels [7]; M. bovis BCG can affect the blood pressure.

Conclusion. So programmed bacterial cell-death is a source of peptide regulators of macroorganism physiological functions.

References

01. O. A. Koksharova, Bacteria and phenoptosis, Biochemistry (Mosc) 2013, 78(9), 963.
02. A. A. Zamyatnin, O. L. Voronina, Antimicrobial and other oligopeptides of grapes, Biochemistry (Mosc) 2010, 75(2), 214.
03. A. A. Zamyatnin, O. L. Voronina, Food protein fragments are regulatory oligopeptides, Biochemistry (Mosc) 2012, 77(5), 502.
04. A.A. Zamyatnin, A.S. Borchikov, M.G. Vladimirov, O.L. Voronina. Nucl. Acids Res. 2006, 34, D261.
05. EROP-Moscow, http://erop.inbi.ras.ru/ (accessed: September 2016).
06. M. Loferer-Krössbacher, J. Klima, R. Psenner, Determination of bacterial cell dry mass by transmission electron microscopy and densitometric image analysis, Appl Environ Microbiol. 1998, 64(2), 688.
07. H. Polena, F. Boudou, S. Tilleul, N. Dubois-Colas, C. Lecointe, N. Rakotosamimanana, M. Pelizzola, S. F. Andriamandimby, V. Raharimanga, P. Charles, J. L. Herrmann, P. Ricciardi-Castagnoli, V. Rasolofo, B. Gicquel, L. Tailleux, Mycobacterium tuberculosis exploits the formation of new blood vessels for its dissemination, Sci Rep. 2016, 6, 33162.

EFFECT OF HUMAN QRFPR LIGANDS IN THE INSULIN TOLERANCE TEST IN MICE

Alim Karima, Picot Marie, Lefranc Benjamin, Prévost Gaëtan, Chuquet Julien, Vaudry David, Vaudry Hubert, Chartrel Nicolas, Leprince Jérôme
Inserm U982, University of Rouen-Normandy, F-76000 Rouen, France

PP IX - 211

Introduction

The term RFamide-related peptides (RFRPs) designates a family of biologically active peptides gathering neuropeptides FF and AF, prolactin-releasing peptide, RFRP-1 and -3, and metastin/kisspeptins that possess the common signature Arg-Phe-NH_2 at their C-terminal extremity. 26RFa is the last member of this regulatory peptide family characterized in our laboratory [1], and is recognized as the endogenous ligand of the former orphan receptor GPR103, now renamed QRFPR. Tissue distribution of 26RFa and QRFPR evince the involvement of this peptide system in several physiological and pathophysiological processes such as regulation of energy homeostasis and bone mineralization [2]. 26RFa also stimulates the gonadotrope axis, increases locomotor activity, induces analgesia and modulates glucose-evoked insulin secretion suggesting that QRFPR ligands should be amenable to drug development [2]. Structure-activity relationship studies conducted on human 26RFa (TSGPLGNLAEELNGYSRKKGGFSFRF-NH_2) reveal that despite a 100-folds decrease in potency, 26RFa$_{(20\text{-}26)}$ (EC_{50} = 10.4 ± 1.5 vs. 739 ± 149 nM) represents a relevant scaffold to design low-molecular weight QRFPR ligands [3]. In this context, we have recently designed three 26RFa analogues (LV-2172, LV-2185 and LV-2186) displaying interesting pharmacological in vitro profiles on human QRFPR-transfected cells. The aim of the present study is to evaluate, in vivo, the effect of these three peptide compounds during an insulin tolerance test in mice.

Results and Discussion

As previously described, intraperitoneal (ip) administration of 26RFa (500 µg/kg) potentiates insulin-evoked hypoglycemia 30 minutes after the insulin load (0.75 U/kg ip) [4]. Surprisingly, LV-2172 (150-750 µg/kg ip), a pseudopeptide more potent, affine and stable than 26RFa and exhibiting a long-lasting orexigenic effect in mice [5] and LV-2186 (150-750 µg/kg ip), a brand-new agonist with a EC_{50} = 67 ± 5 nM, did not exacerbate the insulin effect on glucose level (Fig. 1A et B). Conversely, LV-2185 (1500 µg/kg ip), which significantly reduces the 26RFa-evoked intracellular calcium increase in human QRFPR-transfected cells [6], did not reduce the pro-hypoglycemic activity of 26RFa in mice but perfectly mimicked its effect (Fig. 1C). Thus, it appears that there was a discrepancy between the in vitro effects of these three analogues on human QRFPR and their in vivo effects in mice. Indeed, both human QRFPR agonists LV-2172 and LV-2186 were devoid of agonistic activity in mice whereas the human QRFPR antagonist LV-2185 behaved as a mouse QRFPR agonist. We can speculate that either, our human QRFPR-directed agonisitic analogues are specific of QRFPR but that another receptor mediates the effect of 26RFa on insulin-induced hypoglycemia or that our compounds behave as biased QRFPR ligands for activating differently G protein-coupled QRFPRs. Further experiments are required to decipher these options, as two QRFPR isoforms exist in rodents whereas only one gene encodes human QRFPR [7, 8].

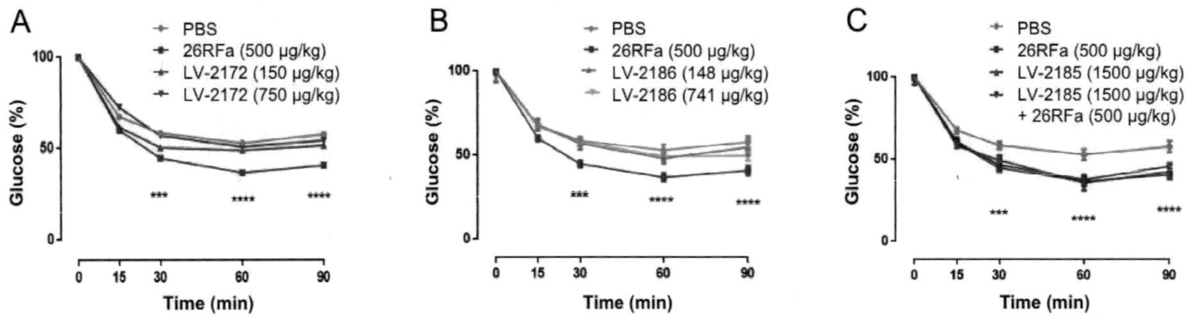

Fig. 1. Effects of intraperitoneal administration of LV-2172 (A), LV-2186 (B) and LV-2185 (C) during the insulin tolerance test. In this test, 26RFa potentiates hypoglycemia induced by insulin 30 to 90 minutes after their intraperitoneal co-injection (Figure 1 A.B.C). LV-2172, described as potent hQRFPR agonist and LV-2186, a new agonist more potent than LV-2172 do not alter insulin-induced hypoglycemia (Figure 1 A.B). By contrast, LV-2185, the first human QRFPR peptide antagonist that blocks the effect of 26RFa on $[Ca^{2+}]_i$ in human QRFPR-transfected cells, does not antagonise but exerts a similar pro-hypoglycemic effect to that of 26RFa (Figure 1 C).

Acknowledgements:

This study was supported by Inserm (U982) and the Normandy Regional Council. The authors wish to thank the Cell Imaging Platform of Normandy PRIMACEN for excellent technical assistance.

References:

01. Chartrel N., Dujardin C., Anouar Y., Leprince J., Decker A., Clerens S., Do-Régo J.-C., Vandesande F., Llorens-Cortes C., Costentin J., Beauvillain J.-C., Vaudry H. (2003) Proc. Natl. Acad. Sci. USA 100, 15247-1252
02. Chartrel N., Alonzeau J., Alexandre D., Jeandel L., Alvear-Perez R., Leprince J., Boutin J.A., Vaudry H., Anouar Y., Llorens-Cortes C. (2011) Front. Neuroendocrinol. 32, 387-397.
03. Le Marec O., Neveu C., Lefranc B., Dubessy C., Boutin J.A., Do Régo J.-C., Costentin J., Tonon M.-C., Tena-Sempere M., Vaudry H., Leprince J. (2011) J. Med. Chem. 54, 4808-4814.
04. Prévost G., Jeandel L., Arabo A., Coëffier M., Alexandre D., Picot M., Gobet F., Leprince J., Berrahmoune H., Dechelotte P., Malagon M.M., Bonner C., Kerr-Conte J., Lefebvre H., Anouar Y., Chartrel N. (2015) Diabetes 64, 2805-2816.
05. Neveu C., Lefranc B., Tasseau O., Do Régo J.-C., Bourmaud A., Chan P., Bauchat P., Le Marec O., Chuquet J., Guilhaudis L., Boutin J.A., Ségalas-Milazzo I., Costentin J., Vaudry H., Baudy-Floc'h M., Vaudry D., Leprince J. (2012) J. Med. Chem. 55, 7516-7524.
06. Neveu C., Dulin F., Lefranc B., Galas L., Calbrix C., Bureau R., Rault S., Chuquet J., Boutin J.A. Guilhaudis L., Ségalas-Milazzo I., Vaudry D., Vaudry H., Sopkova-de Oliveira Santos J., Leprince J. (2014) Br. J. Pharmacology 171, 4425-4439.
07. Kampe J., Wiedmer P., Pfluger P.T., Castaneda T.R., Burget L., Mondala H., Kerr J., Liaw C., Oldfield B.J., Tschöp M.H., Bagnol D. (2006) Brain Res. 1119, 133-149.
08. Takayasu S., Sakurai T., Iwasaki S., Teranishi H., Yamanaka A., Williams S.C., Iguchi H., Kawasawa Y.I., Ikeda Y., Sakakibara I., Ohno K., Ioka R.X., Murakami S., Dohmae N., Xie J., Suda T., Motoike

S.M. Balasanyants[1], T.D. Volkova[1], A.V. Kamynina[1], D.O. Koroev[1], I.J.Aleksandrova[2], I.V. Nesterova[2], A.N. Samokhin[2], N.V. Bobkova[2], O.M. Volpina[1]
[1] Shemyakin-Ovchinnikov Institute of Bioorganic Chemistry, Russian Academy of Sciences, ul. Miklukho-Maklaya, 16/10, 117997, Moscow, Russia. E-mail: balasanjanz@gmail.com
[2] Institute of Cell Biophysics, Russian Academy of Sciences, ul. Institutskaya, 3. 142290, Pushchino, Russia

Introduction

The receptor for advanced glycation end products (RAGE) is a member of the immunoglobulin protein superfamily. RAGE is a multiligand receptor and consists of three extracellular domains (V, C1 and C2), a transmembrane region and a short cytoplasmic region. Most of ligands, such as beta-amyloid and S100B protein bind V- and C1-domains. This receptor is synthesized in different cells, including neurons and glial cells. Activation of RAGE causes brain inflammation, oxidative stress and secretion of beta-amyloid that has been recognized as an essential phase in the development of Alzheimer's disease. It is known that the receptor soluble isoform (sRAGE), which lacks the transmembrane and cytosolic domains, binds to ligands and prevents negative effects of the receptor activation in in vivo and in vitro experiments [1]. We proposed that potential ligand-binding peptide fragments from sRAGE would demonstrate similar to sRAGE biological activity.

Results and discussion

We have selected and synthesized 10 peptide fragments from unstructured surface exposed regions of RAGE using standard Fmoc/Bu^t protocol of solid-phase peptide synthesis. The choice of fragments was based on the data from the X-ray analysis of this protein [2]. Peptides were characterized by the methods of analytical reversed-phase HPLC, MALDI mass spectrometry, and amino acid analysis. Purity of the peptides was estimated >98%, as indicated by the data of HPLC.

Table 1. Synthetic RAGE fragments

	RAGE fragment	Sequence
I	28-32	ARIGE
II	38-45	CKGAPKKP
III	60-76	AWKVLSPQGGGPWDSVA
IV	134-141	SELTAGVPNKV
V	162-169	KPLVPNEK
VI	163-175	PLVPNEKGVSVKE
VII	179-186	RHPETGLF
VIII	195-206	TPARGGDPRPTF

Synthetic peptides were intranasally administrated into olfactory bulbectomized (OBX) mice which developed behavioral, morphological and biochemical signs of the Alzheimer's type degeneration [3]. 2 weeks after olfactory bulb removal, peptide was administrated in a dose of 20 mg per mouse during 15 days. On the 21st day after the operation, the training in the Morris water maze was carried out, and after 5 days the state of the spatial memory was tested. A physiological solution was administered into control sham-operated animals and OBX animals. We have found that administration of RAGE fragment (60-76) only significantly prevents the OBX murine memory from impairment, leads to decrease of beta-amyloid level and blocks the development of neuronal pathology in the brain of the experimental mice.

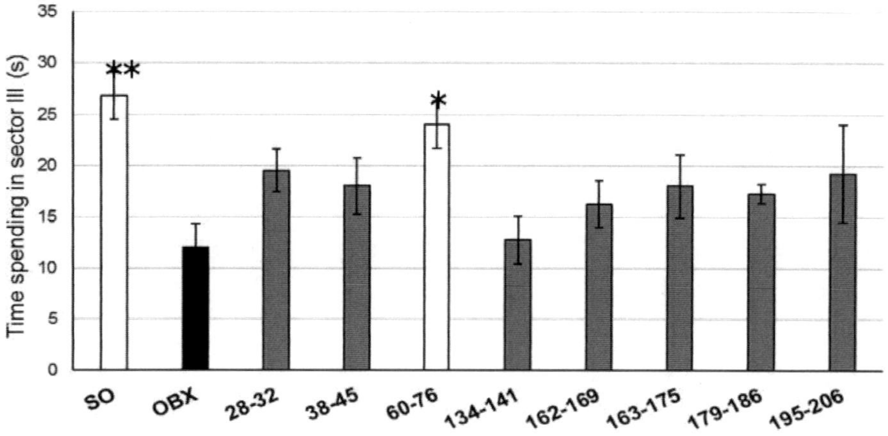

*Figure 1. The occupancy time in the target sector of Morris water maze after the intranasal administration of RAGE fragments. SO - sham-operated; OBX - bulbectomized. * $p < 0.05$: ** $p < 0.01$*

Six overlapping fragments of RAGE (60-76) peptide were synthesized in order to reveal a site, responsible for the therapeutic effect. Tests in OBX mice carried out with these fragments showed that only the N-terminal part of the molecule is responsible for preventing the memory in OBX mice from impairment. Fragment (60-70) was the shortest peptide with statistically significant activity in those experiments.

Peptide (60-76) corresponds to a fragment from V-domain ligand-binding site. We suppose that the discovered active peptides demonstrate biological activity similar to sRAGE - bind with RAGE ligands, such as S100B, β-amyloid and others [4]. This interaction can prevent activation of RAGE-dependent neuroinflammation and neurodegeneration. Our proposal is currently under investigation.

Acknowledgments
This study was financially supported by RFBR (15-04-01360).

References
01.Deane, R., Du, Yan S., Submamaryan, R.K., et al. Nature Med. 9, 907–913 (2003).
02. Koch, M., Chitayat, S., Dattilo, B., et al. Structure. 18, 1342-1352 (2010).
03. Bobkova, N.V., Medvinskaya, N.I., Kamynina, A.N., et al. Neurobiology of Learning and Memory. 107 50–64 (2014).
04. Yan S.F., Ramasamy R., Schmidt A.M., Biochem Pharmacol. 79(10): 1379–1386 (2010).

Daniel Martinez, Cyf Ramos-Colon, James P. Cain, Elizabeth Restituyo-Rosario
Protein Technologies, Inc. 4675 South Coach Drive, Tucson, Arizona, 85714, U.S.A. Tel: +1-520-629-9626, Website: www.ptipep.com, Email: info@ptipep.com

Introduction

Peptides are highly selective and generally well tolerated drug candidates, resulting in increased demand for rapid peptide development in order to accelerate their evaluation as potential therapeutics. Automated solid phase peptide synthesis (SPPS) has proven to be the most effective way to meet the increased need for peptides in industry and academia. However, peptides also typically have unfavorable characteristics for a drug, like poor oral bioavailability and reduced stability, leading to the exploration of complex and cyclic peptides that may offer an improvement in these properties. Owing to steric and conformational factors, the synthesis of complex peptides can be challenging; thus enabling technologies like rapid heating methods have become important. Induction heating has recently been introduced on the Prelude® X, which allows for independent, simultaneous and rapid heating of multiple reactors with increased efficiency.

In order to demonstrate the efficacy of this new technology, the difficult peptide sequences Jung-Redemann (JR) 10-mer[1] and Aib-Enkephalin (Leu-enkephalin with Aib replacing both glycines) have been synthesized using heating during every cycle.

Just as heat-assisted synthesis has been shown to reduce the time necessary to produce high-purity linear peptides, its application in the synthesis of cyclic peptides can provide similar advantages. The utility of heating in the preparation of a cyclic peptide has been illustrated by the synthesis of the potent melanocortin receptor agonist Melanotan II (MT-II)[2]. Multiple temperature profiles were tested in parallel for the optimization of the cyclization reaction.

Jung-Redemann Sequence: **H-Trp-Phe-Thr-Thr-Leu-Ile-Ser-Thr-Ile-Met-NH_2**
Aib Enkephalin (Aib-enk) Sequence: **H-Tyr-Aib-Aib-Phe-Leu-NH_2**
Melanotan II (MT-II) Sequence: **Ac-Nle-cyclo[Asp-His-D-Phe-Arg-Trp-Lys]-NH_2**

Experimental

JR10 was synthesized by protocol A and B, Aib-enk was synthesized by protocol B and the linear MT-II was synthesized by protocol C (Table 1). Alloc and Allyl side chain protection was used for Lys and Asp residues on MT-II.

MT-II cyclization: Following Pd-mediated removal of the side chain protecting groups and washing, a solution of PyClock (50 mM, 5 eq) and DIEA (100 mM, 10 eq) in DMF was added to the resin. After cyclization, the resin was washed with DMF and DCM.

Table 1. Descriptions of different protocols followed for the synthesis of difficult peptide sequences.

Protocols	Deprotection	Coupling	Final Cleavage
A: 50 μmol scale Rink Amide MBHA PS resin (0.32 mmol/g)	1 min, 25°C 20% Piperidine in DMF	2 min at 25°C, 60°C or 90°C 250 mM AA/HCTU, 500 mM NMM	
B: 50 μmol scale Rink Amide ChemMatrix resin (0.47 mmol/g)	2 min at 25°C, 60°C or 90°C 20% Piperidine in DMF	3 min at 25°C, 60°C or 90°C 100 mM AA/ Activators:Additives, 200 mM DIEA	2 h at 25°C TFA:TIS:EDT:Water (95:1:2.5:2.5)
C: at 200 μmol scale Rink Amide MBHA PS resin (0.32 mmol/g)	2 x 3 min, 25°C 20% Piperidine in DMF	30 min at 25°C 200 mM AA/HCTU, 400 mM NMM	

Analysis: MT-II and JR10 prepared using Protocol A were analyzed on a Varian ProStar HPLC using a Polaris C18, 180 Å, 5 μm, 250 x 4.6 mm column, over 60 min with a flow rate of 1 mL/min, and using a gradient of 5-95%B, where Buffer A is 0.1% TFA in water, and Buffer B is 0.1% TFA in acetonitrile. JR10 and Aib enk prepared using Protocol B were analyzed using a Varian Microsorb MW 300-5 C18 50 x 4.6 mm column with a gradient of 5-95%B in 15 min at 1 mL/min. Detection was at 214 nm.

Mass analysis of MT-II and JR10 (Protocol A) was performed on a Shimadzu LCMS-2020 Single-Quad mass spectrometer, equipped with a C18, 100 Å, 2.6 μm, 50 x 2.1 mm column (Phenomenex Kinetex), over 7 min with a flow rate of 1 mL/min and using a gradient of 5-50% B where Buffer A is 0.1% formic acid in water and Buffer B is 0.1% formic acid in acetonitrile. Mass of peptides synthesized by Protocol B was determined with a gradient of 5-95%B in 15 min for JR10 and 10-20%B in 9 min for Aib-Enkephalin at 1 mL/min.

Results and Discussion

Across all conditions tested an increase in reaction temperature up to 90°C led to improved purity of JR10 (Table 2). The same trend was observed for Aib-enk, with reactions at 90°C showing the best purity results.

Table 2. Effect on peptide crude purity of different temperature protocols during coupling for JR 10-mer.

	Reagent	25°C	60°C	90°C
Protocol A:	HCTU	15.0%	48.4%	65.6%
Protocol B:	HDMC/Oxyma Pure	28.7%	50.1%	55.0%
	HDMC	29.6%	50.2%	55.4%
	COMU	25.9%	53.9%	60.2%

Table 3. Effect on peptide crude purity of different temperature protocols during MT-II cyclization.

Cyclization Times	Temp °C	Purity
1 min	55	68.7%
	85	69.5%
5 min	55	67.8%
	85	72.6%

MT-II cyclization was done at two different temperatures for two different coupling lengths. Induction heating resulted in reduced MT-II cyclization times with cyclization occurring after 1 min at both temperatures, with the highest purity achieved by heating at 85°C for 5 min.

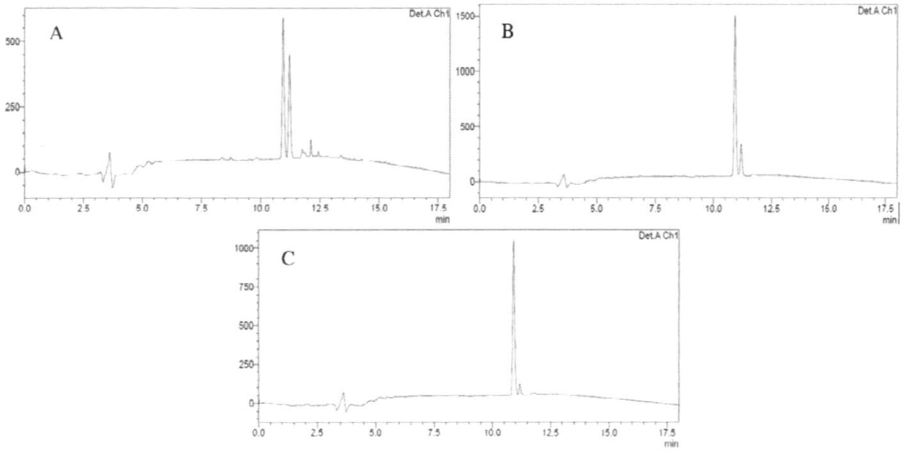

Figure 1. Aib enkephalin synthesized with COMU HPLC chromatogram: A. at 25°C, B. at 60°C, C. at 90°C.

Conclusions

Difficult peptide sequences, JR10 and Aib-enkephalin, along with a cyclic peptide, MT-II, were successfully synthesized with short synthesis times using the Prelude® X. As observed for the difficult peptides, the crude purities improved with an increase in temperature up to 90°C. By using induction heating the cyclization time of MT-II was significantly reduced without compromising crude purity, with an increased purity when run at 85°C for 5 min. In conclusion, multi-variable conditions were tested in parallel for the process optimization of JR10, Aib-enkephalin and MT-II.

References

01. Redemann, T.; Jung, G. In Peptides 1996, Proc. of the 24th European Peptide Symposium; Ramage, R., Epton, R., Eds.; Mayflower Scientific Ltd: Kingswinford, UK, 1998; 749.
02. Al-Obeidi, F., A.M., Hadley, M.E., Pettitt, B.M., and Hruby, V.J. J. Am. Chem.Soc. 111, 3413-3416 (1989).

Daniel Martinez, Cyf Ramos-Colon, James P. Cain, Elizabeth Restituyo-Rosario
Protein Technologies, Inc. 4675 South Coach Drive, Tucson, Arizona, 85714, U.S.A. Tel: +1-520-629-9626, Website: www.ptipep.com, Email: info@ptipep.com

PP IX - 220

OP
PP I
PP II
PP VI
PP VII
PP VIII
PP IX
PP X
PP XI
PP XII
PP XIII
PP XIV

Introduction

Efficient and thorough process development is crucial for scaling up the manufacture of peptides from an economic and regulatory perspective. By screening multiple methods, optimal synthesis conditions are found that minimize impurities from amino acid deletions and side reactions while balancing cost and time parameters.

The Symphony® X peptide synthesizer offers the most flexibility and independent protocols of any instrument on the market, making it the industry standard for high-throughput process development. In a previously reported example, this system has been used to synthesize C-Peptide under 22 independent conditions, producing crude peptide with purities as high as 92%.[1] Here we show the result of simultaneous screens done on the Symphony® X, including variations in the resin, coupling reagents, and reaction time, for the optimization of the 25-mer ziconotide synthesis.

Originally isolated from the venom of the marine snail Conus magus[2] and developed as SNX-111[3], ziconotide (Prialt®) acts as a potent and selective antagonist of N-type calcium channels and has been approved for the treatment of severe pain, particularly for patients with morphine tolerance.[4,5] This ω-conotoxin contains 25 amino acid residues, including six cysteines which in the native peptide form three specific disulfide bonds.[6]

Ziconotide Sequence: **H-CKGKGAKCSRLMYDCCTGSCRSGKC-NH_2**

Experimental

The ziconotide linear peptide (H-CKGKGAKCSRLMYDCCTGSCRSGKC-NH2) was synthesized on a Symphony X using Rink Amide ChemMatrix® Resin (0.47 mmol/g substitution) and TentaGel® R RAM Resin (0.19 mmol/g substitution) at a 20μmol scale. Deprotection time was 5 min (2X) at 25°C with 20% Piperidine in DMF. Coupling used 3 mL of solution at 25°C with final concentrations of 100 mM:100 mM:200 mM for Amino Acids:Activators:DIPEA with a coupling time of either 5 min, 10 min, or 20 min. DMF washes followed both deprotect and coupling steps with 3 repetitions of 30 s each. Final cleave used TFA:TIS:EDT:Water (95:1:2.5:2.5) for 2 h at 25°C followed by cold Ethyl Ether precipitation, centrifugation, and overnight drying. Analysis was performed with a 6 mg/mL solution on a Polaris 180 Å, C18, 5 μm, 250 x 4.6 mm column for HPLC analysis with a gradient of 15-30%B in 60 min using Water(0.1%TFA):ACN(0.1%TFA) at 1 mL/min. A 1:10 dilution of a standard sample of 3 mg/mL was run on a Phenomenex Kinetex 2.6 μ, C18, 100 A, 50 x 2.1 mm column for LCMS with a gradient of 5-25%B in 10 min using Water (0.1%FA):ACN(0.1%FA) at 1 mL/min.

Results

Linear ziconotide was synthesized with crude purities ranging from 28.5 to 44% (Table 1 and 2). The highest purities were observed with ChemMatrix resin. This was true whether HCTU or HDMC was used as the coupling reagent. In contrast, Tentagel produced ~20% higher yields under both activator conditions.

Table 1. Effect of different coupling times and resins on crude purity and percent yield for ziconotide synthesized with HDMC.

Coupling Times	ChemMatrix		Tentagel	
	Purity (%)	Yield (%)	Purity (%)	Yield (%)
5 min	42.7	69.8	30.2	86.8
10 min	40.4	66.9	35.3	97.2
20 min	43.4	64.2	35.6	90.6

Table 2. Effect of different coupling times and resins on crude purity and percent yield for ziconotide synthesized with HCTU.

Coupling Times	ChemMatrix		Tentagel	
	Purity (%)	Yield (%)	Purity (%)	Yield (%)
5 min	36.9	73.6	28.5	94.4
10 min	35.8	69.8	31.9	97.2
20 min	44.0	67.9	34.7	93.4

Increasing the coupling time with the combination of HDMC and ChemMatrix resin did not produce significant changes in the purity, while small increases were observed under the other conditions tested.

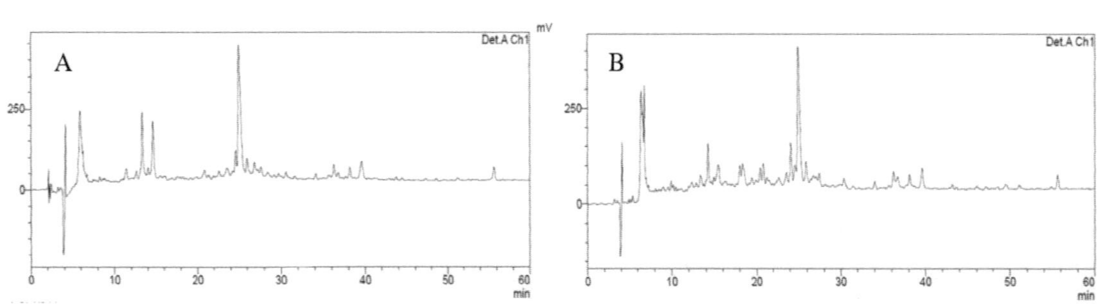

Figure 1. HPLC chromatograms of Ziconotide synthesized with HCTU on Tentagel at A) 5 min coupling and B) 20 min coupling.

Conclusion

The synthetically challenging ziconotide sequence was successfully synthesized on the Symphony X. Independent protocols allowed for the screening of 12 different conditions simultaneously. Synthesis on Tentagel resin produced the highest crude yields, while ChemMatrix consistently gave the highest crude purities under both activator conditions. The coupling reagents used are relatively stable and effective for the reaction times used here and much shorter or longer times could also be examined to clearly identify the optimal operating range for a demanding sequence such as ziconotide.

Finally, there appears to be an interaction effect between the resin type and coupling reagent, suggesting that this should also be considered along with resin/sequence and reagent/sequence compatibility.

References

01. Martinez D. et al. Proc. 24th Am. Pept. Sym. 2015, 259.
02. Olivera BM. et al. Biochemistry 1987; 26: 2086-2090.
03. Nadasdi L. et al. Biochemistry 1995; 34: 8076-8081.
04. Schmidtko A, Lotsch J, Freynhagen R, Geisslinger G. Lancet 2010; 375: 1569-1577.
05. McGivern JG. Neuropsych. Dis. Treatment 2007; 3: 69-85.
06. Chung D, Gaur S, Bell JR, Ramachandran J, Nadasdi L. Int. J. Peptide Protein Res. 1995; 46: 320-325.

NTS1 AND NTS2 DIVERSITY DELINEATED BY MOLECULAR MODELLING DRIVES THE SYNTHESIS OF NEW SELECTIVE NEUROTENSIN ANALOGUES

Roberto Fanelli[1], Nicolas Floquet[1], Mélanie Vivancos[2], Bartholomé Delort[1], Élie Besserer-Offroy[2], Jean-Michel Longpré[2], Jean Martinez[1], Philippe Sarret[2], Florine Cavelier[1*]
[1] Institut des Biomolécules Max Mousseron, IBMM, UMR-5247, CNRS, Université de Montpellier, ENSCM, Place Eugène Bataillon, 34095 Montpellier cedex 5, FRANCE.
[2] Department of pharmacology and physiology, Faculty of medicine and health sciences, Université de Sherbrooke, CANADA.
* florine.cavelier@umontpellier.fr
Keywords: Neurotensin, NTS2 selectivity, binding, peptide synthesis, modelling.

Introduction

Neurotensin is a tridecapeptide first isolated by Caraway and Leeman (1973) from bovine hypotalami.[(1)] NT exerts a wide range of biological functions including hypothermic[(2)], analgesic[(3)] and antipsychotic properties[(4)]. Its activity is related to the binding with different receptors (NTSs) belonging either to the superfamily of the G-protein-coupled receptors (NTS1 and NTS2) or to the family of sortilin receptors (NTS3)[(5)]. NTS2 receptor is an important target for the analgesic effect of NT analogues since it has been demonstrated recently its implication in pain modulation[(6)]. For this reason and for a better understanding of its physiological role, there is an urgent need to provide further information on the structure activity relationship of the receptor-ligand interaction to develop new selective analogues.

Results and Discussion

The recently crystallized rNTS1 receptor bound to its agonist peptide Neurotensin 8-13[(7)] gave the structural basis for NTS1 targeting. In this work, starting from these structural data, we built models for both hNTS1 and hNTS2 receptors bound to the NT[8-13] peptide and we were able to observe some discrepancies in terms of interaction. Interestingly, the positively charged Arg^{212} and the negatively charged Glu^{179} residues were aligned on the initial sequence alignment used for the homology modelling step, and were therefore located at the same position of the extracellular loop 2 of the two receptors, at the entry of the binding site (Figure 1). Based on this observation we chose to replace the Tyr^{11} with the basic amino acid lysine for the preparation of compound 1 (JMV 5836) and with aspartic acid for compound 2 (JMV 5839 and glutamic acid for compound 3 (JMV 5963). We also prepared compound 4 (JMV 5965) in which a lysine replaces the Tyr^{11} and the TMSAla residue replaces the C-terminal leucine since we have recently demonstrated that this modification not only increased the binding affinity compared to the NT native peptide but it produced analgesia in vivo in experimental models of acute and persistent pain.[(8)]

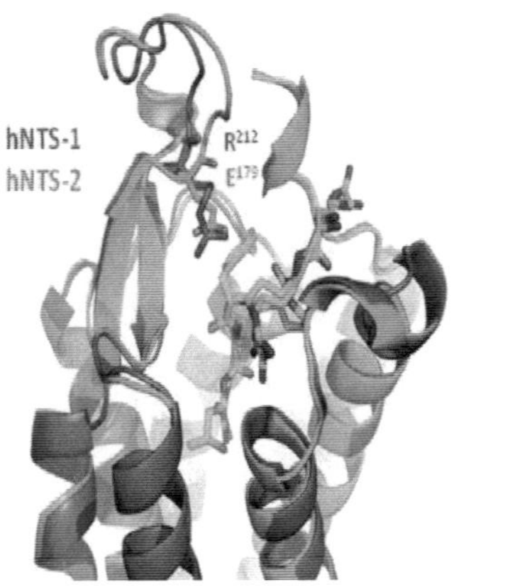

Figure 1. Localization of the two R212 and E179 residues in the extra-cellular loop 2 of hNTS1 (in magenta) and hNTS2 (in green) receptors, respectively.

We evaluated the ability of the NT[8-13] derivatives to inhibit the binding of ^{125}I-Tyr^{3}-NT on membranes prepared from cells stably expressing either hNTS1 or hNTS2 receptors and results are shown in table 1. The replacement of the residue at position 11 resulted in a decreased binding affinity for both hNTS1 and hNTS2, as compared to the native NT[8-13] peptide but in the case of compounds 1 and 4, a gain of selectivity toward hNTS2 was observed as it was expected from molecular modelling results.

Table 1. Binding potencies of the reference compound NT[8-13] and NT analogues.

compd	sequence	IC_{50} binding (nM) hNTS1	IC_{50} binding (nM) hNTS2	selectivity hNTS1/hNTS2
NT[8-13]	H-Arg-Arg-Pro-Tyr-Ile-Leu-OH	1.21 ± 0.06	7.46 ± 2.47	0.16
1	H-Lys-Lys-Pro-Lys-Ile-Leu-OH	6426 ± 858	297 ± 82.6	21.6
2	H-Lys-Lys-Pro-Asp-Ile-Leu-OH	>10,000	4753 ± 840	nd[a]
3	H-Lys-Lys-Pro-Glu-Ile-Leu-OH	>10,000	1824 ± 420	nd[a]
4	H-Lys-Lys-Pro-Lys-Ile-TMSAla-OH	752 ± 90.8	76 ± 29.5	9.89

[a] *Selectivity for these compounds is not determinable since binding on hNTS1 is greater than 10,000 nM.*

Once the binding results were determined, in order to confirm our hypothesis, we decided to express a mutated version of NTS1 receptor in order to verify if the affinity of our ligand could be restored. We then evaluated the affinity of compounds 1 and 4 on the hNTS1-R212E mutant by competition with ^{125}I-Tyr^{3}-NT. As shown in table 2, the binding affinity was increased for compounds 1 and 4 when tested on hNTS1-R212E. Compound 1 showed the most important gain in binding affinity with a 28-folds increase, whereas compound 4 displays a moderate affinity gain of 5-folds.

Table 2. Binding potencies of 1 and 4 toward the wild-type hNTS1 and mutated hNTS1-R212E receptors.

		IC_{50} binding (nM)		Affinity gain
compd	sequence	hNTS1-WT	hNTS1-R212E	R212E / WT
NT[8-13]	H-Arg-Arg-Pro-Tyr-Ile-Leu-OH	1.21 ± 0.06	5.32 ± 0.85	- 4.4
1	H-Lys-Lys-Pro-Lys-Ile-Leu-OH	6426 ± 858	226 ± 35.9	28.5
4	H-Lys-Lys-Pro-Lys-Ile-TMSAla-OH	752 ± 90.8	144 ± 19.4	5.22

These results indicate that the ionic interaction influence the binding affinity and plays a very important role in the selectivity towards the hNTS2 receptor.

References

01. Carraway R., Leeman S.E., J. Biol. Chem., 1973, 248(19), 6854-6861.
02. Dubuc I, Costentin, J., Doulut, S., Rodriguez, M., Martinez, J., Kitabgi, P. Eur. J. Pharmacol., 1992, 219(2), 327-329.
03. Nemeroff C.B., Osbahr A.J., Ervin G.N., Prange A.J., Jr. Proc. Natl. Acad. Sci. 1979, 76(10), 5368-5371.
04. Skoog K.M., Cain S.T., Nemeroff C.B. Neuropharmacology, 1986, 25(7), 777-782.
05. Kleczkowska P., Lipkowski A.W., Eur. J. Pharmacol., 2013, 716(1-3), 54-60.
06. Tetreault P., et al. FASEB J, 2013, 27(9), 3741-3752.
07. White J.F., et al. Nature, 2012, 490(7421), 508-513.
08. Fanelli R, et al. J. Med. Chem. 2015, 58(19), 7785-7795.

SYNTHESIS AND IN VITRO EVALUATION OF POTENT NEUROMEDIN U RECEPTOR AGONISTS

An De Prins[1,2], Charlotte Martin[1], Dirk Tourwé[1], Vicky Caveliers[3], Ann Van Eeckhaut[2], Csaba Tömböly[4], Mette M. Rosenkilde[5], Birgitte Holst[5], Ilse Smolders[2], Steven Ballet[1].
[1] Research Group of Organic Chemistry, Vrije Universiteit Brussel, Brussels, Belgium
[2] Department of Pharmaceutical Chemistry and Drug Analysis, Vrije Universiteit Brussel, Brussels, Belgium
[3] Department of In vivo Cellular and Molecular Imaging, Vrije Universiteit Brussel, Brussels, Belgium
[4] Biological Research Centre, Laboratory of Chemical Biology, Szeged, Hungary
[5] Department of Neuroscience and Pharmacology, University of Copenhagen, Copenhagen, Denmark

Introduction

Neuromedin U (NMU) is a structurally highly conserved neuropeptide which is ubiquitously distributed through the body. This neuropeptide exerts its biological effects via two G protein-coupled receptors, namely NMU receptor 1 (NMUR1) and NMU receptor 2 (NMUR2). NMUR1 is predominantly found in the periphery whereas NMUR2 is most abundant in the central nervous system. NMU is involved in different physiological processes such as contraction of smooth muscles, feeding and energy homeostasis, regulation of the blood pressure, nociception and regulation of the stress response [1]. A remarkable homology is described between the different forms of NMU in different species, in particular at the C-terminus of the peptide. The amidated C-terminal heptapeptide is conserved in all mammalian species, which indicates that this segment of the peptide is crucial for receptor activation [1].

The anorexigenic characteristics of NMU have led to a great interest in this peptide as potential therapeutic for the treatment of diseases such as diabetes and obesity. Therefore, several NMU derived ligands, including long acting peptides, have been developed over the past decade with the objective to find novel candidates for these increasing health concerns. Our aim is to design short NMU derived peptides with improved characteristics such as increased enzymatic stability, higher potency or improved selectivity.

Results and Discussion

In our study, NMU-8 (Figure 1) is taken as the lead molecule for the synthesis of novel NMU derived peptides. We selected NMU-8 since it is a natural occurring form of NMU in certain species and it represents the conserved C-terminal part of the more extended NMU sequence.

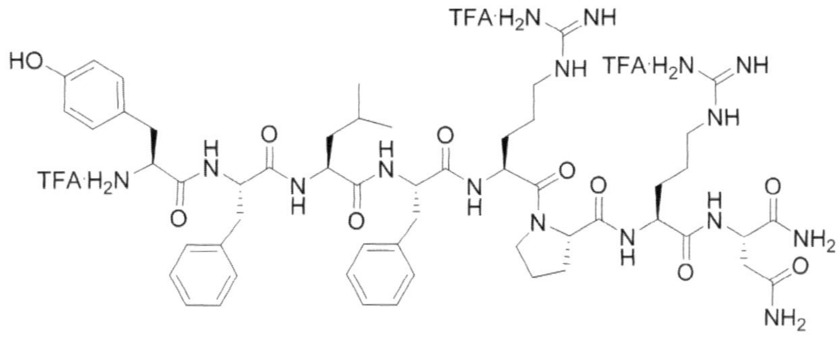

Figure 1: Structure of NMU-8.

Within a first generation of NMU analogs we tried mainly to confirm the existing literature by introducing D-amino acids and different N-capping groups in the NMU-8 sequence. In vitro characterization was performed on HEK293 cells, transiently transfected with NMUR1 or NMUR2, by means of an inositol phosphate accumulation assay, for determination of agonistic properties, and a competitive binding experiment. These results demonstrated that we could confirm the literature. Moreover, previously described weak non-competitive antagonists, such as H-Tyr-Phe-Leu-Phe-Arg-D-Pro-Arg-Asn-NH_2[2], did not show any agonistic activity on the NMURs, as expected, but interestingly they possessed in fact a markedly decreased affinity for the receptors. Another key finding of this first generation of NMU analogs is that acetylation of the N-terminus in general leads to an increase of the relative activity, compared to the non-acetylated analogs.
The second generation of NMU derived peptides consisted of sequences with more advanced modifications such as the use of unnatural and constrained amino acids and the introduction of N-alkylated glycines (peptoid approach) in NMU-8. The results of the in vitro characterization revealed that all peptoid analogs are less potent on the NMURs, as compared to NMU-8. Substitution of Tyr^1 by 7-OH-L-Tic gave rise to a peptide with an increased activity and affinity (a selection of highly active analogs is presented in Figure 2). Additionally, some peptides with an improved activity on both NMURs were discovered.

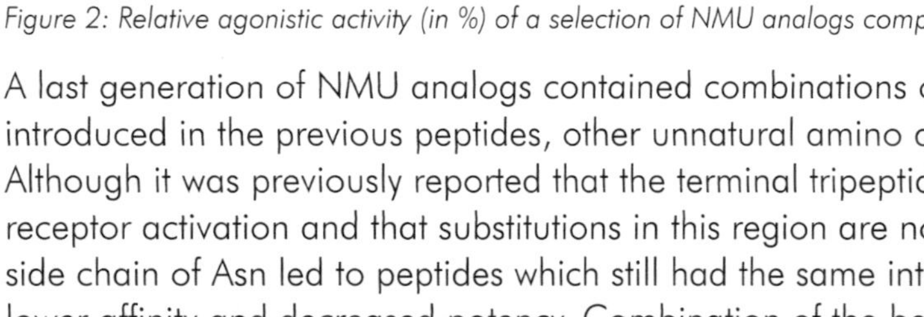

ADP40B: Ac-Dmt-Phe-Leu-Phe-Arg-Pro-Arg-Asn-NH_2
ADP42: Ac-Tyr-Phe-Leu-2'Nal-Arg-Pro-Arg-Asn-NH_2
ADP43B: Ac-7-OH-Tic-Phe-Leu-Phe-Arg-Pro-Arg-Asn-NH_2

Figure 2: Relative agonistic activity (in %) of a selection of NMU analogs compared to NMU-8 at a concentration of 1000 nM.

A last generation of NMU analogs contained combinations of the most promising modifications introduced in the previous peptides, other unnatural amino acids and C-terminal modifications. Although it was previously reported that the terminal tripeptide -Pro-Arg-Asn-NH_2 is crucial for receptor activation and that substitutions in this region are not tolerated [3], modifications in the side chain of Asn led to peptides which still had the same intrinsic activity as NMU-8, but a slightly lower affinity and decreased potency. Combination of the best modifications of two potent agonists, namely ADP42 and ADP43B (EC_{50} = 4,2 10^{-9} M and 5,1 10^{-9} M respectively for NMUR1), gave

rise to a NMU analog with an even higher potency (EC_{50} = 2,9 10^{-9} M on NMUR1) and increased intrinsic activity on the NMURs.
In conclusion, we have designed a series of novel NMU derived peptides which are potent agonists for both NMUR1 and NMUR2. These peptides can be possible new therapeutic targets for the treatment of obesity and diabetes.

Acknowledgements

An De Prins is funded by a Ph.D. fellowship of the Agency for Innovation by Science and Technology in Flanders. We would like to thank the Research Foundation - Flanders (FWO Vlaanderen) for providing two travel grants.

References

01. Brighton et al, Pharmacol. Rev., 2004; 56:231-48
02. Hashimoto et al, Chem. Pharm. Bull., 1991; 39(9): 2319 - 22
03. Takayama et al, J. Med. Chem., 2014; 57(15): 6583-93

PP IX - 229

OP
PP I
PP II
PP VI
PP VII
PP VIII
PP IX
PP X
PP XI
PP XII
PP XIII
PP XIV

THE MECHANISM OF ANXIOLYTIC-LIKE EFFECT OF GD-23, THE DIPEPTIDE TS

Deeva O.A., Yarkova M. A., Gudasheva T.A., Seredenin S.B.

The 18 kDa translocator protein (TSPO) is a five transmembrane protein that mainly located in the outer mitochondrial membrane. The main function of TSPO is cholesterol transport from the outer to the inner mitochondrial membrane that is the rate-limiting step of neurosteroids biosynthesis. Neurosteroids interact with non benzodiazepine site of $GABA_a$ receptor causing an anxiolytic effect without the side effects of benzodiazepines [Nothdurfter C. e., J. Neuroendocrinol., 2012].

A new original dipeptide TSPO ligand was designed using the of drug-based peptide design method [Gudasheva T.A. et al., J. Med. Chem., 1998]. We used Alpidem as non-peptide prototype. This structure contains all the necessary TSPO ligands pharmacophore elements including two aromatic nuclei and branched aliphatic chain. These groups can be simulated by N-protected tryptophan's and isoleucine's side radicals, respectively. Due to these arguments the dipeptide N-carbobenzoxy-L-tryptophanyl-L-isoleucine amide (GD-23) was designed and synthesized. Docking was performed using Autodock Vina and Glide-3 software. The results showed that the GD-23 possesses a high TSPO affinity ($Ki = 7*10^{-7}$ M). GD-23 revealed good superposition with Alpidem in active pocket of the receptor. Glide data predicts that the compound GD-23 is perfectly adapted in the mainly hydrophobic binding pocket of TSPO. The anxiolytic activity was investigated in two behavioral tests: the illuminated open-field test in Balb/C mice [Seredenin S.B. et al., Biull .Eksp. Biol. Med. (Russia), 1979] and elevated plus-maze test in CD-1 mice [Pellow S. et al., J. of Neurosci. methods, 1985]. The activating effect on locomotor activity of animals was taken as a measurement of the anxiolytic activity of the compound. The total locomotor activity was calculated as sum of number of squares crossed in the periphery, in the central regions , number of entries into the center and number of rearing. GD-23 in the dosage range 0.05-1.0 mg/kg i.p. significantly ($p<0.005$) increased total locomotor activity of mice compared with control groups. GD-23 increased the number of open arms entries and time spent in open arms in elevated plus maze test in dose range 0.1-0.5 mg/kg i.p.

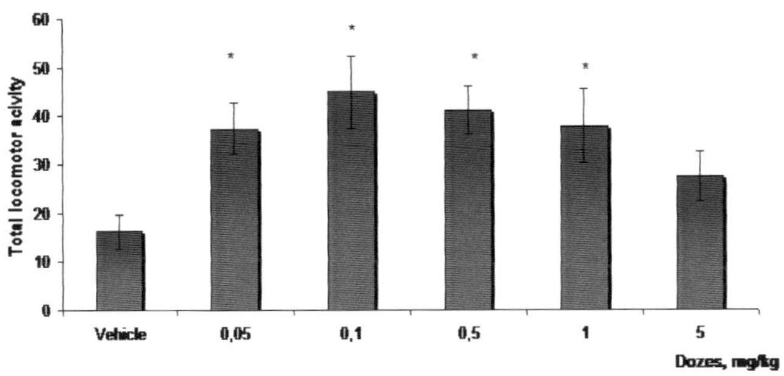

Fig.1. Influence of GD-23 on total locomotor activity of mice in open field test

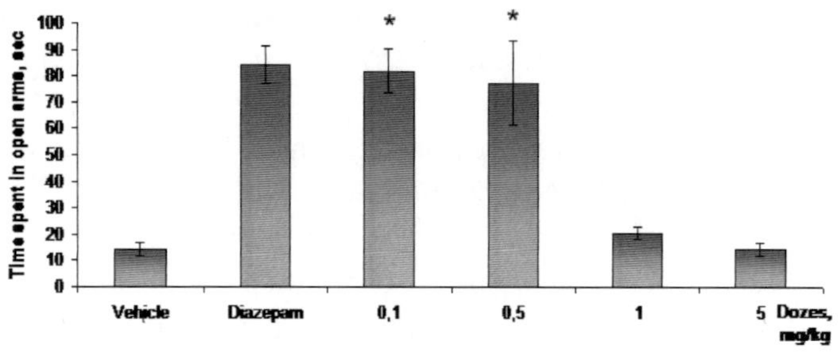

Fig.2. Influence of GD-23 on time spent in open arms in elevated plus maze test

Involvement of TSPO receptor in mechanism of anxiolytic activity of GD-23 was proved using two experiments. We used an elevated plus-maze behavioral test in CD1 mice. We used GD-23 in dose of 0.5 mg/kg that was chosen from the previous experiments. Preliminary administration of PK11195, selective TSPO antagonist, completely abolished the anxiolytic effect of GD-23. The data reveal that the anxiolytic effect of diazepam, that is agonist of $GABA_A$ receptor benzodiazepine site, is not removed by PK11195.

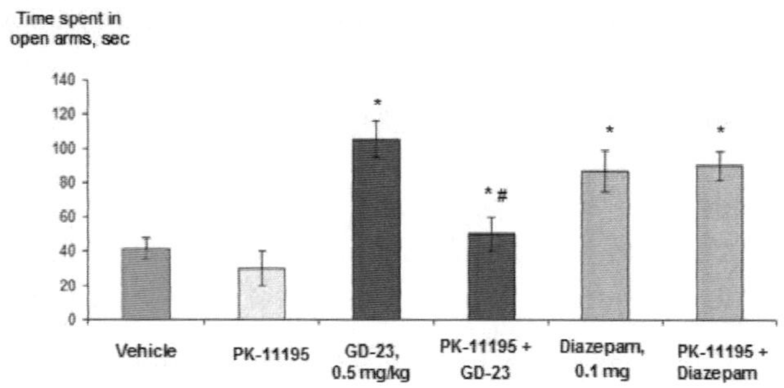

Fig.3. Influence of PK-11195 on anxiolytic effect of GD-23

At the second experiment we used two inhibitors of enzymes which are involved in the biosynthesis of neurosteroids: trilostane and finasteride in the doses of 10 mg/kg. These inhibitors did not exhibit any effects in comparison with the control by themselves, but they completely blocked the anxiolytic effect of GD-23. In the same time, trilostane and finasteride did not influence on the anxiolytic activity of diazepam.

OP
PP I
PP II
PP VI
PP VII
PP VIII
PP IX
PP X
PP XI
PP XII
PP XIII
PP XIV

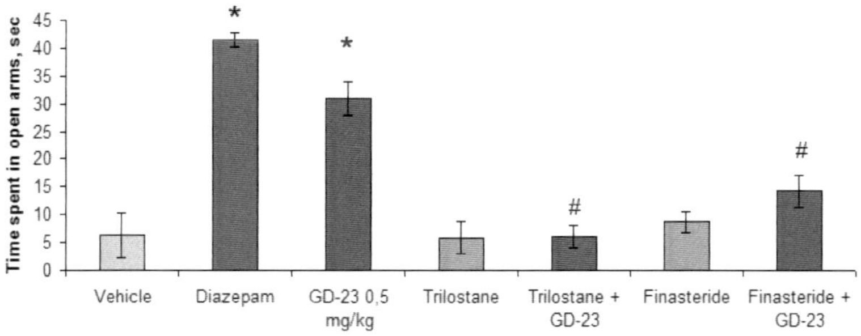

Fig.4. Influence of steroid biosynthesis inhibitors on anxiolytic effect of GD-23

Evaluation of acute toxicity was investigated in male outbred mice. LD_{50} is over 1000 mg/kg ip.

The obtained results demonstrate that the anxiolytic effect is mediated by interaction of the compound GD-23 with TSPO receptor. Hence GD-23 can provide a basis for a new peptide class of fast anxiolytics without side effects of benzodiazepines. This work was partially supported by the grant of Russian Foundation for Basic Research № 17-04-00861-A.

Kata Nóra Enyedi[1], Szilárd Tóth[2], Gergely Szakács[2], Gábor Mező[1]
[1] MTA-ELTE Research Group of Peptide Chemistry, Hungarian Academy of Sciences, Eötvös L. University, 1117 Budapest, Hungary
[2] Institute of Enzimology, Research Center for Natural Sciences, Hungarian Academy of Sciences, 1117 Budapest, Hungary

Introduction

NGR (Asn-Gly-Arg) peptides received particular interest when phage display libraries were used to identify non-RGD integrin binding motifs. Among the non-RGD peptides the NGR motif was the most frequent that showed integrin binding properties.[1] However, it has been found that peptides containing the NGR motif, are not integrin ligands, but specifically recognize Aminopeptidase N (CD13) receptor isoforms that are overexpressed in tumor vasculature and on some tumor cells.[2] Furthermore, these peptides can spontaneously decompose through succinimid ring formation to aspartyl and isoaspartyl (DGR and isoDGR) derivatives, which is strongly influenced by their structure. Studies have shown that the resulting isoDGR derivatives might be a potent ligand for RGD-integrins that are essential for tumor cell invasion and metastasis.[3] Hence NGR-peptide-drug conjugates, through their NGR-to-isoDGR rearrangement and CD13/RGD-integrin switching, can serve as ideal chemotherapeutic candidates, due to their potential in dual drug targeting.

Results and discussion

Based on the literature and on our previous work, in this study four small cyclic NGR-peptides with different stability characteristics were selected for drug targeting: c[KNGRE]-NH_2 and Ac-c[CN-GRC]-NH_2 as the most stable, the thioether bond containing c[CH_2-CO-KNGRC]-NH_2 and the least stable c[CH_2-CO-NGRC]-NH_2.[4] Using these small cyclic NGR derivatives novel cyclic NGR peptide-drug conjugates with oxime-linked daunomycin (Dau) were developed. Cathepsin-B labile spacer (GFLG) and a MMP-2 cleavable spacer (GPLGVRG) were incorporated between the homing peptides and drug molecule to ensure the effective drug release. The following conjugates (compound 1-12) were prepared.

Linear peptides were synthesized on Rink Amide MBHA resin by Fmoc chemistry. To avoid side reactions the aminooxyacetyl group was blocked with isopropyliden protection. The cyclization through thioether bond formation was carried out in TRIS buffer (0.1 M, pH: 8.1), followed by deprotection of the aminooxyacetyl group with 0.1 M methoxyamine in NH_4OAc buffer (0.2 M, pH: 5.0). The cyclization of the KNGRE motif containing peptides was done in solution, through in situ active ester formation. Finally, the daunomycin was conjugated via oxime bond formation in NH_4OAc-buffer in all cases.
The stability of the conjugates was determined in serum containing DMEM cell medium at 37°C. The in vitro cytotoxicity of the conjugates was tested both on HT1080 (human fibrosarcoma) cells that selectively express CD13 receptors and on CD13 negative HT-29 (human colorectal adenocarcinoma) cells as control using MTT assay. Both cell lines express RGD integrins. The results suggest-

ed that the RGD integrin receptor selectivity correlates with the quantity of the resulted isoAsp form. Furthermore, it seems, that the type of spacer and the conjugation site of the drug in the peptide sequence have influence on the in vitro cytostatic effect. This is supported by the fact, that conjugates in which the drug-spacer was attached to their C terminal had different biological activity and specificity for CD13, compared to the conjugates which were elongated through the side chain of Lys (5-7, 6-8, 9-11, 10-12). From the results, it could be assumed, that compound 1 and 6 act through RGD integrin receptors. In turn Compound 7 and 11 shows CD13 receptor selectivity. The most active conjugates were Compound 9 and 12.

Stability		Compounds	6 h IC$_{50}$ (μM)		72 h IC$_{50}$ (μM)	
6 h	72 h		HT1080 (CD13+)	HT-29 (CD13-)	HT1080 (CD13+)	HT-29 (CD13-)
(Asn/isoAsp/Asp)						
-	-	**Daunomycin**	0.9±0.4	1.4±0.4	0.3±0.3	0.2±0.1
0/83/17	0/83/17	**1**	>>50	>>50	23.8±10.7	20.6±5.6
100/0/0	74/0/26	**3**	32.3 ±11.2	46.2±3.5	12.4±5.6	27.4±1.1
0/57/43	0/57/43	**5**	8.0 ±1.9	10.3±2.2	3.3±0.6	5.7±0.9
30/48/22	0/62/38	**7**	>>50	>>50	14.5±2.0	>>50
100/0/0	100/0/0	**9**	5.4±0.05	8.7±1.7	1.2±0.66	2.8±0.6
100/0/0	54/0/46	**11**	43.7±4.1	>>50	15.8±4.4	>>50
0/0/100	0/0/100	**2**	>>50	>>50	>>50	>>50
0/0/100	0/0/100	**4**	>>50	>>50	30.5 ± 3.5	>>50
0/100/0	0/100/0	**6**	13.5±0.3	11.2±4.0	4.8±0.3	5.8±0.8
46/0/54	0/0/100	**8**	>>50	30.4± 1.5	>>50	>>50
100/0/0	100/0/0	**10**	>>50	>>50	>>50	>>50
100/0/0	52/0/48	**12**	1.8±0.1	0.8±0.1	0.6±0.1	0.4±0.1

Acknowledgments

This project has received funding from the European Union's Horizon 2020 research and innovation program under the Marie Sklodowska-Curie grant agreement No 642004, and from the Hungarian National Science Fund (OTKA 104045).

References

01. Koivunen, E., et al. J. Cell Biol. 124, 373 (1994)
02. Pasqualini, M., et al. Cancer Res. 60, 722 (2000)
03. Hildenbrand, T., et al. Int. J. Oncol.. 34, 15 (2009)
04. Enyedi et al.: J. Med. Chem. 58, 1086 (2015)

METABOLIC PATHWAY MONITORING OF TRIAZOLOPEPTIDIC ANALOGUES OF THE ANTIANGIOGENIC PEPTIDE A7R IN HUMAN PLASMA WITH HPLC-MS

Bartlomiej Fedorczyk[1], Beata Wilenska[1], Anna Puszko[1], Dagmara Tymecka[1], Anna Starzec[2], Gerard Y. Perret[2], Aleksandra Misicka[1]
[1] Faculty of Chemistry, Biological and Chemical Research Center, University of Warsaw, Warsaw, Poland
[2] Université Paris 13, Sorbonne Paris Cité, INSERM U1125, France

OP
PP I
PP II
PP VI
PP VII
PP VIII
PP IX
PP X
PP XI
PP XII
PP XIII
PP XIV

Introduction

One of the major challenges for peptide-based therapeutic development is the susceptibility of peptides to proteases. Peptides are vulnerable to the proteolytic enzymes that are present in the blood. Nonspecific proteolysis is considered to be the major elimination pathway for peptide and protein-based drugs from bloodstream. Therefore determination of peptide stability in blood plasma constitutes a powerful and important screening assay for the elimination of unstable peptides in the pipeline of drug development [1]. Hereby we present this approach in a case of our peptidomimetics derived from antiangiogenic peptide A7R, which was described in literature earlier [2]. Our intension was to investigate stability of our triazolopeptidic analogues (2), (3) in human plasma and compare this parameter therewith obtained for our lead compound (1) [3].

Materials and Methods

The stability of peptidomimetics was tested in plasma samples obtained from healthy donors. Endomorphin-2 was used as a control sample, because it is known to be a very unstable peptide in human plasma. Each experiment was repeated three times. Stock solutions of each compound (1-3) were prepared by dissolving its TFA salt in water to achieve concentration 1mg/ml. In an Eppendorf tube, samples of human plasma were temperature-equilibrated at 37°C for a few minutes before adding the equal volume of proper peptide stock solution. The time of degradation (t_d) was recorded at different intervals. After incubation time a portion of 96 % ethanol was added to the samples in order to precipitate the plasma proteins. Cloudy mixtures were shaken by vortex, left at 4°C for a few minutes and subsequently centrifuged. The reaction supernatant was taken, lyophilized, dissolved in phase A and analyzed.

All presented analysis were proceeded on HPLC (Shimadzu) coupled with mass spectrometer (LC-MS2010EV, Shimadzu, ESI as ion source, single quadrupole as mass detector). For chromatografic separation a column Jupiter Proteo was used (2.0 mm x 250 mm, beads 4 microns, C-12), which was termostated at 30 °C. As detector a PDA diode array detector was used and chromatografic data was analyzed at 210 nm. Method (1): phase A – water + 0.05% TFA, phase B – methanol + 0.05% TFA, linear gradient from 0% to 21% of phase B in 44 min, flow 0.3 ml/min. Method (2), (3): phase A – water + 0.05% TFA, phase B – acetonitrile + 0.05% TFA, linear gradient from 0% to 15% of phase B in 30 min, flow 0.2 ml/min

Results

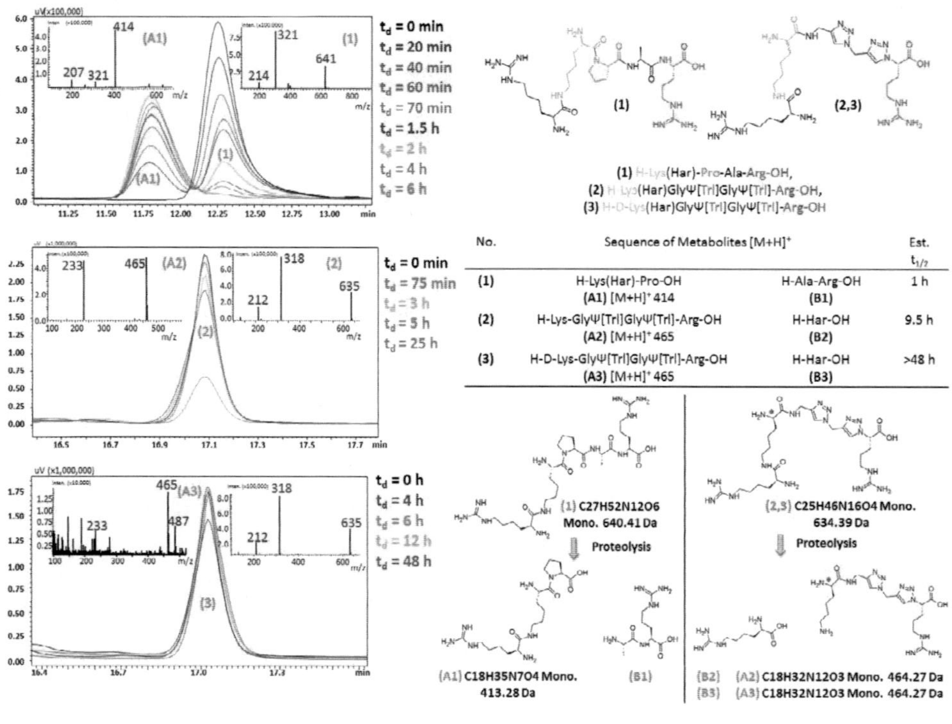

Figure 1. Summary of HPLC-MS experiments. Left side - overlapped chromatograms for compounds (1), (2) and (3) in different time intervals with representative MS spectra recorded for starting compounds and metabolites. Right - enzymatic degradation pathway proposed in the frame of current studies with estimated half-life ($t_{1/2}$) values.

Conclusions

As we have shown on figure 1., degradation of our lead compound (1) proceeds with first enzymatic cleavage between Pro and Ala and estimated half-life $t_{1/2}$ varies c.a. 1 h. Chemical modification in the middle site of (1) has led us to obtain a family of triazolopeptides (2) and (3), which exhibit significantly longer $t_{1/2}$. Moreover, we have not observed degradation of triazole moieties during the experiment, thus triazoles might be used as very stable peptide bond non-classical isosteres.

References

01. Jambunathan K., Galande A.K., Protein Pept. Lett 2014, 21, 32–38.
02. Starzec A. et al., Life Sci 2006;79: 2370–2381.
03. Misicka A. et al., Inter. Pat. WO2015026251 A1

BILE ACID AS AN EFFECTIVE ABSORPTION ENHANCER FOR ORAL DELIVERY OF HYBRID PEPTIDE

Arong Gaowa, Tomohisa Horibe, Masayuki Kohno, and Koji Kawakami
Department of Pharmacoepidemiology, Graduate School of Medicine and Public Health, Kyoto University, Yoshidakonoe-cho, Sakyo-ku, Kyoto, 606-8501, Japan

Abstract
The aim of this study is to enhance oral absorption of hybrid peptide using hydrophilic bile acid as an absorption enhancer. The oral formulation of peptide was formed through the electrostatic interaction between the cationic peptide and anionic bile acid. In Caco-2 cell monolayers, absorption permeability of peptide from the peptide formulation was increased 5-fold compared with that of peptide alone. Furthermore, oral administration of peptide formulation to xenograft nude mouse showed significantly improved anti-tumor activity compared to free peptide. These results suggested that the bile acid is an effective absorption enhancer for improving the oral bioavailability and bioactivity of hybrid peptide.
Keywords: Hybrid peptide; Cell permeability; Absorption enhancer; Anti-tumor activity

Introduction
We have previously reported that the EGFR2R-lytic hybrid peptide has cytotoxic and anti-tumor activities against EGFR over-expressing cancers both in vitro and in vivo [1, 2]. Moreover, our further studies have shown that the intravenous injection of this hybrid peptide is also a potential treatment option for patients with colorectal cancer metastases in the liver [3]. Oral drug delivery is the most preferred route to deliver therapeutic agents due to ease of administration and patient acceptance. However, oral delivery of peptide drugs faces many hurdles such as poor absorption, poor permeability and rapid degradation. To overcome these problems several approaches have been investigated. Among them, the use of absorption enhancers is a simplest approach to enhance the oral absorption of peptides across the epithelial intestinal membrane [4]. Because of biocompatibility, bile acids have been widely used as absorption enhancers for drug delivery [5]. Therefore, in this study we used the bile acid to enhance the oral absorption of EGFR2R-lytic hybrid peptide.

Results and Discussion
For the development of the transwell assay system for permeability experiments, Caco-2 cells monolayers were used. The integrity of the Caco-2 cell monolayer was measured by the transepithelial electrical resistance (TEER). Peptide and peptide oral formulations were labelled with fluorescein isothiocyanate (FITC) for quantitative detection by spectrofluorimetry in permeability studies. In Caco-2 cells, the values of TEER were decreased and the absorption permeability of peptide was markedly increased in cells incubated with peptide formulation compared to that of free peptide (Fig. 1A and 1B). The in vitro dissolution test was carried out under various pH conditions in the range of 1.5 to 9.0. Stability investigations showed that the peptide formulations were reasonably stable at gastric acid pH (< 3.5), but were relatively fast release at intestinal fluid pH (data not shown). In vivo antitumor activities of peptide and peptide formulations were performed using the xenograft nude mouse model of human gastric cancer cell line. In animal experiments, the anti-

tumor activity of peptide formulation was higher to that of peptide alone (Fig. 1C). These results suggested that the bile acid is an effective absorption enhancer for improving the oral bioavailability and bioactivity of hybrid peptide.

This work was supported by Grants-in-Aid for Young Scientists (A) (grant No. 23680089) and Young Scientist (B) (grant No. 16K18937) from the Japan Society for the Promotion of Science.

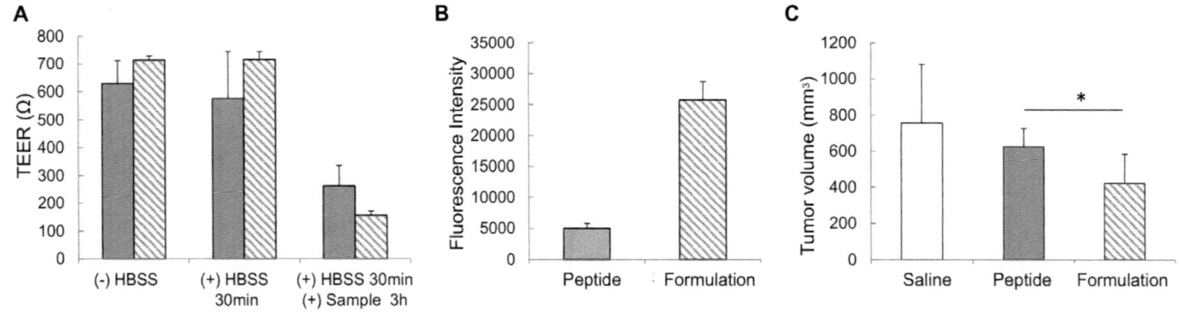

*Fig. 1. (A) The transepithelial electrical resistance (TEER) values of Caco-2 cell monolayer pre-incubation with Hank's balanced salt solution (HBSS) for 30 min, then followed by further incubation with peptide or formulation for 3 hours at 37°C. (B) The fluorescence intensity of FITC-labeled peptide and FITC-labeled peptide formulation across Caco-2 cell monolayers (Apical→Basolateral). The columns represent the means ± SD., n=3. (C) The tumor volumes of MKN45 human gastric-carcinoma xenografts treated orally with saline (control), peptide (5mg/kg, twice/day), and equivalent peptide-loading formulation (5mg/kg, twice/day). The columns represent the means ± SD., n=6 for each group.*P<0.05.*

References

01. Kohno, M et al. (2011) Eur. J. Cancer, 47, 773-783.
02. Tada, N et al. (2011) Biochem. Biophys. Res. Commun., 407, 383-388.
03. Gaowa, A et al. (2016) Clin. Exp. Metastasis, 33, 87-95.
04. Kramer, W et al. (1994) J. Biol. Chem., 269, 10621-10627.
05. Moghimipour, E et al. (2015) Molecules, 20, 14451-14473.

DESIGNING AND CHEMICAL SYNTHESES OF SELECTIVE MATRIPTASE-2 INHIBITORS BASED ON TRYPSIN INHIBITOR SFTI-1 ISOLATED FROM SUNFLOWER SEEDS

Agata Gitlin-Domagalska[1], Dawid Debowski[1], Natalia Ptaszynska[1], Anna Łegowska[1], Marit Stirnberg[2], Michael Gütschow[2], Krzysztof Rolka[1]
[1] Faculty of Chemistry, Department of Molecular Biochemistry, University of Gdansk, Poland
[2] Faculty of Mathematics and Natural Sciences, Pharmaceutical Chemistry, University of Bonn, Germany

Introduction

Matriptase-2, member of type II transmembrane serine proteases (TTSPs), was first identified in 2002 [1]. TTSPs are anchored in cell membrane thus localized at the cell surface, what makes them perfectly positioned to interact with other proteins and mediate signal transduction between the cell and its extracellular environment [2]. What focused our attention, is a link between matriptase-2 and iron-refractory iron-deficiency anemia (IRIDA) [3]. Matriptase-2 is one of many precise, highly specified proteins responsible for keeping iron concentrations in a narrow physiological range. In short: matriptase-2 degrades hemojuveline in cell membranes and subsequently inhibits hepcidin expression (negative regulator of iron absorption) leading to high iron concentrations [3].

Results and discussion

The aim of our work was to obtain selective matriptase-2 inhibitors which might be helpful in determining the exact role of this enzyme in iron homeostasis. Hepcidin expression and its regulation is the only biological mechanism to dispose excess of iron. Inhibitors of this protease may become adjuvant way to control iron concentration. Close structural similarity between matriptase-1 and matriptase-2 makes it particular challenge to develop potent and selective peptide inhibitors of one protease and we are first to take it up. Recently we published [4] a series of 17 SFTI-1 analogues that were designed and examined for their inhibitory activity towards matriptase-1 and matriptase-2. Based on our results and previous studies [5-7] we synthesized next 23 SFTI-1 analogues and examined all 40 peptides against selected serine proteases displaying trypsin-like activity. The most potent inhibitors of matriptase-2 were analogues: 6[4], 7[4], 11[4], 12 [4], 21-24, 28-30 and 34-37 and both wild and monocyclic SFTI-1[4], that were chosen for further, detailed research. Inhibitory constants (K_i) values against matriptase-1, matriptase-2, trypsin, plazmin and thrombin were determined for aforementioned peptides. The most interesting results are shown in Table 1. Our results confirm that native SFTI-1 and monocyclic SFTI-1 are able to inhibit matriptase-2 and SFTI-1 is good lead structure for designing inhibitors of this protease. In effect we obtained 14 potent inhibitors of matriptase-2 with K_i values lower than 1 μm. Two of them revealed the highest inhibitory potency towards matriptase-2 described to date: analogue 11[4] with K_i = 19 nm and analogue 29 with K_i = 15 nm. Both demonstrated highest proteolytic resistance, and were stable for at least 3 hours of incubation with enzyme. Importantly, we observed that introduction of D-arginine instead of L-arginine in position 2 resulted in dramatic decrease of inhibitory potency of examined analogues against matriptase-1 and plazmin. Peptides 7 and 12 were 176 and 228 times respectively stronger inhibitors of matriptase-2 than matriptase-1 [4], their inhibitory potency towards plasmin was marginal and they activated thrombin. Additionally, five analogues (28, 29, 34, 35, 36) revealed remarkable inhibitory potency towards matriptase-1 with single-digit nanomolar K_i values, that were 40 to 100-fold weaker inhibitors of matriptase-2, over 1000-fold less potent inhibitors of plazmin and did not influence activity of thrombin

Table 1. Inhibition constants (K_i) of selected analogues against selected proteases. Compounds with additional 'head-to-tail' cyclisation are marked by * and these with 'side chain-to-tail' cyclisation are marked by ^.

Analogue number and its structure	K_i ± SEM (μm)				
	Matriptase-1	Matriptase-2	Plazmin	Throm-bin	Trypsin
Monocyclic SFTI-1	0.061 ± 0.004	1.365 ± 0329	0.028 ± 0.010		0.009 ± 0.001
Native SFTI-1	0.102 ± 0.007	0.218 ± 0.021	0.025 ± 0.005		0.003 ± 0.001
6 [Arg^{5}] SFTI-1	0.091 ± 0.011	0.115 ± 0.016	2.260 ± 0.405		0.015 ± 0.001
7 [d-Arg^{2}, Arg^{5}] SFTI-1	76.320 ± 27.220	0.433 ± 0.062	>100		0.191 ± 0.013
11 [Arg^{5}] SFTI-1 *	0.269 ± 0.270	0.019 ± 0.003	0.365 ± 0.097		0.013 ± 0.001
12 [d-Arg^{2}, Arg^{5}] SFTI-1 *	63.365 ± 10.483	0.278 ± 0.019	>100		0.045 ± 0.003
21 [Lys^{1}, Arg^{5}] SFTI-1	0.224 ± 0.024	0.274 ± 0.024	-	-	-
22 [Lys^{1}, Arg^{5}] SFTI-1 *	0.532 ± 0.053	0.127 ± 0.009	0.488 ± 0.099		0.011 ± 0.001
23 [Lys^{1}, Arg^{5}] SFTI-1 ^	0.179 ± 0.018	0.370 ± 0.027	-	-	-
24 [Lys^{1}, d-Arg^{2}, Arg^{5}] SFTI-1 *	4.109 ± 0.515	2.641 ± 0.339	>100		0.017 ± 0.001
28 [Arg^{5}, Arg^{10},His^{12}] SFTI-1	0.006 ± 0.001	0.288 ± 0.017	7.760 ± 0.884	-	0.010 ± 0.005
29 [Arg^{5}, Arg^{10},His^{12}] SFTI-1 *	0.005 ± 0.001	0.015 ± 0.001	-	-	-
30 [d-Arg^{2},Arg^{5}, Arg^{10},His^{12}] SFTI-1 *	0.241 ± 0.021	0.848 ± 0.043	-	-	-
34 [Lys^{1},Arg^{5}, Arg^{10},His^{12}] SFTI-1	0.004 ± 0.001	0.318 ± 0.025	5.789 ± 0.772	«	0.005 ± 0.001
35 [Lys^{1},Arg^{5}, Arg^{10},His^{12}] SFTI-1 *	0.008 ± 0.001	0.102 ± 0.008	3.047 ± 0.567	«	0.009 ± 0.003
36 [Lys^{1},Arg^{5}, Arg^{10},His^{12}] SFTI-1 ^	0.003 ± 0.000	0.257 ± 0.012	3.308 ± 0.359	«	0.005 ± 0.001
37 [Lys^{1},d-Arg^{2},Arg^{5}, Arg^{10},His^{12}] SFTI-1 *	0.028 ± 0.002	0.607 ± 0.041	>100	«	0.021 ± 0.011

Acknowledgments

This work was supported by the National Science Centre (NCN) in Poland under grant UMO-2014/13/N/ST5/01299.

References

01. Velasco G., Cal S., Quesada V., Sanchez L.M., Lopez-Otin C.,J. Biol. Chem., 2002, 277, 37637-37646.
02. Antalis T. M., Bugge T., Wu Q., Prog. Mol. Biol. Transl. Sci., 2011, 99, 1-50.
03. Folgueras A.R., de Lara F.M., Pendas A.M., Garabaya C., Rodriguez F., Astudillo A., Bernal T., Cabanillas R., Lopez-Otin C., Velasco G, Blood, 2008, 112, 2539-2545.
04. A. Gitlin , D. Debowski, N. Karna, A. Łegowska, M. Stirnberg, M. Gütschow, K. Rolka, ChemBioChem, 2015, 16, 1601-1607.
05. Fittler H., Avrutina O., Empting M., Kolmar H., J. Pept. Sci., 2014, 20, 415-420.
06. Wysocka M., Gruba N., Miecznikowska A., Popow-Stellmaszyk J., Gütschow M., Strinberg M., Lesner A., Rolka K., Biochimie., 2014, 97, 121-127.
07. Sisay M.T., Steinmetzer T., Stirnberg M., Maurer E., Hammami M., Bajorath J., Gütschow M., J. Med. Chem., 2010, 53, 5523-5535.

Karolina Grabowska[1], Anna K. Puszko[1], Piotr F. J Lipinski[2], Anna K. Laskowska[2], Beata Wilenska[1], Ewa Witkowska[1], Gerard Y. Perret[3], Aleksandra Misicka[1]
[1] Faculty of Chemistry, University of Warsaw, 02-093 Warsaw, Poland
[2] Department of Neuropeptides, Mossakowski Medical Research Centre Polish Academy of Sciences, 02-106 Warsaw, Poland
[3] Université Paris 13, Sorbonne Paris Cité, INSERM U1125, 74 rue Marcel Cachin, 93017 Bobigny, France

Introduction

During pathological form of angiogenesis tumour growth relies on the development of new vasculature that delivers oxygen and nutrients to proliferating cells and allows simultaneous removal of metabolic waste. One of the most important signalling molecule in angiogenesis is Vascular Endothelial Growth Factor-165 ($VEGF_{165}$). Recently, many reports have suggested that at least part of VEGF signalling in tumour growth may be mediated by interaction with NRP-1.[1] Therefore compounds which are able to selectively block the interaction $VEGF_{165}$/NRP-1 could become antiangiogenic and antitumor drugs. Heptapeptide ATWLPPR (A7R) which are peptidic inhibitor of $VEGF_{165}$/NRP-1 has been identified by screening a mutated library.[2] This peptide shows in vitro and also in vivo antiangiogenic activity.[3] The shortest part of this peptide which is crucial in antiangiogenic activity is LPPR. Based on this tetrapeptide we synthesized a cyclic tetrapeptide with exocyclic arginine, which is important for interaction with NRP-1.[4] To be able to make a ring we changed Leu in position 1 into Lys residue and Pro in the position 3 into Glu residue. Here we present structure-activity relationship study (SAR) on peptide 1. We synthesized analogues 2-6 with different ring size (14-,15- and 30- membered ring), and replacement of some L- into D-amino acids.

Cyclic peptides were synthesized manually on Merrifield resin (including cyclization). The final cyclic peptides were cleaved from the resin using HF, purified by RP-HPLC and analyzed by LC-MS. Pure cyclic peptides (purity > 95%) were examined by enzyme-linked immunosorbent assay (ELISA). Structures and inhibitory effect of the obtained cyclic compounds are presented in table 1. Peptides 2 and 6 show high inhibitory effect, peptides 3 and 5 containing one D-amino acid residue show low inhibitory effect.

Table 1. Structure and inhibitory effect of the obtained cyclic peptides

Cmpd	Structure	% inhibition of $VEGF_{165}$ /NRP-1 binding			
		10 μM	3 μM	1 μM	0.3 μM
1	H-c[Lys-Pro-Glu]-Arg-OH	94.7±0.7	86.7±1.3	76.5± 0.5	61.7±0.5
2	H-c[Glu-Pro-Lys]-Arg-OH	89.0±1.1	85.4±1.4	79.0±2.0	68.3±2.2
3	H-c[Lys-Pro-Asp]-Arg-OH	14.7±5.6	12.5±1.5	11.8±1.2	ND
4	H-c[Orn-Pro-Glu]-Arg-OH	46.6±7.9	35.4±5.5	27.8±7.9	14.6±5.1
5	H-c[D-Lys-Pro-Glu]-Arg-OH	4.0±0.1	2.3±3.0	1.1±0.1	ND
6	H-c[Lys-D-Pro-Glu(Arg-OH)-Lys-D-Pro-Glu]-Arg-OH	86.8±1.3	72.2±1.8	46.6±3.3	30.1±3.0

We measured also stability of our parent peptide 1 in human plasma. The first metabolite product (cyclic peptide formed after cleavage of egzocyclic Arg) was observed after 1 hour of incubation (Fig. 1, ii). Our preliminary stability study of compound 1 performed in human plasma showed that the half time ($t_{1/2}$) of this peptide is about 6 h.

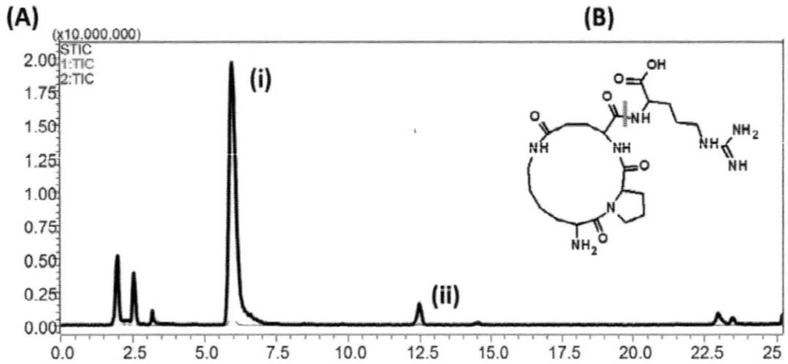

Figure 1. (A): MS chromatogram of 1 after one hour incubation with human plasma (i) starting compound, (ii) metabolite ; (B): Structure of 1 indicating first place of degradation (green line).

According to the results obtained by ELISA assay and molecular modeling (not presented here), both the ring size and configuration of amino acids present in a structure are important for high inhibition of $VEGF_{165}$/NRP-1 interaction. We plan to extend our SAR study by preparing more analogues.

Acknowledgments

This work was supported by National Science Centre (NCN) grant N204 350940 and co-financed by the EU from the European Regional Development Fund under the Operational Programme Innovative Economy, 2007-2013, and with the use of CePT infrastructure financed by the same EU program and a grant from the University of Warsaw for young researchers, no. 120000-501/86-DSM-110200.

References

01. Grun, D.; Adhikary, G.; Eckert, R. L. Oncogene. 2016, 1
02. Binetruy-Tournaire, R.; Demangel, C.; Malavaud, B.; Vassy, R.; Rouyre, S.; Kreamer. EMBO J, 2000, 19, 1525.
03. Starzec, A.; Vassy, R.; Martin, A.; Lecouvey, M.; Di Benedetto, M.; Crepin, M.; Perret, G.Y. Life Sciences, 2006, 79, 2370.
04. Grabowska, K.; Puszko, A.K.; Lipinski, P.F.J.; Laskowska, A.K.; Wilenska, B.; Witkowska, E.; Misicka, A. Bioorg Med Chem Lett, 2016, 26, 2843.

Povarnina P.Yu., Volkova A.A., Gudasheva T.A., Seredenin S.B.

PP IX - 248

Stroke is one of the leading causes of death and long-term disability worldwide, and no drugs are available for promoting recovery after a stroke has occurred.

The dimeric dipeptide, bis(N-succinyl-L-glutamyl-L-lysine) hexamethylenediamide (GK-2), was designed based on the most exposed outside fragment of NGF loop 4 β-turn sequence [1]. It has been shown in vitro, using both immortalized and primary cell cultures under conditions of H_2O_2, glutamate or MPTP-induced toxicity, that GK-2 exerts NGF-like neuroprotective activity ($10^{-5} – 10^{-9}$ M) [1].

It was revealed by western blot analysis that GK-2 elevated the level of TrkA receptor phosphorylation and selectively increased the level of AKT phosphorylation [2].

In vivo GK-2 exhibited therapeutic effects in models of Parkinson's disease, Alzheimer's disease, cerebral ischemia and diabetes mellitus in doses 0.01 - 5 mg/kg intraperitoneally and 5-10 mg/kg per os [3]. It was also found that GK-2 has no side effects accompanying NGF treatment namely hyperalgesia and weight loss [2].
I
The aim of the present work was to study the pharmacological effects of GK-2 in a rat model of transient middle cerebral artery occlusion at therapeutic time windows of 4 and 6 h.

It was established that GK-2 reduced the cerebral infarct volume by about 40% when treatment was begin at 4 h after surgery and this effect was saved when the therapeutic time window was increased to 6 hours.

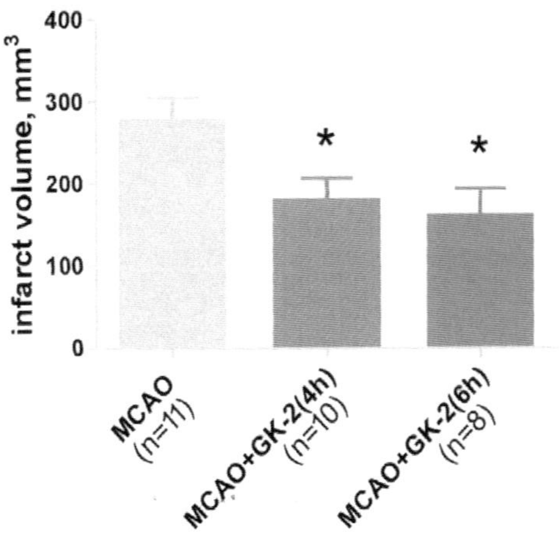

The results obtained suggest a potential role for the dipeptide as a therapeutic agent useful in the treatment of stroke.

This work was supported by the Russian Science Foundation (Project №14-15-00596) and by Russian Federal programme "Pharma2020" (Contract №14.N08.12.0051).

References

01. Gudasheva TA, Antipova TA, Seredenin SB. Novel low-molecular-weight mimetics of the nerve growth factor. Dokl. Biochem. Biophys. 2010; 434: 262–5.
02. Gudasheva TA, Povarnina PY, Antipova TA, Firsova YN, Konstantinopolsky MA, Seredenin SB. Dimeric dipeptide mimetics of the nerve growth factor Loop 4 and Loop 1 activate TRKA with different patterns of intracellular signal transduction. J Biomed Sci. 2015; 22: 106.
03. Gudasheva TA, Povarnina PY, Antipova TA, Seredenin SB. A Novel dimeric dipeptide mimetic of the nerve growth factor exhibits pharmacological effects upon systemic administration and has no side effects accompanying the neurotrophin treatment. Neuroscience & Medicine. 2014; 2:101-108.

Keiko Hojo[1] Mohammad Akhter Hossain[2, 3] K Johan Rosengren[4] Sherie Ma[2] Andrew L Gundlach[2] Yuko Tsuda[1] Koushi Hidaka[1] Ross AD Bathgate[2] John D Wade[2, 3]
[1] Kobe Gakuin University, Faculty of Pharmaceutical Sciences, Kobe, Japan
[2] University of Melbourne, Florey Institute for Neuroscience & Mental Health, Melbourne, Australia
[3] University of Melbourne, School of Chemistry, Melbounre, Australia
[4] University of Queensland, School of Biomedical Sciences, Brisbane, Australia

PP IX - 253

Introduction

Relaxin-3 is a two-chain neuropeptide that plays a key role in stress responses, arousal and affective behaviors through interaction with its G protein-coupled receptor RXFP3 that is highly expressed in the brain [1]. From detailed structure-activity relationship studies together with the known tertiary structure of relaxin-3 and molecular modeling of the RXFP3-relaxin-3 complex, the central B-chain a-helical region of relaxin-3 is essential for the binding to RXFP3 as it contains several key residues [2, 3]. The C-terminal residues are critical for activation of RXFP3. Although all the critical residues are located within the B-chain, the native relaxin-3 B-chain alone displays only weak agonistic activity at RXFP3 probably due to loss of the native helical binding conformation when separated from the stabilizing A-chain. In this study we prepared a series of stapled single B-chain analogues that mimic the binding conformation of relaxin-3 for RXFP3. We instead examined the utility of hydrocarbon stapling, which has gained increasing popularity given the development of improved ring closure metathesis methods [4] and the recognition of its effectiveness in helical induction. Additional stapling chemistries were also assessed for comparison. Such acquisition of an active simplified RXFP3-selective analogue will be an important molecular probe of relaxin-3 function [5].

Results and Discussion

In the present study, we undertook to develop a selective single-chain agonist for RXFP3 based on stabilization of the helical domain of the B chain using olefin stapling methods based on RCM [4]. From structure-activity relationship studies using Ala scanning, the primary binding site, Arg^{B12}, Ile^{B15}, Arg^{B16}, and Phe^{B20}, is located in the same surface of the helical domain in the relaxin-3 B-chain. In addition to these residues, relaxin-3 requires Arg^{B26} and Trp^{B27} located toward C-terminus of B-chain, to activate RXFP3. RXFP3. Examination of the central helical motif in the B-chain via a classical helical wheel representation shows that two faces of helix with the active site (Arg^{B12}, Ile^{B15}, Arg^{B16}, and Phe^{B20}) exposed on one face and the predominantly hydrophobic residues that interact with the A-chain in the native structure exposed on the second face. Based on this analysis we chose to introduce staples at positions 13, 17, and 21 which should be sufficiently distant from the active site. Furthermore, the results from a previous chain truncation study revealed that the deletion of the N-terminal-seven residues of the B-chain did not significantly affect the ligand activity for the RXFP3, indicating that N-terminal residues, from Arg^{B1} to Tyr^{B7}, of native relaxin-3 B-chain are not important for receptor binding and activation. We thus designed a minimized stapled peptide analogue that was truncated at the B-chain N-terminus by up to 10 residues from Arg^{B1} to Leu^{B9}. A series of N-terminal truncated stapled analogues were prepared, the smallest of which (peptide 5 with 13/17 hydrocarbon stapling) is only18 residues length [5].

The series of stapled peptides was tested for their ability to bind to and active the relaxin-3 receptor RXFP3 and the related receptors. In the comprehensive in vitro test, the analogs with 13/17 hydrocarbon stapling showed marked increases its binding affinity to RXFP3 [5]. Interestingly the use of different positions for the staple or the use of different staples showed far less improvement in binding, suggesting that the 13/17 positions and hydrocarbon linkage is ideally suited for this peptide. We also tested the activity of the peptides on the RXFP1 receptor, which is also activated by H3 relaxin. All the peptides demonstrated very poor activity on RXFP1 which is consistent with our previous publications which have demonstrated the importance of the A-chain for RXFP1 activity. Thus single chain peptides are likely to be strongly selective for RXFP3 over RXFP1 [5].
To be able to directly assess the effect of the staple on the peptide structure we studied peptide 5 by solution NMR spectroscopy. This analysis clearly showed that the hydrocarbon staple has a major influence and indeed is able to reconstruct a helical conformation throughout the central part of the relaxin-3 B-chain [5]. The dramatic increase in helical structure and RXFP3 activity of peptides 5 highlight the significant advantage of use of the stapling position B13 and B17 with incorporating α-methyl amino acids.
Next, we also assessed the effect of central administration of analogue 5A on food intake in satiated rats during the light phase. The analogue 5 significantly stimulated food intake, as compared to control vehicle, to a level not statistically different from native relaxin-3 [5]. Analogue 5 therefore demonstrates similar properties to relaxin-3 both in vitro and in vivo and is further evidence that Analogue 5A is a full agonist of both human and rat RXFP3 [5].

Conclusion

Relaxin family peptides have similar structural features to insulin and are difficult to reconstruct their own conformational structures in the reduced sized analogues. We succeeded in the mimicking the native conformation of relaxin-3 B-chain perfectly without A-chain. We prepared stapled single B-chain analogues that mimic the binding conformation of relaxin-3 for RXFP3. These analogues were readily prepared by combination of RCM and solid-phase peptide synthesis using commercially available building blocks. The conformational stapled peptide strategy can dramatically reduce analogue size from total 51 to only 18 amino acid residues. Therefore we have developed single-chain stapled peptide that is a high affinity and highly selective agonist for relaxin-3 receptor RXFP3. These peptides is now being used in our laboratories to characterize the nature of neutral relaxin-3/RXFP3 signaling circuits in vivo and represents a powerful lead molecule for the developing drugs for the modulation of motivation and anxiety [5].

Acknowledgement

This research was partly funded by NHMRC (Australia) project grants (508995) to JDW and RADB, and (1065481 and 1066369) to RADB and KJR. This research was also funded by The Naito Foundation (Japan) Subsidy for Female Researchers to KH. We are grateful to Tania Ferraro and Sharon Layfield for assistance with cell-based assays and to Feng Lin for amino acid analysis. We thank Prof Andrea Robinson and Dr Alessia Belgi (Monash University, Australia) for assistance with the RCM reactions. During these studies, MAH was the recipient of a Florey Foundation Fellowships. RADB is an NHMRC Senior Research Fellow, and JDW is an NHMRC Principal Research Fellow. KJR is an Australian Research Council Future Fellow. Studies at the FNI were supported by the Victorian Government's Operational Infrastructure Support Program.

References

01. C. M. Smith, A. W. Walker, I. T. Hosken, B. E. Chua, C. Zhang, M. Haider, A. L. Gundlach, Front Pharmacpl. 2013, 5, 46 (1-17).
02. C. Liu, J. Chen, C. Kuei, S. Sutton, D. Nepomuceno, P. Bonaventure, T. W. Lovenberg, Mol. Pharmacol 2005, 67, 231-240.
03. K. J. Rosengern, S. Zhang, F. Lin, F. J. Biol. Chem. 2006, 281, 28287-282895.
04. Y. W. Kim, T. M. Grosmann, G. L. Verdine, Nature Protocols 2011, 6, 761-771.
05. K. Hojo, K., A. M. Hossain, J. Tailhades, F. Shabanpoor, L. L. L. Wang, E. E. K. Ong-Palsson, H. E. Kastman, S. Ma, A. L. Gundlach, K. J. Rossengresn, J. D. Wade, R. A. D. Bathgate, R.A.D, J. Med. Chem. 2016, 59, 7445-7456.

R. A. Kirschner, J. Müller, A. Geyer*, G. Klebe**
*Department of Pharmacy, Philipps-University Marburg, Wilhelm-Roser-Str. 2
**Department of Chemistry, Philipps-University Marburg, Hans-Meerwein-Str., D-35032 Marburg

PP IX - 260

We investigated adaptive chemistry for its suitability to assemble substrate analog peptide inhibitors of protein kinase A (PKA). In a first step, Lewis-acid promoted glycosylation of various O- and S-nucleophilic amino acids with O-acetyl protected D-ribopyranose (1) yielded building blocks which bear the high affinity trihydroxy ax-eq-ax motif for spontaneous esterification with boronic acids.[1]

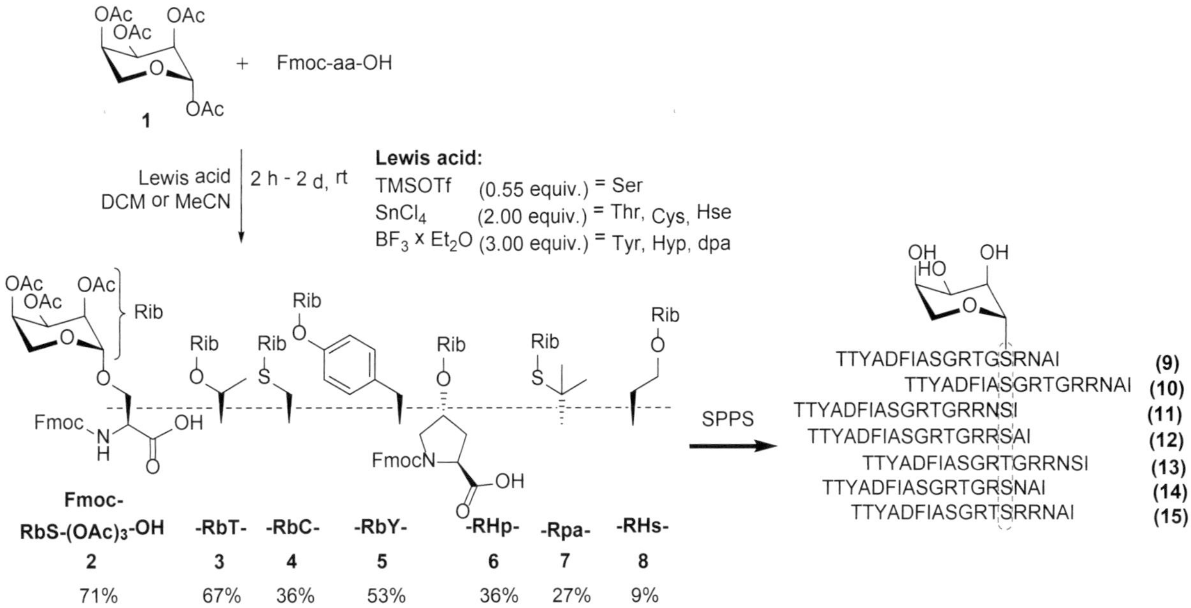

Scheme 1. Lewis-acid promoted glycosylation reaction for various Fmoc-ribopyranosylated amino acids (Raa) and the assembly of glycopeptides containing those Raas by SPPS.

The known protease inhibitor Fasudil was functionalized with boronic acid by reductive amination of the imine formed between the secondary amine of the homopiperazine moiety and 2-formylphenylboronic acid (16).[2] Further structural modifications to regulate the distance of the participating agents in the esterification are shown in Scheme 2.

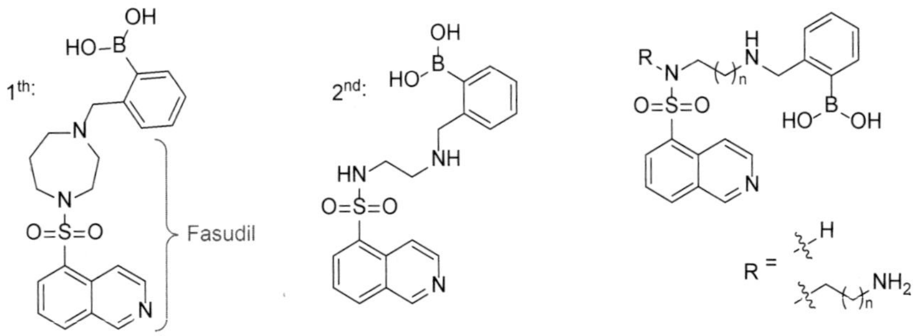

Scheme 2. Different isoquinoline-type inhibitors functionalized with boronic acids. The distance between the iso-quinoline ring and the boronic acid is varied by the incorporation of primary or secondary amines (1th & 2nd generation).

Co-crystallization trials of the enzyme PKA, the peptide (9-15), and the isoquinoline 16 were performed and the complex stability and binding affinities were measured. First insights with respect to the spatial orientation of RbS in the protein-bound state were obtained for peptide 9. Using other peptides, the sugar moiety exhibited large flexibility and no unique orientation could be observed (well-defined electron density until O-linkage of the ribose). Also the position of the modified inhibitor was resolved, which showed that no esterification occurred and the hydroxyl groups at position 2 & 4 are in 2.8 and 5.2 Å distance to the boronic acid.

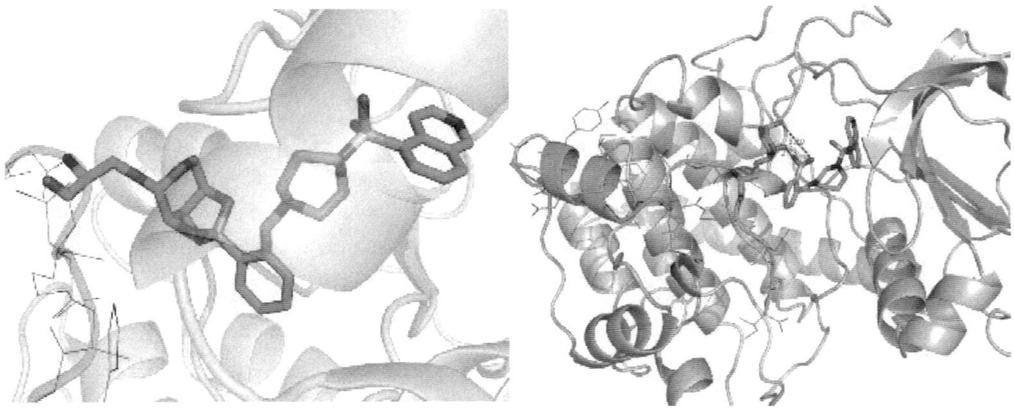

Figure 1. Proposed (left) and crystallographically obtained structure (right) of peptide 9 and inhibitor 16 in PKA.

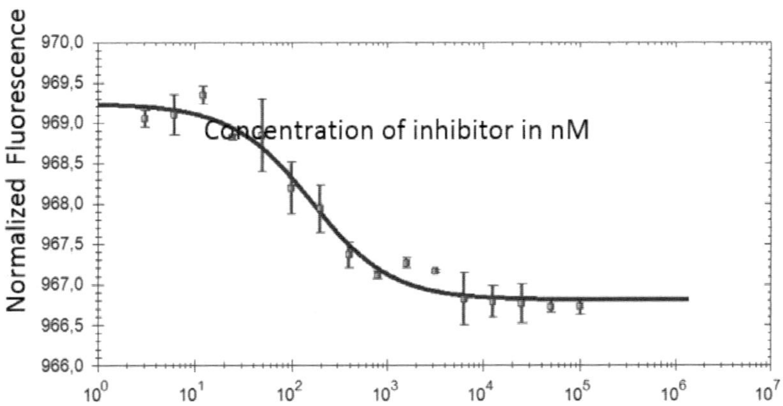

Thermal-shift and micro thermophoresis assays were performed to collect information about the inhibitory activity or the complex stability, which showed reduced but still well-detectable affinity compared to the unmodified parent peptide PKI (µM against nM range – Figure 2).

Inhibitor	K_D
PKI	144 ± 30 nM
Peptide 9	> 400 µM
Peptide 10	2670 ± 350 nM
Peptide 12	29.6 ± 10 µM

Figure 2. Binding curve of the literature known PKI to PKA in presence of ATP measured by micro thermophoresis assay (left). The assay is based on the diffusion of the involved molecules along a temperature gradient. Since diffusion of larger molecules is slower, the formation of the size-increased protein-inhibitor complex (compared to protein itself) depends on the applied inhibitor concentration and can be monitored. Selected inhibitory values are listed in the table on the right.

Further measurements using similar inhibitors and peptide sequences with other mutations (other position as well as other Raas) are in progress.

01. R. A. Kirschner, A. Geyer, Chemistry Select 2016, 1, 4570-4576.
02. C. Breitenlechner, M. Gaßel, V. Kinzel, R. Huber, R. A. Engh, D Bossemeyer, Structure 2003, 11, 1595-1607.

PP IX - 275

MODIFICATION OF PEPTIDE SEQUENCE SELECTED FOR HT-29 COLON CANCE OF CONJUGATES DEVELOPED FOR TARGETED TUMOUR THERAPY

Krisztina Kiss, Rita Oláhné Szabó, Beáta Biri-Kovács, Gábor Mezö
MTA-ELTE Research Group of Peptide Chemistry, Eötvös Loránd University, 1117 Budapest, Hungary

Introduction

Colorectal cancer is the third most common type of cancer worldwide, with nearly 1.4 million new cases diagnosed in 2012 [1]. Therefore, the development of efficient therapeutic strategies is of utmost importance. Peptide-based targeted tumour therapy might be an effective therapeutic approach to cure colon cancer as well. The principle of targeted tumour therapy relies on the structural and/or functional differences between cancer cells and healthy ones. One of the possible targeted chemotherapeutic approaches is based on the attachment of an anticancer drug to a peptide based targeting moiety, which recognizes tumour specific receptors or cell surface structures that are highly expressed on tumour cells. In our previous studies, we applied hormone peptides (GnRH and somatostatin derivatives) as homing devices. Nevertheless, to increase the drug uptake of cancer cells, a combination of conjugates that recognize different receptors on cancer cells should be applied. Phage display is a molecular diversity technology that allows the presentation of a large number of peptides permitting the selection of peptides with high affinity and selectivity for almost any target. A phage display-7 peptide library that contained 10^{11} pfu was applied and phage clones that bind to colon cancers cells were isolated by 3 rounds of positive panning. Approximately 50 phage clones were randomly picked for further analysis. Peptide sequence VHLGYAT showed the highest binding activity for HT-29 colon cancer cell line [2]. Therefore, we chose this heptapeptide as targeting moiety to develop peptide – drug conjugates for targeted tumour therapy.

Results and discussion

Daunomycin (Dau) as an anticancer agent was attached to the N-terminus of the peptide via oxime linkage through a Cathepsin B labile spacer (LRRY) that allows an easy release of the active drug metabolite in lysosomes of cancer cells [3]. The conjugate Dau=Aoa-LRRY-VHLGYAT-NH_2 (where Aoa is aminooxyacetyl moiety) showed moderate cytostatic effect (IC_{50} = 46.9±9.4 mmol). In order to increase its anti-tumour activity, the peptide sequence was modified by Ala-scan. The cytostatic effect of the conjugates was determined by MTT assay. The results indicated that Gly can be replaced by Ala inducing higher cytostatic effect (IC_{50} = 24.1±1.6 mmol) which was related to a significantly higher cellular uptake measured either by flow cytometry or fluorescent microscopy (Figure 1). Replacement of ^{1}Val, ^{3}Leu and ^{5}Tyr in the sequence completely diminished the activity (IC_{50} > 100 mmol). The exchange of Thr by Ala slightly decreased (IC_{50} = 70.9±3.8 mmol), while of His by Ala increased the anti-tumour activity (IC_{50} = 36.8±0.4 mmol) compared to the basic

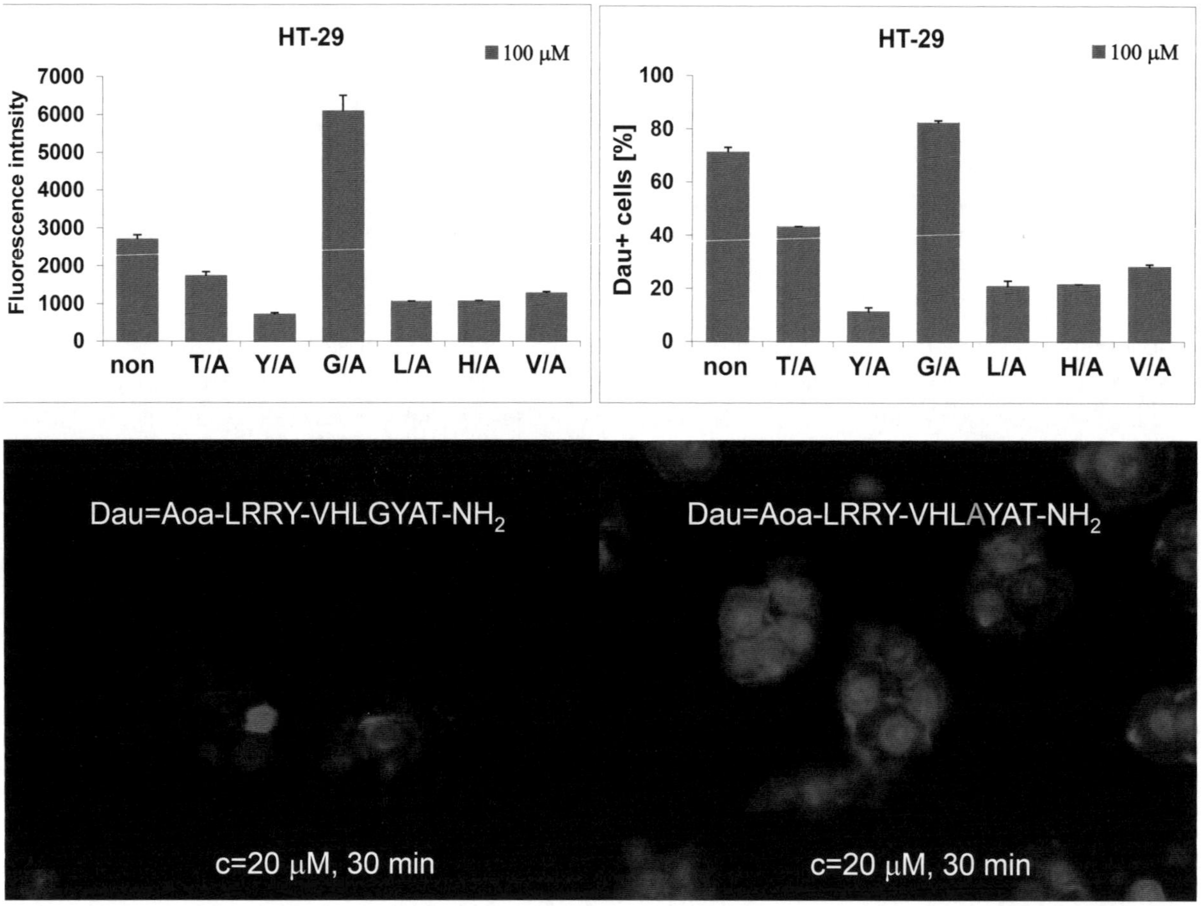

Figure 1. In vitro cellular uptake results measured by flow cytometry and fluorescent microscopy

Since it was suggested that the amino acid in position 5 can be modified, further replacements were done. Thus, amino acids with different characters were incorporated in this position of the conjugates (K, E, N, P, S, T, L, F). The cytostatic effect of the novel bioconjugates gave the following activity rank: F ≅ L > T ≥ S ≅ A ≅ N ≅ E > G > K > P. This observation suggests that the apolar amino acids with bulky side chains in position 5 increase the bioactivity of the conjugates. Ser-containing conjugate has equal antitumor activity as the Ala-containing one, but it has higher solubility. Polar amino acids mainly decrease the effect. Pro-containing conjugate has very low anti-tumour activity that might be because of the change of conformation. This study indicates that modifying peptide sequences selected by phage display technics may result in enhancement of their biological activity.

Acknowledgement

This work was supported by the grant from the Hungarian National Science Fund (OTKA 104045) and by the MTA PostDoc Fellowship for Rita Oláhné Szabó (2014-2016).

References

01. http://www.wcrf.org/cancer_statistics/data_specific_cancers/colorectal_cancer_statistics.php
02. Zhang J. Y., et al. Journal of Biomolecular Screening, 12, 429-435 (2007)
03. Orbán E., et al. Amino Acids, 41, 469-483 (2011)

SYNTHESIS AND ANTI-TUMOR ACTIVITY OF NANO-SIZED PARTICLE WITH AMPHIPHILIC LIPOPEPTIDES CONTAINING TT-232 DERIVATIVES

Anna Miyazaki[1] Koushi Hidaka[1,2] Hiromi Yoshida[3] Yoshiyuki Mizushina[3] Yuko Tsuda[1,2]
[1] Kobe Gakuin University, Cooperative Research Center of Life Sciences, Kobe, Japan
[2] Kobe Gakuin University, Pharmaceutical Sciences, Kobe, Japan
[3] Kobe Gakuin University, Nutrition, Kobe, Japan

PP IX - 279

The discovery and development of novel chemotherapeutic agents against cancer is a field of vigorous study. The goals are to identify, new biological targets and to reduce serious side effects of therapeutics. We have designed and synthesized small cyclic and linear peptide mimetics derived from somatostatin analog, TT-232 [H-D-Phe-c(Cys-Tyr-D-Trp-Lys-Cys)-Thr-NH_2] reported by Keri et al., and demonstrated their anti-tumor activities [1]. The active sequence is -Tyr-D-Trp-Lys-, of which structure-activity relationship studies showed that hydrophobicity and bulkiness on N- and C-terminus of peptide had beneficial effects on the antiproliferative activities [2,3]. On the other hand, the enhancement of hydrophobicity caused the poor solubility, loss of selectivity, poor bioavailability and high toxicity. In order to overcome their problems, various approaches and methods were developed on a field of tumor targeting therapy and reduction of side effects, recently. As our approach, we designed amphiphilic lipopeptides conjugated with a palmitic or stearic acid through various linkers to TT-232 or its active sequence, -Tyr-D-Trp-Lys- to good biocompatibility with keeping high antitumor activity. We chose linkers; miniPEG chain, $V_2A_2E_2$ and $A_2L_3K_3$ containing both hydrophobic and acidic/basic amino acid. Then prepared lipopeptides were measured their antitumor activities on HCT116 cells. Furthermore, we prepared a nano-sized liposome by insertion of a lipopeptide to liposome membrane in order to improve their efficacy in vivo.

Table. Antitumor activities of prepared lipopeptides on HCT116 cells.

Sequences		IC_{50} (μM)
YO-136	stearoyl-(miniPEG)$_2$-G-TT-232/TFA salt	6.1
YO-137	palmitoyl-(miniPEG)$_2$-G-TT-232/TFA salt	0.8
YO-138	palmitoyl-(miniPEG)$_2$-G-YwK-NH_2/TFA salt	55
YO-139	palmitoyl-$V_2A_2E_2$-YwK-NH_2/TFA salt	>100
YO-140	palmitoyl-$A_2L_3K_3$-G-YwK-NH_2/4TFA salt	19
YO-141	palmitoyl-$A_2L_3K_3$-G-TT-232/4TFA salt	43
TT-232	H-fc(CYwKC)T-NH_2/2TFA salt	26

Some lipopeptides exhibited more potent antitumor activities than TT-232 on HCT116 cells (Table). The result was supported that palmitoyl group and $A_2L_3K_3$ linker was beneficial to enhance activity. We supposed that the difference on antitumor activities between lipopeptides containing active sequences, TT-232 and -Tyr-D-Trp-Lys-, was due to effects of their original chemical property. The balance on amiphiphilicity as whole molecule was effected on antitumor activity. Furthermore a lipopeptide, YO-140 containing an active sequence, -Tyr-D-Trp-Lys-, exhibited more potent activity than YO-141 containing TT-232, although only tripeptide exhibited almost no activity. On the other hand, a lipopeptide YO-140, was successfully inserted to liposome membrane consisted by 1, 2-dioleoyl-sn-glycero-3-phosphocholine (DOPC), with confirmation by measurement of zeta potential and fluorescence (diameter: 82.36 nm, zeta potential: 58.3 mV). The prepared liposome was stable for 1 week at least. The moderated liposome was easily prepared by mix of a various concentrated lipopeptide and liposome, and left to stand at room temperature for 1 h. The reaction mixture was purified by centrifugation at 4,000 g for 40-60 min using Amicon Ultra-50. For next study, we would determine its antitumor activity. Further, also other efficient lipopeptides would be similarly examined. Finally, most useful compound would study about hemolysis and efficiency in vivo.

References

01. Keri, G. et al. (1996) Proc. Natl. Acad. Sci. USA, 93, 12513-12518.
02. Miyazaki, A. et al. (2008) J. Med Chem., 51(16), 5121-5124.
03. Kuriyama, I. et al. (2010) ANTICANCER RESEARCH, 30, 4841-4850.

A FORMYLGLYCINE-PEPTIDE USED FOR THE IDENTIFICATION OF NOVEL PHOSPHOTYROSINE MIMETICS

E. Nawrotzky[1], E. Burda[2], V. Martos Riaño[3], J. Rademann[1]
[1] Institut für Pharmazie Freie Universität Berlin, Königin-Luise-Str. 2+4, 14195 Berlin
[2] Institut für Pharmazie Universität Leipzig, Brüderstraße 34, 04103 Leipzig
[3] Leibniz-Institut für Molekulare Pharmakologie, Robert-Rössle-Str. 10, 13125 Berlin

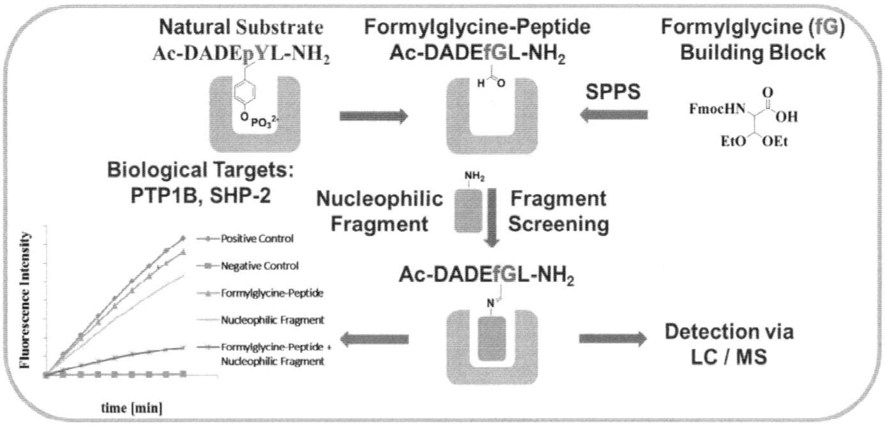

Mimetics of phosphotyrosine (pTyr, pY) residues can be starting points for the development of potent and specific inhibitors of protein tyrosine phosphatases (PTPs).[1,2,3]

We reasoned that replacement of the bulky side chain of pY in a peptide substrate of PTP by a small formyl residue, should enable the identification of pTyr-mimetic fragments in a fragment ligation assay (Figure 1).

Firstly, a facilitated protocol of a formylglycine (fG) building block suitable for Fmoc-based solid-phase peptide synthesis (SPPS) was devised (Figure 2a). Next, the formylglycine-peptide Ac-DADEfGL-NH_2 was derived from a reported peptide substrate of PTP1B,[4] DADEpYL-NH_2, and prepared. Novel phosphotyrosine (pTyr) mimetics were detected in a dynamic fragment ligation assay from a collection of nucleophilic fragments.[5,6]

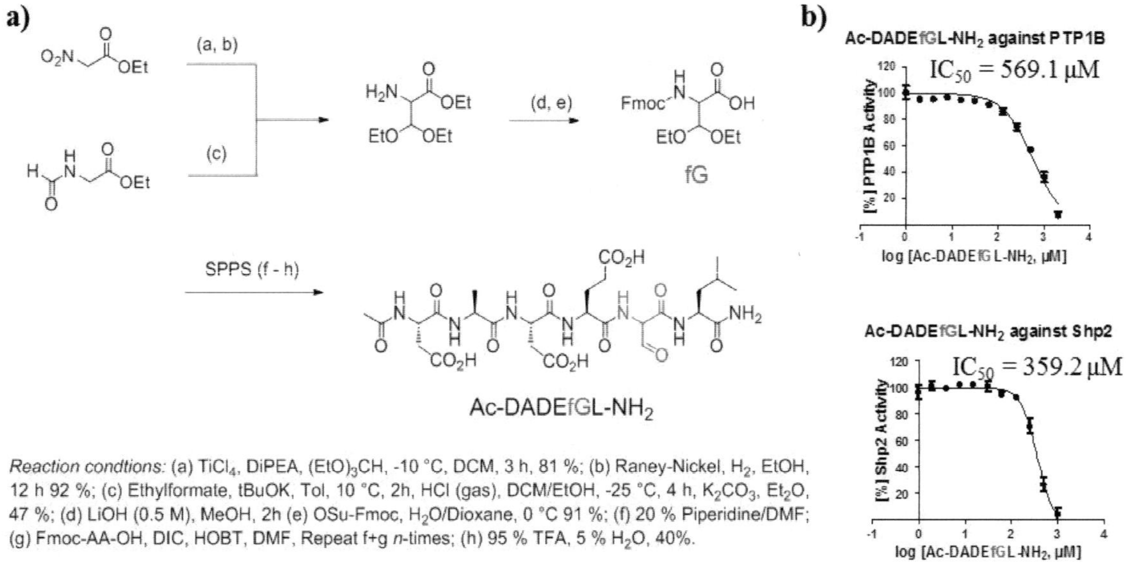

Figure 2: a) *Scheme of synthesis Formylglycine-Peptide; b) Determination of IC_{50} values of Formylglycine-Peptide against PTP1B and Shp2.*

Novel phosphotyrosine (pTyr) mimetics were detected in a dynamic fragment ligation assay from a collection of nucleophilic fragments (Figure 3). Nucleophilic fragments displaying the over-additive enhancement of inhibition of PTPs in presence of the formylglycine peptide were further validated by LC/MS analysis indicating formation of a covalent fragment addition product. Finally, peptides containing the pY-mimetic fragment in position of the formyl residue were prepared and confirmed in the enzyme assay as PTP inhibitors (Figure 4).

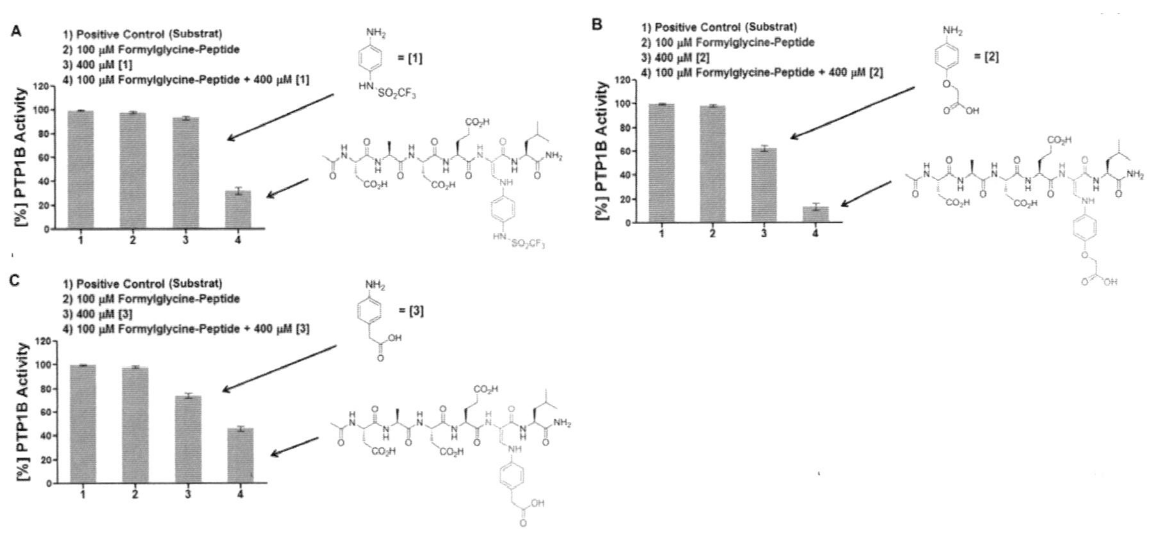

Figure 3: Nucleophilic Fragments [1], [2] and [3] were identified as novel phosphotyrosine mimetics. The formed ligation product displaying the over-additive enhancement of inhibition of PTP1B (A, B and C) and also Shp2 in presence of the formylglycine-peptide and were further validated by LC/MS analysis indicating formation of a covalent fragment addition product, which was not isolated.

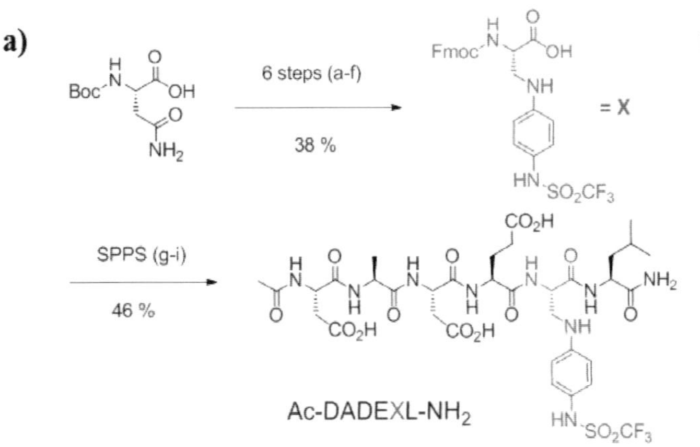

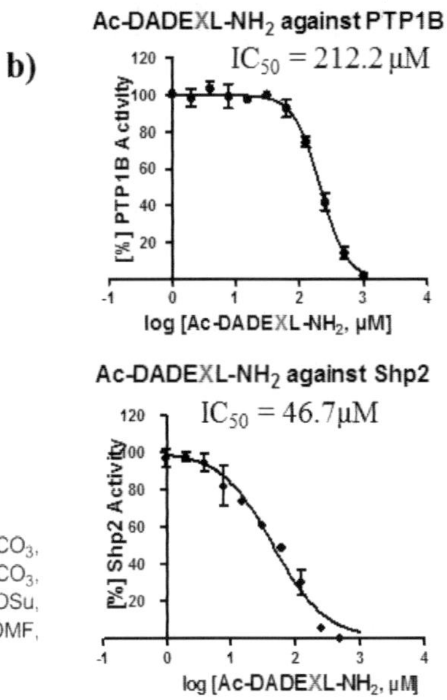

Reaction condtions: (a) $PhI(OAc)_2$, H_2O/MeCN/EtOAc), 0 °C -> r.t., 4h; (b) K_2CO_3, EtOH, 90 °C, 48h; (c) Pd/C/H_2, EtOH, r.t., 24h; (d+e) Tricflic anhydride, $NaHCO_3$, CH_2Cl_2, 0 °C -> r.t., Ar-Atmosphere; (f) 10 % Na_2CO_3 (aq.), 1-4-dioxane, Fmoc-OSu, 0 °C -> r.t., 4h; (g) 20 % Piperidine/DMF; (h) Fmoc-AA-OH, DIC, HOBT, DMF, Repeat g+h *n*-times; (i) 95 % TFA, 5 % H_2O.

Figure 4: a) *Scheme of synthesis pY-mimetic [1] containing Peptide; b) Determination of IC_{50} values of pY-mimetic [1] containing Peptide against PTP1B and Shp2.*

Conclusion: We developed a novel protocol for the synthesis of an formylglycine (fG) building block, which was integrated in Fmoc-based solid-phase peptide synthesis (SPPS).The affinity of AC-DADEfGL-NH_2 was in micromolar range and therefor suitable for the fragment ligation assay (Figure 2b). Nucleophilic fragments were identified as novel phosphotyrosine mimetics via fragment ligation assay and LC/MS analysis. Finally Peptide containing pY-mimetic [1] fragment in position of the formyl residue were prepared, shows increasing affinity against PTP1B and Shp2 and therefore confirmed in the enzyme assay as PTP inhibitors (Figure 4b)

References

01. S. Großkopf, et al., ChemMedChem 2015, 10, 815-826.
02. L. Lan, et al., EMBO J. 2015, 34, 1493-1508.
03. K. Hellmuth, et al., PNAS 2008, 105, 7275-7280.
04. Z. Jia, et al., Science 1995, 268, 1754-1758.
05. M. F. Schmidt et al., Angew. Chem. Int. Ed. 2008, 47, 3275-3278.
06. M. F. Schmidt, et al., ChemBioChem 2011, 12, 2640-2646.

Neubauer D.[1], Jaskiewicz M.[1], Sikora K.[1], Baranska-Rybak W.[2], Kamysz W.[1]
[1] Medical University of Gdansk, Chair & Department of Inorganic Chemistry, Gdansk, Poland
[2] Medical University of Gdansk, Chair & Clinic of Dermatology, Venereology and Allergology, Gdansk, Poland

PP IX - 285

Discussion

Treatment of staphylococcal infections becomes more and more difficult nowadays. The major reason for this situation is the rapid spread of resistance to antibiotics among those strains. As a result, Staphylococcus aureus is currently the leading pathogen, responsible for a variety of life threatening infections [1]. As many antimicrobial peptides (AMPs) have been found to be effective against S. aureus, research in this field is on the raise [2]. Purification of peptides by preparative high-performance liquid chromatography
(RP-HPLC) requires the use of trifluoroacetic acid (TFA) to afford trifluoroacetates as the final products. These, however, have been found to be cytotoxic. Hence the final activity of synthetic peptides may be affected [3, 4]. In this study, we synthesized the following peptides : CAMEL (KWKLFKKIGAVLKVL-NH_2), Citropin 1.1. (GLFDVIKK-VASVIGGL-NH_2), LL-37 (LLGDFFRKSKEKIG-KEFKRIVQRIKDFLRNLVPRTES), Pexiganan (GIGKFLKKAKKFGKAFVKILKK-NH_2) and Temporin A (LIGSLVRGL-IPLF-NH_2), substituted their counterions for biocompatible chlorides, and determined their antimicrobial activity against the reference and clinical strains of S aureus. Our study has shown that there are no significant differences in antimicrobial activity of the majority of the tested peptides. Moreover, the chloride anion in the CAMEL and Citropin 1.1. salts seems to be essential for antistaphylococcal activity. In addition, an enhanced activity of the chlorides has been found to be strain-dependent.

Results

Table 1. MIC values [μg/mL] against reference strains of S. aureus

Compound	*S. aureus* ATCC *25923*	*S. aureus* ATCC *6538*	*S. aureus* ATCC *6538/P*	*S. aureus* ATCC *9144*	*S. aureus* ATCC *12598*
CAMEL	8	2	4	8	4
CAMEL	2	≤0.5	0.5	1	1
Citropin 1.1	32	16	16	32	16
Citropin 1.1	16	4	8	16	8
LL-37	>256	128	>256	>256	>256

LL-37	>256	64	>256	>256	>256
Pexiganan	8	4	8	8	16
Pexiganan	8	4	8	8	8
Temporin A	4	4	8	8	4
Temporin A	4	4	8	8	4

□ - TFA^- ; ■ Cl^-

Table 2. MIC values [µg/mL] against clinical strains of S. aureus

Compound	*S. aureus* MRSA 001 N	*S. aureus* MRSA 001 S	*S. aureus* MSSA 002 N	*S. aureus* MSSA 002 S	*S. aureus* MSSA K19 N
CAMEL	4	4	4	4	4
CAMEL	2	2	≤0.5	1	1
Citropin 1.1	16	16	16	16	16
Citropin 1.1	16	16	16	8	16
LL-37	>256	>256	>256	>256	>256
LL-37	>256	>256	>256	>256	>256
Pexiganan	16	8	8	16	8
Pexiganan	8	8	16	16	16
Temporin A	8	4	4	4	8
Temporin A	8	4	4	4	8

□ - TFA^- ; ■ Cl^-

Acknowledgments

This study was supported by a grant from the Polish National Science Centre (Project No. 2011/03/B/NZ7/00548).

References

01. B. C. Kahl and K. Becker, "Clinical Significance and Pathogenesis of Staphylococcal Small Colony Variants in Persistent Infections," vol. 29, no. 2, pp. 401–427, 2016.
02. M. Dawgul, W. Baranska-Rybak, E. Kamysz, A. Karafova, R. Nowicki, and W. Kamysz, "Activity of short lipopeptides and conventional antimicrobials against planktonic cells and biofilms formed by clinical strains of Staphylococcus aureus," Future Med. Chem., vol. 4, no. 12, pp. 1541–1551, 2012.
03. M. J. Little, N. Aubry, M.-E. Beaudoin, N. Goudreau, and S. R. LaPlante, "Quantifying trifluoroacetic acid as a counterion in drug discovery by 19F NMR and capillary electrophoresis," J. Pharm. Biomed. Anal., vol. 43, no. 4, pp. 1324–1330, 2007.
04. W. Mrozik, A. Markowska, L. Guzik, B. Kraska, and W. Kamysz, "Determination of counter-ions in synthetic peptides by ion chromatography, capillary isotachophoresis and capillary electrophoresis," J. Pept. Sci., vol. 18, no. 3, pp. 192–198, Mar. 2012.

DESIGN AND SYNTHESIS OF A BIOACTIVE PEPTIDE CONJUGATED WITH ANTHRAQUINONE: TARGETING SELECTIVE IMMUNOSUPPRESSION

Anthi Tapeinou[1], Agathi Nteli[1], Carmen Simal[1], Efstathia Giannopoulou[2], Haralabos P. Kalofonos[2], Alexios Vlamis-Gardikas[1], Theodore V. Tselios[1]
[1] Department of Chemistry, University of Patras, Rion Patras, 26504, Greece
[2] Clinical Oncology Laboratory, Department of Medicine, University of Patras, Rion Patras, 26504, Greece

Introduction

Multiple Sclerosis (MS) is a serious autoimmune disease of the Central Nervous System (CNS) [1]. The disease is triggered by the stimulation of encephalitogenic T cells via the formation of a trimolecular complex between the human leukocyte antigen (HLA), an immunodominant epitope of the myelin proteins and the T cell Receptor (TCR) [2, 3]. In this study, the design and synthesis of the immunodominant myelin basic protein 85-99 epitope (MBP_{85-99}) conjugated to an immunosuppressive anthraquinone derivative was carried out [4] to give analogue 5. In this analogue, the MBP_{85-99} epitope is expected to participate in the formation of the trimolecular complex and carry the anthraquinone moiety in close proximity to the surface of encephalitogenic T cells. Thus, the T cells that recognize the MBP epitope and are responsible for the induction of MS could be selectively suppressed. The synthesized analogue was evaluated in vitro using Jurkat cells.

Results and Discussion

Chemistry: The MBP_{85-99} epitope was synthesized in solid phase using the 2-chlorotrityl chloride resin (CLTR-Cl) in combination with the Fmoc/tBu methodology, utilizing N,N'-Diisopropylcarbodiimide (DIC) and 1-Hydroxybenzotriazole (HOBt) as coupling reagents. The side chains of used amino acids were protected as follows: Trt for His; Pbf for Arg; tBu for Ser, Thr, Asp, Glu; and Boc for Lys. The $(Ahx)_6$ linker and SPDP [Succinimidyl 3-2(Pyridyldithio)Propionate] were coupled at the N terminal of the amino acid sequence. Cleavage was accomplished with Dichloromethane/2,2,2-Trifluoroethanol (DCM/TFE) and the final deprotection was achieved by treatment of the linear protected peptide with Trifluoroacetic acid/Dichloromethane (TFA/DCM) in presence of scavengers. Purification and identification of the peptide were accomplished with Reverse Phase-High Performance Liquid Chromatography (RP-HPLC) and Electrospray Ionisation Mass Spectrometry (ESI-MS) respectively. The anthraquinone type analogue was synthesized using the commercially available leucoquinizarin.

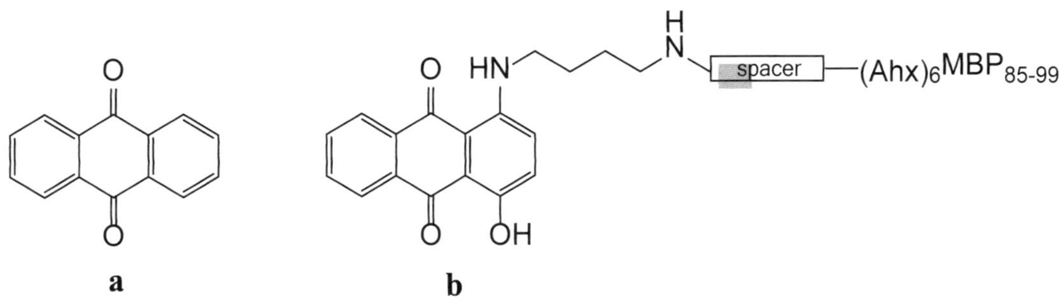

Figure 1: Chemical structures of a) Anthraquinone, b) Analogue 5.

Biological assay: The detection of analogue 5 in Jurkat cells was visible by microscopy (fluorescence). The Hoechst stain was used for DNA staining. Analogue 5 was found in the cytoplasm and nucleus, 10 and 20 min after cells treatment, respectively. Pre-treatment of cells with cisplatin, a known inhibitor of thioredoxin reductase, inhibited the entry of analogue 5 into cells suggesting involment of thiol disulfide interchange in the entrance. The analogue 5 caused apoptosis through reduction of Bcl-2 protein levels (data not shown).

Perspectives
Future studies will examine the immunosuppression potential of analogue 5 against encephalitogenic T cells, after its binding to specific HLA-DR2 tetramers.

Acknowledgment
This work was financially supported by the "Cooperation" program 09SYN-609-21, (O. P. Competitiveness & Entrepreneurship (EPAN II), ROP Macedonia - Thrace, ROP Crete and Aegean Islands, ROP Thessaly – Mainland Greece – Epirus, ROP Attica).

References
01. Mantzourani, E.D.; Mavromoustakos, T.M.; Platts, J.A.; Matsoukas, J.M.; Tselios, T.V.; Curr Med Chem, 2005, 12, 1521-1535.
02. Hahn, M.; Nicholson, M.J.; Pyrdol, J.; Wucherpfennig, K.W.; 2005 Nat Immunol 6(5): 490-496
03. Mantzourani, E.D.; Platts, J.A.; Brancale, A.; Mavromoustakos, T.M.; Tselios, T.V.; J Mol Graph Model, 2007, 26(2): 471-481
04. Watson, C.M.; Davison A.N.; Baker, D.; O'Neill, J.K.; Turk, J.L.; Int J Immunopharmacol, 1991, 13(7): 923-930

SYNTHESIS AND IN VITRO CYTOTOXIC ACTIVITY OF EGF RECEPTOR TARGETING DRUG-PEPTIDE-POLYMER CONJUGATES

Lilla Pethő[1], György Kasza[2], Orsolya Láng[3], Eszter Lajkó[3], Béla Iván[2], László Kőhidai[3], Gábor Mező[1]
[1] MTA-ELTE Research Group of Peptide Chemistry, Hungarian Academy of Sciences, Eötvös L. University, Budapest, Hungary
[2] Polymer Chemistry Research Group, Institute of Materials and Environmental Chemistry, Research Centre for Natural Sciences, Hungarian Academy of Sciences, Budapest, Hungary
[3] Department of Genetics, Cell- and Immunobiology, Semmelweis University, Budapest, Hungary

Introduction

Cancer is a leading cause of death worldwide. Targeted tumor therapy is a very perspective research area, because antitumor drugs can be selectively allocated into tumor cells. The drug moiety is usually attached to a peptide that can specifically bind to a receptor overexpressed on tumor cells. The conjugate enters the cell by receptor mediated endocytosis, gets metabolized in the lysosomes, where the active drug molecule or its active metabolite gets released, resulting in selective antitumor effect.[1]

Epidermal growth factor receptor (EGFR) overexpression was found by more than 60% of human tumor cells, therefore it is a promising target for drug delivery systems. EGFR plays a crucial role in tumorigenesis, its signaling enhances proliferation, cell survival, angiogenesis, invasion and metastasis, while it inhibits apoptosis.[2-4] GE11 (YHWYGYTPQNVI) was identified by Li et al. by using a phage display peptide library.[5] It binds to the EGF receptor as an antagonist, avoiding the activation of the signaling pathways,[6] furthermore it has no significant mitogenic and neoangiogenic activity.[5] A hexapeptide, D4 (LARLLT) with specific bound to the EGF receptor was identified by computer-aided design by Song et al. D4 binds to a binding pocket on the surface near the top, which is different to the EGF binding pocket. D4 containing liposomes were tested efficiently in in vitro and in vivo tests, respectively.[7]

Results and discussion

In this work EGFR targeting conjugates were prepared for targeted tumor therapy. GE11 and D4 peptides were used as targeting moieties and daunomycin (Dau), an anthracyclin antibiotic was used as anticancer agent. The peptides were synthesized on Rink Amide MBHA resin manually by SPPS using Fmoc/tBu strategy. Two enzyme-labile spacers (GFLG, YRRL) were also incorporated into the sequences on the N-termini that can be cleaved in the lysosomes by Cathepsin B providing a more effective drug release. The conjugation of the daunomycin to the aminooxyacetylated peptides via oxime linkage was performed in solution (0,2 M ammonium acetate buffer, pH 5,1). Solubility tests were carried out before the in vitro biological tests. Unfortunately, in some cases precipitation of the conjugate in serum free medium (diluted from DMSO containing stock solution), furthermore colloid formation of Dau=Aoa-GFLG-D4 conjugate in complete medium was observed. The in vitro cellular uptake tests were performed using BioRad ZOETM fluorescent microscope on HT-29 human colon carcinoma cells using DMSO containing stock solutions and complete medium. The conjugates with an enzyme-labile spacer are already internalized after 30 minutes and localization in the nucleus was also observed after 1 hour. To avoid high concentration of DMSO and to increase the solubility of the conjugates a highly soluble polymer was incorporated into the construct. Hiperbranched polyglycerol (HbPG) is a polyether-polyol that was selected for the conjugation. Its water solubility is extraordinary, it does not induce the immune system and its biocompatibility was approved by FDA. The worst soluble (GFLG spacer containing) peptides were synthesized on Wang resin using Fmoc/tBu strategy. Pentaglycine linker containing analogs were synthesized as well to study the significance of the peptide-polymer distance. The conjugation of the polymer was performed in solution using BOP reagent as coupling agent. After the cleavage of the isopropylidene protecting group used for protection of aminooxyacetyl moiety, daunomycin was conjugated via oxime linkage. Conjugates containing PEG were also used as controls. Solubility problems could be solved by the conjugation of polymers to the peptide-drug conjugates. HbPG could increase the solubility more than PEG (clear solution vs. colloid in $5 \cdot 10^{-3}$ M). Due to the branched structure the hydrodynamic volume (V_h) of HbPG is lower than the V_h of PEG with similar molecular weight[8] that can maybe cause lower receptor binding hindrance. The in vitro cytotoxicity of the prepared conjugates was determined on HT-29 cells using xCELLigence SP (ACEA Biosciences), IC_{50} values were determined after 48 hours (Table 1).

Table 1 IC_{50} values of the synthesized drug-peptide-polymer conjugates

Conjugate	IC_{50} / μM	Conjugate	IC_{50} / μM
Dau=Aoa-GFLG-GE11-HbPG	0.26	Dau=Aoa-GFLG-D4-HbPG	6.47
Dau=Aoa-GFLG-GE11-G_5-HbPG	0.96	Dau=Aoa-GFLG-D4-G_5-HbPG	1.17
Dau=Aoa-GFLG-GE11-PEG	1.08	Dau=Aoa-GFLG-D4-PEG	6.32
Dau=Aoa-GFLG-GE11-G_5-PEG	0.12	Dau=Aoa-GFLG-D4-G_5-PEG	26.00

The in vitro internalization of the conjugates was determined on HT-29 cells in suspension (10^{-5} M, 30 min). All conjugates had low IC_{50} values but GE11 containing conjugates were more effective. There was one outstanding conjugate from each group in the cytotoxicity measurements that correlated well with the internalization studies. Further biological investigations are in progress.

Acknowledgements
This work was supported by grants from the Hungarian National Research Fund (OTKA K104045), the MedInProt Protein Science Research Synergy Program (MedInProt) and the by the Hungarian Templeton Program (a grant from Templeton World Charity Foundation, Inc.).

References
01. G. Mezö, M. Manea, Expert Opin. Drug Deliv., 2010, 7, 79-96.
02. R. Bianco et al., Int. J. Biochem. Cell Biol., 2007, 39, 1416-1431.
03. R. Roskoski, Biochem. Biophys . Res. Commun., 2004, 319, 1-11.
04. E.K. Rowinsky, Annu. Rev. Med., 2004, 55, 433-457.
05. Z. Li et al., FASEB J., 2005, 19, 1978-1985.
06. F. M. Mickler et al., Nano Lett., 2012, 12, 3417-3423.
07. S. Song et al., FASEB J., 2009, 23, 1396-1404.
08. K. Knop et al., Angew. Chem. Int. Ed., 2010, 49, 6288-6308.

Zhanna Potetinova[1], Hilary Groom[1], Teresa Peart[1], Cornelia Tolg[1], Eva A. Turley[1,2], Leonard G. Luyt[1,2]
[1]London Health Sciences Centre, London, ON N6A 4L6, Canada
[2]University of Western Ontario, London, ON N6A 3K7, Canada

Introduction

Receptor for Hyaluronan Mediated Motility (RHAMM) is a multifunctional protein involved in wound repair that is overexpressed in disease processes [1] making it a promising therapeutic target. RHAMM interacts with hyaluronan (HA), a naturally occurring polysaccharide, and microtubules [2], a cytoskeleton component formed from alpha/beta-tubulin heterodimers. Tubulin-derived peptides were discovered to selectively block RHAMM interactions with HA fragments without binding to CD44, another predominant HA receptor [3]. In this work, we studied analogues of one of these potential therapeutic and diagnostic agents, human beta-tubulinIII(408-419) (bTUB) which binds to RHAMM with a nanomolar K_d [3]. However, the hydrophobic nature of this compound complicates its use in biological assays. In order to optimize biophysical parameters of this ligand, Ala- and D-amino acid scanning bTUB libraries as well as non-acetylated and acetylated bTUB fragments were synthesized, characterized and tested using biological approaches.

Results and Discussion

The bTUB fragment and scanning peptides were prepared in an amide form by parallel synthesis on Rink amide MBHA resin and purified by HPLC-MS. The acetylated beta-tubulin analogue (Ac-bTUB) was obtained with Ac_2O/DIPEA [4] and purified by flash chromatography with a C_{18} reverse-phase column due to its high hydrophobicity. Peptide identity of obtained compounds was confirmed by UPLC-MS.

Comparative analysis of UPLC retention times of scanning peptides showed that Ala substitutions of N- and C-terminal hydrophobic residues (Phe408, Leu418 and Val419) and Met415 decrease retention times indicative of higher hydrophilicity and structural changes (Figure 1A). These UPLC data fully correlate with consensus LogP values. In the D-amino acid scan, the modification effect on UPLC retention times was less remarkable. However, analogues with substitutions between Glu412 and Asp417 did have shorter retention times.

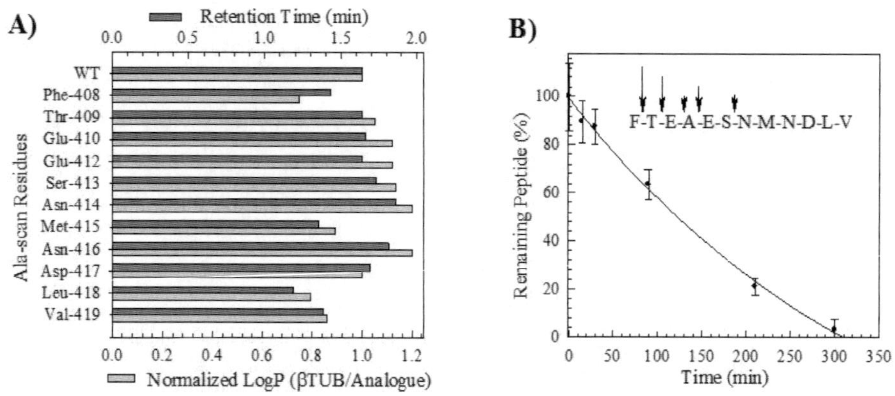

Fig. 1. Hydrophobicity of Ala-scan beta-tubulin analogues evaluated by UPLC (A).

Table 1. Stability in human serum and bioactivity of the beta-tubulin fragment and its acetylated analogue.

Compound	*Half-Life (h)*	*Inhibition (%)*	
		Cell Migration	*Cell Invasion*
βTUB(408-419)	**2**	**55**	**18**

Stability of all synthesized compounds was checked in 25% human serum. The half-life of the native bTUB fragment was two hours, and degradation started from the first two N-terminal peptide bonds being the main targets in human serum (Figure 1B). Replacement of Phe408 or Thr409 with D-amino acids and N-terminal acetylation greatly increased the serum stability of this peptide, and degradation was not detected even after 24 h incubation in serum. Furthermore, the D-Glu410 modification increased the half-life to 11 hours, and this construct had only one degradation point between Phe408 and Thr409, as identified by UPLC-MS.
The native beta-tubulin fragment and the stable acetylated analogue were next tested for bioactivity in cell migration and invasion assays (Table 1). Migration and invasion of RHAMM transfected 10T1/2 mesenchymal progenitor cells were assessed in the presence and absence of the peptides using Chemicon Cell Migration and Matrigel Invasion Assay Kits (Millipore). 30% fetal bovine serum in DMEM was added into the lower chamber, stimulating chemotactic migration (chemotaxis). For invasion, the membrane between upper and low chambers was coated with Matrigel. The number

of cells that had migrated through the membrane and adhered to the lower face of the membrane was quantified by CyQUANT GR dye. Both non-acetylated and acetylated beta-tubulin fragments inhibited cell migration and invasion. The bioactivity of the acetylated analogue was slightly increased compared to the native fragment.
In conclusion, analysis of scanning beta-tubulin libraries identified key amino acid residues in the hydrophobic structure and peptide bonds that are subject to degradation in human serum. Changes at these positions are able to improve biophysical characteristics and enzymatic stability of the tubulin fragment. N-terminal modifications significantly stabilize the RHAMM ligand structure in human serum and do not diminish bioactivity as indicated by bioassay data.

Acknowledgments
This work was supported by Novare Pharmaceuticals and the Canadian Breast Cancer Foundation.

References
01. Tolg, C., McCarthy, J.B., Yazdani, A., Turley, E.A. Biomed. Res. Int. 2014, 1-18 (2014).
02. Assmann, V., Jenkinson, D., Marshall, J.F., Hart, I.R. J. Cell. Sci. 112, 3943-3954 (1999).
03. Esguerra, K.V., Tolg, C., Akentieva, N., Price, M., Cho, C.F., Lewis, J.D., McCarthy, J.B., Turley, E.A., Luyt, L.G. Integr. Biol. 7, 1547-1560 (2015).
04. Pennington, M.W. Methods in Molecular Biology 35, 173-174 (1994).

ANTIANGIOGENIC ACTIVITY AND PLASMA STABILITY STUDY OF PEPTIDOMIMETICS CONTAINING UNNATURAL PROLINE ANALOGS

Anna K. Puszko[1], Piotr Sosnowski[2], Karolina Pułka-Ziach[1], Anna Laskowska[2], Adam Mieczkowski[3], Gerard Y. Perret[4], Aleksandra Misicka[1]
[1] University of Warsaw, Faculty of Chemistry, Warsaw, Poland
[2] Mossakowski Medical Research Centre Polish Academy of Science, Department of Neuropeptides, Warsaw, Poland
[3] Institute of Biochemistry and Biophysics Polish Academy of Science, Department of Biophysics, Warsaw, Poland
[4] Université Paris 13, Sorbonne Paris Cité, INSERUM EA4222, Bobigny, France

Introduction

Peptides having biological activity are often considered as potential drugs. Unfortunately, they are also subjected to fairly rapid enzymatic degradation in the plasma. To overcome this, on the basis of the active peptide sequence, new analogs are designed with a similar backbone but different pharmacological properties. The structure can be modified by different amino acid substitutions or the peptide bond modification to form a so-called peptidomimetics [1]. In general, such compounds are similar to the parent peptide (to preserve biological activity), but structural changes ensure higher degradation resistance due to the enzyme failure of recognizing cleavage site. Peptide K(hR)PPR, synthesized in our laboratory, exhibits potent anti-angiogenic properties. This compound is an analogue of C-terminal fragment of heptapeptide A7R (ATWLPPR) which inhibits $VEGF_{165}$ binding to NRP-1 and decreases breast cancer angiogenesis and growth in vivo [2].
We report the synthesis, biological activity and in vitro plasma stability studies of a series of peptidomimetics derived from K(hR)PPR, in which proline has been replaced by its analogs such as hydroxyproline (Hyp). Enzyme-Linked Immunosorbent Assay (ELISA) was used to test how proline replacement influence the activity.

Results and Discussion

The synthesis of peptides was carried out manually according to standard Fmoc strategy using Wang resin pre-loaded with Fmoc-Arg(Pbf) in the presence of HATU and DIPEA. Guanidinylation of lysine was done using DMPCN and DIPEA. Subsequently compounds were characterized by LC-MS and purified by preparative RP-HPLC. Afterwards, pure compounds were analyzed for their biological activity. ELISA is characterized by a relatively high sensitivity, which allows analyzing the activity of designed compounds at very low concentrations. For the experiment, a variant of this test was adapted – competitive "sandwich" ELISA. The assay was performed with several concentrations of peptidomimetics to determine the minimal effective amount of the compound.

Tab.1 The results of biological activity of selected peptidomimetics and their parent sequences (*):

Sequence	% Inhibition of $VEGF_{165}$ to NRP-1 binding in a different concentration			
	10μM	3μM	1μM	0.3μM
H_2N-Lys(Har)-Pro-Pro-Arg-H*	92,2	79,0	41,4	4,7
H_2N-Lys(Har)-Hyp-Hyp-Arg-H	89,8	76,6	42,1	30,5
H_2N-Lys(Har)-Pro-Ala-Arg-H*	94,4	89,4	77,5	54,2
H_2N-Lys(Har)-Hyp-Ala-Arg-H	86,6	62,0	25,8	21,6

Synthesized analogues were also tested for stability in human plasma. At specific time intervals, samples were taken and the plasma proteins were precipitated with ethanol. The supernatant obtained after centrifugation was lyophilized and redissolved in water prior to analyzing by RP-HPLC and LC-MS. Gradual decrease in the signal intensity of degraded peptidomimetics was observed. On the basis of obtained chromatogram of the compound, half-life ($t_{1/2}$) in human plasma was determined.

Tab.2 Half-life in human plasma of selected peptidomimetic and its parent sequence (*):

Sequence	$t_{1/2}$ [h]
H_2N-Lys(Har)-Pro-Ala-Arg-H*	1
H_2N-Lys(Har)-Hyp-Ala-Arg-H	2.8

Synthesized peptidomimetics have a slightly lower binding activity to NRP-1 compared to parent sequence. However, compounds with proline analogs seem to be much more stable in human plasma, probably due to the fact that the obtained derivatives are not easily recognized by proteolytic enzymes. Based on obtained results, new structures of further peptidomimetics will be designed.

Acknowledgements

This work was supported by National Science Centre (NCN) grant N204 350940 and co-financed by the EU from the European Regional Development Fund under the Operational Programme Innovative Economy, 2007-2013, and with the use of CePT infrastructure financed by the same EU program and a grant from the University of Warsaw for young researchers, no. 120000-501/86-DSM-110200.

References

01. Liskamp R.M., Rijkers D.T., Kruijtzer J.A., Kemmink J., ChemBioChem., 12 (2011) 1626-53.
02. Starzec A., Vassy R., Martin A., Lecouvey M., Di Benedetto M., Crépin M, Perret G.Y., Life Sciences, 79 (2006) 2370–2381

SIMULTANEOUS OPTIMIZATION OF THE SYNTHESIS OF DIFFICULT PEPTIDES IN THE PRELUDE® X AUTOMATED SYNTHESIZER USING A NOVEL REAGENT COMBINATION

Daniel Martinez, Cyf Ramos-Colon, Elizabeth Restituyo-Rosario, James P. Cain, Beatriz G. De la Torre, Fernando Albericio
Protein Technologies, Inc. 4675 South Coach Drive, Tucson, Arizona, 85714, U.S.A. Tel: +1-520-629-9626, Website: www.ptipep.com, Email: info@ptipep.com
Department of Organic Chemistry, University of Barcelona, Spain
School of Chemistry, University of ZwaZulu-Natal, South Africa

Introduction

The coupling reagent COMU has been demonstrated to be highly efficient in the synthesis of a variety of peptides[1,2]. Unfortunately, the high reactivity of COMU leads to decreased stability in the presence of solvents like DMF, making it problematic for use over multiple days on automated peptide synthesizers without preparing new solutions.

Herein we performed preliminary tests of the ability to, in essence, make COMU in situ in the reaction vessel by adding HDMC[3] and Oxyma Pure[4] separately. In this way the coupling efficiency of COMU might be achieved without losing reactivity over the course of a longer synthesis, or multiple syntheses. The use of the Prelude® X peptide synthesizer facilitated the screening of different sequences and multiple temperatures simultaneously, utilizing the independent, parallel heating capacity of the instrument.

Two difficult sequences were chosen for these experiments: the Jung-Redemann (JR) 10-mer and Aib -Leu enkephalin (where Aib replaces both glycines).

JR-10: H-WFTTLISTIM-NH_2
Aib Enkephalin amide: H-Tyr-Aib-Aib-Phe-Leu-NH_2

Experimental

JR10 (H-WFTTLISTIM-NH_2) and Aib-enkephalin (H-Y-Aib-Aib-F-L-NH_2) were synthesized on a Prelude® X using either Rink Amide ChemMatrix resin (0.47 mmol/g substitution), high loaded Rink Amide-MBHA resin (0.78 mmol/g), or low-loaded Rink Amide-MBHA resin (0.22 mmol/g) at a 50 μmol scale in duplicates. Deprotection time was 2 min at 25°C, 60°C, or 90°C using 20% Piperidine in DMF. Amino acids were coupled using a six-fold excess and final concentration of 100 mM for Amino Acids and Activators/Additive and 200 mM for DIEA. Coupling time was 3 min at 25°C, 60°C, or 90°C. After each deprotection and coupling step, three 30 s DMF washes were performed. The final cleave used TFA:TIS:EDT:Water (95:1:2.5:2.5) for 2 h at 25°C.

The JR10 peptides underwent cold Ethyl Ether precipitation and centrifugation followed by overnight drying. Prior to precipitation of Aib-Enkephalin, the majority of the cleavage solution was removed by rotary evaporation, followed by cold Ethyl Ether precipitation and drying overnight. HPLC Analysis consisted of 3 mg/ml solution on a Varian Microsorb MW 300-5 C18 50x4.6 mm column with a gradient of 5-95%B in 15 min using Water (0.1%TFA):ACN(0.1%TFA) at 1 mL/min. A 1:10 dilution of the standard sample was run on a Phenomenex Kinetex 2.6 μ C18 100A 50x2.1 mm column for LCMS with a gradient of 5-95%B in 15 min for JR10 and 10-20%B in 9 min for Aib-Enkephalin using Water (0.1%FA):ACN(0.1%FA) at 1 mL/min.

Results and Discussion

COMU, HDMC, and HDMC/Oxyma Pure produced similar purities in the synthesis of the Jung-Redemann sequence at three different temperatures using short coupling times on Rink Amide ChemMatrix resin (Table 1). For Aib-enkephalin, the highest purities were observed with COMU, but addition of Oxyma Pure significantly improved the coupling with HDMC compared to HDMC alone (Table 1).

	JR 10-mer			Aib-enkephalin		
Reagents	25°C	60°C	90°C	25°C	60°C	90°C
HDMC/Oxyma Pure	28.7%	50.1%	55.0%	29.2%	54.2%	71.3%
HDMC	29.6%	50.2%	55.4%	21.3%	38.1%	56.9%
COMU	25.9%	53.9%	60.2%	49.9%	78.5%	89.3%

Table 1. Effect on crude purity of different reagent and temperature protocols during coupling for JR 10-mer and Aib-enkephalin on Rink Amide ChemMatrix resin.

Crude purities resulting from synthesis using Rink MBHA polystyrene resin with different loading capacities were also evaluated. JR 10-mer was synthesized on high (0.78 mmol/g) and low loading (0.22 mmol/g) Rink Amide MBHA resin under the same conditions as before. Interestingly, at room temperature the difference in resin loading had only a minor and inconsistent effect, while at elevated temperature COMU, HDMC, and HDMC/Oxyma Pure all produced higher purities when low loading Rink Amide MBHA resin was used, with the best results obtained at 90 °C (Table 2). Similar to Aib-enkephalin results using ChemMatrix resin, COMU showed the best crude purities, and Oxyma Pure improved the coupling with HDMC when synthesized using the low loaded Rink MBHA PS resin.

Table 2. Effect on crude purity of different reagent and temperature protocols during coupling for JR 10-mer on Rink Amide MBHA polystyrene resins of different loading.

JR 10-mer	MBHA – High Loading			MBHA – Low Loading		
Reagents	25°C	60°C	90°C	25°C	60°C	90°C
HDMC/Oxyma Pure	20.9%	41.6%	45.7%	20.8%	52.9%	59.8%
HDMC	23.5%	41.1%	47.2%	21.6%	44.9%	51.3%
COMU	18.5%	43.7%	47.3%	23.5%	55.7%	66.9%

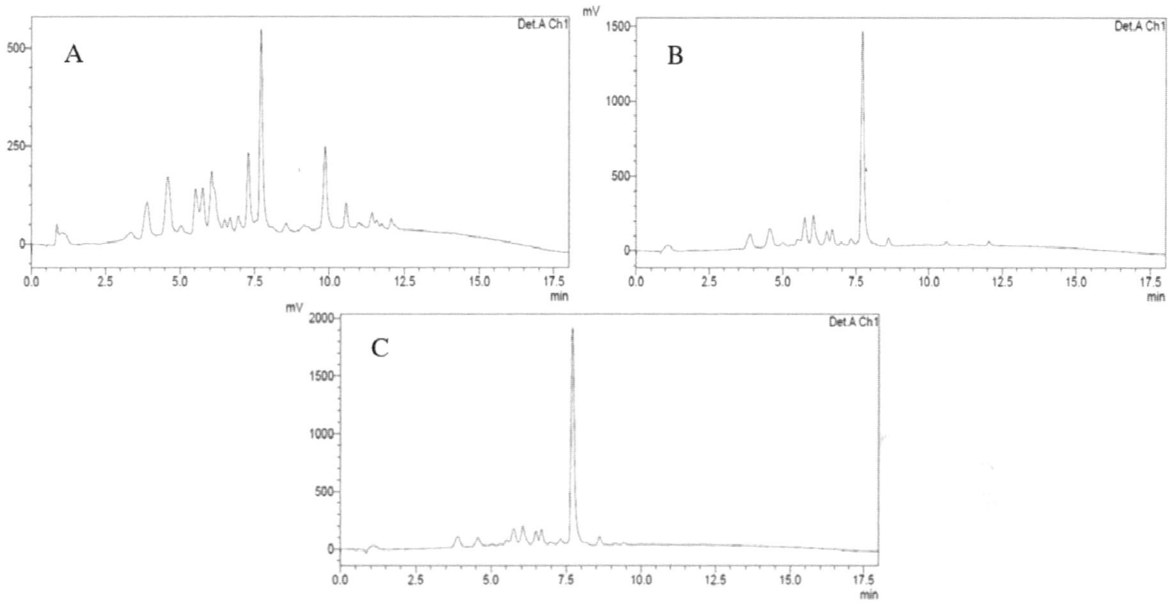

Figure 1. HPLC chromatogram of JR10 synthesized on low loading Rink Amide MBHA with HDMC/Oxyma Pure at A) 25°C B) 60°C C) 90°C.

Conclusions

Jung-Redemann 10-mer was successfully synthesized and HPLC results showed that HDMC and Oxyma Pure can produce purities similar to COMU when using a Rink Amide ChemMatrix or polystyrene resin.

Aib-enkephalin results, and to a lesser extent those for JR 10-mer synthesized on low-loaded polystyrene, suggest the possibility of competing active species (cholorobenzotriazolyl and Oxyma derivatives) and differences in associated coupling kinetics.

Purities of both difficult sequences improved with an increase in temperature up to 90°C. Additional synthesis conditions are currently under examination, including the effect of pre-activation.

References

01. El-Faham, A.; Funosas, R.S.; Prohens, R.; Albericio, F. Chemistry; 2009; 15, 9404.
02. Chantell, C.A.; Onaiyekan, M.A.; Menakuru, M. J. Peptide Sci.; 2012; 18, 88.
03. El-Faham, A.; Albericio, F. J. Org. Chem.; 2008; 73, 2731.
04. Funosas, R.S.; Prohens, R.; Barbas, R.; El-Faham, A.; Albericio, F. Chemistry; 2009; 15, 9394.

Sabine Schuster, Beáta Biri-Kovács, Gábor Mező
MTA-ELTE Research Group of Peptide Chemistry, Eötvös Loránd University, 1117 Budapest, Hungary

PP IX - 302

Introduction

Targeted tumour therapy has become an extremely important approach for the treatment of cancer. In comparison to classical chemotherapy, Drug Delivery Systems (DDS) provide a selective application of chemotherapeutics to tumour cells. Gonadotropin-releasing hormone-III (GnRH-III, <EHWSHDWKPG-NH_2; <E is pyroglutamic acid) is a native isoform of the human GnRH isolated from sea lamprey and offers different benefits like binding to human GnRH-receptors on the surface of various cancer cells, antiproliferative activity on several tumour cell lines and lower endocrine effect in mammals.[1,2] Due to these observations, GnRH-III analogues can be used as targeting moieties for anticancer drugs e.g. daunorubicin (Dau).[3] In order to improve their in vitro cytostatic effect, we developed different Dau-GnRH-III compounds in which various unnatural amino acids were incorporated in the sequence of GnRH-III.[4]

Results and discussion

Here, we report on the synthesis, lysosomal digestion, in vitro cytostatic effect and cellular uptake by HT-29-colon cancer and MCF-7 breast cancer cells of novel Dau-GnRH-III bioconjugates containing 4Ser or 4Lys(Bu) and 6D-Asp, 6D-Glu or 6D-Trp (G1-G6). Besides, we synthesized Dau-GnRH-III analogues with D-Trp in positions 3 and/or 7 as well as compounds bearing a C-terminal N-ethylamide (G7-G16) and studied their in vitro cytostatic effect on MCF-7 breast cancer cells (Table 2). These amino acid substitutions are based on the previously reported antiproliferative activity studies of GnRH-III analogues and on the fact that highly effective agonists can be produced by incorporation of D-amino acids in position 6 (6D-Aaa) of human GnRH.[4,5] All GnRH-III derivatives were prepared by SPPS. Dau was conjugated in solution to an aminooxyacetic acid linker at the side chain of 8Lys by formation of an oxime bond. The cytostatic effect was determined by Alamar blue® assay. The previously analyzed bioconjugates K1 and K2 were used as positive controls.[3,6] All Dau-GnRH-III-derivatives displayed an in vitro cytostatic effect (IC50 values are displayed in Table 1 and 2). The cellular uptake of K1, K2, G1, G2, G4 and G5 by HT-29 (Fig 1A) and MCF-7 (data not shown) cells was determined by flow cytometry. On both cell lines, the bioconjugates were uptaken similarly. To gain insight into the release of the drug or the smallest drug-containing metabolite (H-K(Dau=Aoa)-OH), the degradation of the compounds by rat liver lysosomal homogenate was determined at 37°C. All Dau-GnRH-III derivatives were digested by lysosomal enzymes revealing various cleavage sites (Fig 1B). However, the degradation level and the cleavage sites within the GnRH-III sequence differ substantially depending on the incorporated amino acids. For instance, the active metabolite H-K(Dau=Aoa)-OH was released most efficiently and much faster, in case of the two ^{6}L-Asp containing bioconjugates K1 and K2, whereas the metabolite was not detected in case of the ^{6}D-Asp analogues. Moreover, compound G1 showed the highest cellular uptake of the new derivatives, but the IC_{50} values were in the same range or even higher than the

IC_{50} value of the other [6]D-Aaa compounds. Considering all these data, we can conclude that the cytostatic effect is not only influenced by the cellular uptake, but also the release of the effective metabolite plays an important role.

Table 1. *In vitro* cytostatic effect of GnRH-III bioconjugates **G1**-**G6** on HT-29 cancer cell line (24 h treatment)

Code	Compound [^{4}Ser]	HT-29 IC_{50} [µM]	MCF-7 IC_{50} [µM]	Code	Compound [^{4}Lys(Bu)]	HT-29 IC_{50} [µM]	MCF-7 IC_{50} [µM]
K1	[^{6}D,^{8}K(Dau=Aoa)]	1.5±0.5	3.2±0.1	**K2**	[^{8}K(Dau=Aoa)]	1.9±0.7	2.7±0.1
G1	[^{6}d,^{8}K(Dau=Aoa)]	8.9±1.3	13.0±0.5	**G4**	[^{6}d,^{8}K(Dau=Aoa)]	9.3±1.1	6.2±0.2
G2	[^{6}e,^{8}K(Dau=Aoa)]	10.1±1.4	6.8±1.0	**G5**	[^{6}e,^{8}K(Dau=Aoa)]	13.7±2.6	7.0±1.2
G3	[^{6}w,^{8}K(Dau=Aoa)]	n.d.	n.d.	**G6**	[^{6}w,^{8}K(Dau=Aoa)]	n.d.	n.d.

n.d. no data (**G3** and **G6** precipitated in medium at concentrations higher than 20 µM – no dose response)

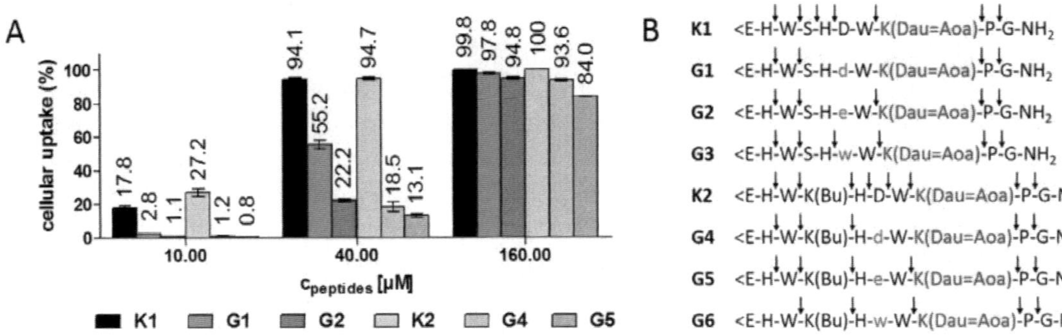

Figure 1. A) Cellular uptake of the GnRH-III bioconjugates at different concentrations by HT-29 cells after 6h determined by flow cytometry. B) Cleavage sites produced by the proteolysis of GnRH-III bioconjugates in the presence of rat liver lysosomal homogenate after 24 h of incubation (full-line arrows).

Table 2. *In vitro* cytostatic effect of GnRH-III bioconjugates **G7-G16** on MCF-7 cancer cells (24 h treatment)

Code	Compound [^{4}Ser]	MCF-7 IC_{50} [μM]	Code	Compound [^{4}Lys(Bu)]	MCF-7 IC_{50} [μM]
G7	[^{3}w,^{8}K(Dau=Aoa)]	3.6±0.3	**G12**	[^{3}w,^{8}K(Dau=Aoa)]	6.6±1.6
G8	[^{3}D-Tic,^{8}K(Dau=Aoa)]	2.9±0.6	**G13**	[^{3}D-Tic,^{8}K(Dau=Aoa)]	2.6±0.5
G9	[^{3}D-Tic,^{7}w,^{8}K(Dau=Aoa)]	3.4±0.4	**G14**	[^{3}D-Tic,^{7}w,^{8}K(Dau=Aoa)]	2.6±0.5
G10	[^{6}D(OMe),^{8}K(Dau=Aoa)]	4.8±0.7	**G15**	[^{6}D(OMe),^{8}K(Dau=Aoa)]	3.4±0.5
G11	[^{8}K(Dau=Aoa),^{10}ΔGly-NHEt]	4.9±0.1	**G16**	[^{8}K(Dau=Aoa),^{10}ΔGly-NHEt]	2.2±0.4

Acknowledgement

This project has received funding from the European Union's Horizon 2020 research and innovation program under the Marie Sklodowska-Curie grant agreement No 642004, and from the Hungarian National Science Fund (OTKA 104045).

References

01. S. Lovas et al. J Pept Res (1998), 52, 384
02. M. Kovács et al. J Neuroendocrinol (2002), 14, 647
03. R. Hegedüs et al. Eur J Med Chem (2012), 56, 155
04. E.V. Pappa et al. Biopolymers (Pept Sci) (2012), 98, 525
05. M. Padula Anim Reprod Sci (2005), 88, 115
06. M. Manea et al.Bioconjug. Chem. (2011), 22, 1320

Monika Swiontek, Patrycja Król, Mateusz Pawlaczyk, Justyna Fraczyk, Zbigniew J. Kaminski, Beata Kolesinska
Institute of Organic Chemistry, Lodz University of Technology, Zeromskiego 116, 90-924 Lodz, Poland.

Introduction

Insulin, in solution exists as a mixture of different oligomeric states, including hexamers, dimers and monomers [1]. Due to its propensity to undergo stress induced conformational changes insulin aggregate forming amyloid fibrils [2, 3]. It has been identified that segments of insulin formed the spine of amyloid fibrils are fragments A13-A19 H-LYQLENY-OH and B12-B17 H-VEALYL-OH [4]. The investigation and development of fibrillogenesis inhibitors is an important scientific and therapeutic goal.

Results and discussion

As part of our search on inhibitors of insulin aggregation process it was used the strategy of applying analogues containing α,α-disubstituted amino acids. The choice of α-methylalanine is based on the observation that peptides containing α,α-disubstituted amino acid residues are prone to formation stable helical structures, which should interfere with or even prevent the formation of stable β-sheet structure characteristic for amyloid fibers. Herein synthesis of analogues of ^{13}LYQLENY^{19} (chain A) – part of amyloidogenic core of human insulin is presented. The first stage of the studies was α-methylalanine scan of ^{13}LYQLENY^{19}, wherein each of the amino acid residue have been replaced with α-methylalanine. Eight analogues have been synthesized and their susceptibility to aggregation were tested. Studies on the aggregation of the analogues have been done by using three independent techniques: spectrophotometric with Congo red, fluorescent with Thioflavin T and microscope examination (Figure 1). Fluorescence and spectrophotometric measurements were carried out after incubation of analogues ^{13}LYQLENY^{19} in phosphorus buffer (pH = 7.2 in the case of Congo Red and pH = 6.0 in the case of Thioflavin T) for 7 days in temperature 37,2 °C.

Figure 1. The results of spectrophotometric and microscopic studies on the aggregation of modified with α-methylalanine (Aib) A13-A19 human insulin fragment.

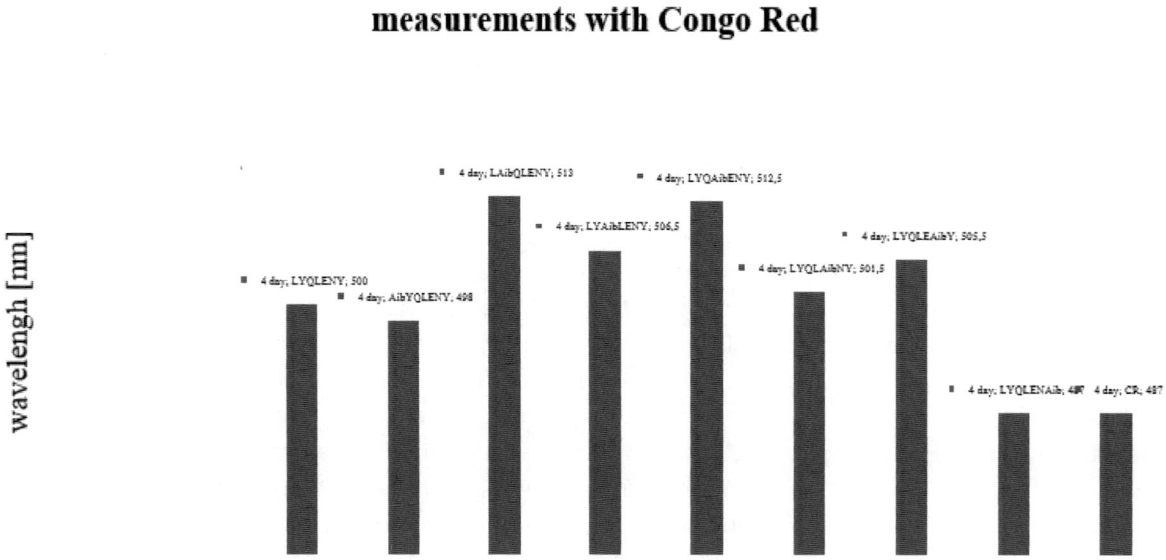

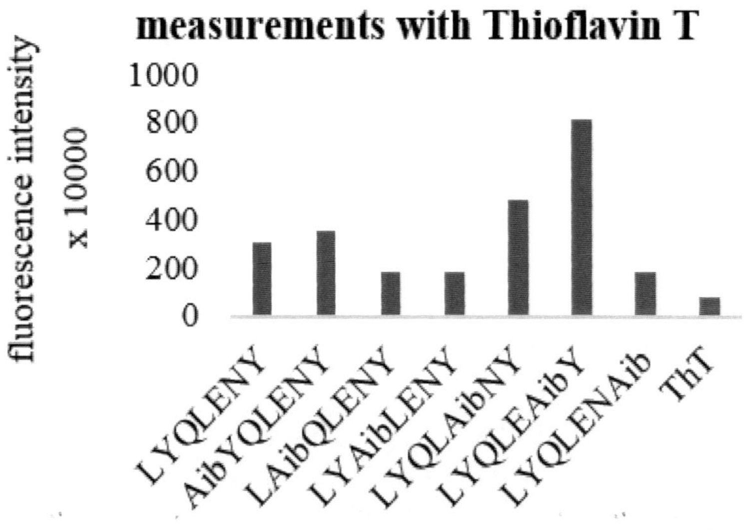

microscopic assays

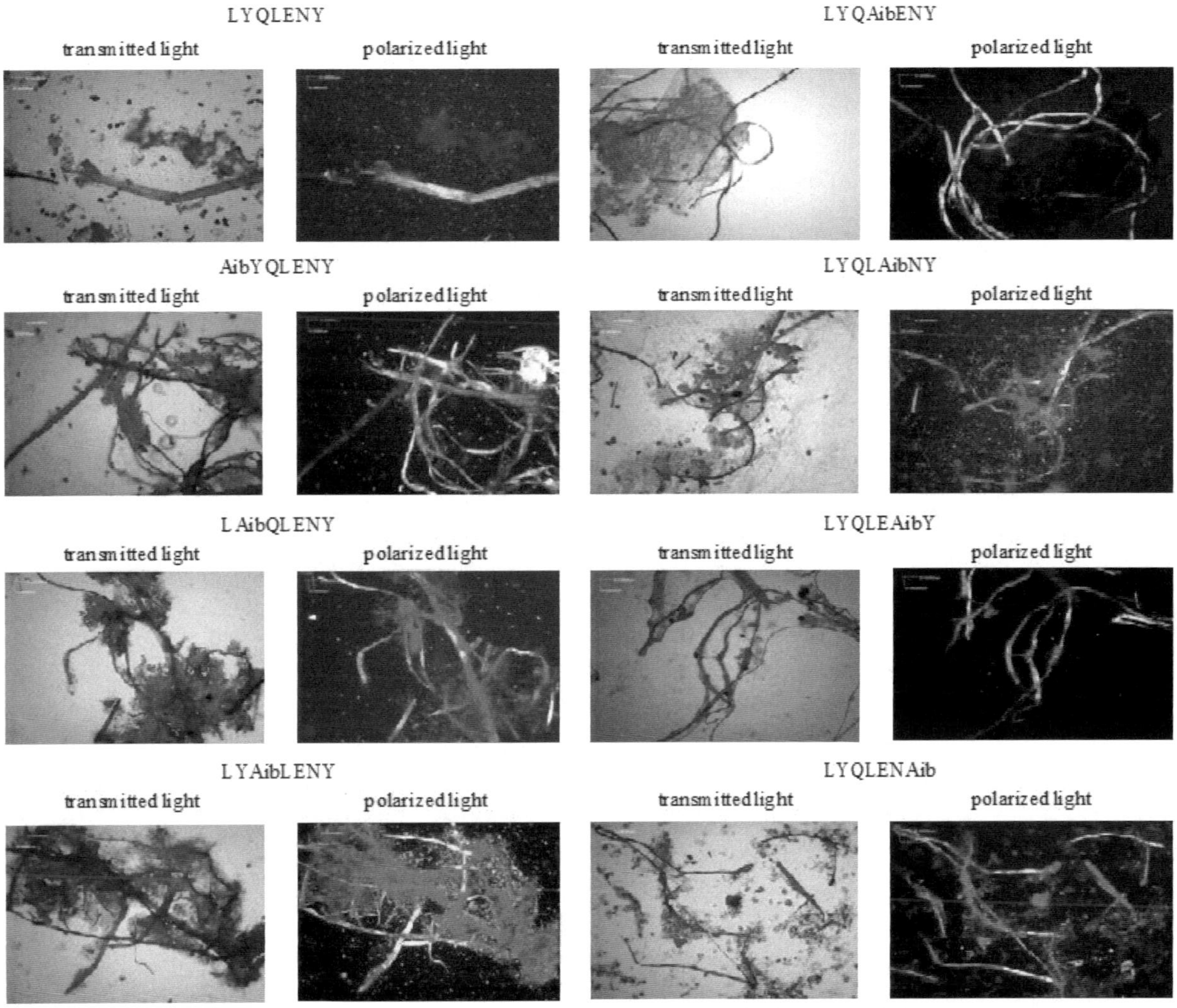

Three independent tests made it possible to determine which analogues of fragment A13-A19 showed less tendency to aggregate as a result of incorporation of α-methyl alanine residues. Results obtained with Congo red showed that all of analogues undergo aggregation process (shift of the absorbance maximum approx. 510 nm). However, a study based on measurements of fluorescence (Thioflavin T) and microscopic tests clearly indicated that two analogs of A13-A19: H-LAibQLENY-OH and H-LYAibLENY-OH were characterized by a significantly reduced potential for aggregation. Both these peptides showed the smallest increase in the intensity of fluorescence, and microscopy images showed amorphous structure in advantage.

This results might be a starting point for the rational design of inhibitors of insulin aggregation process comprising at least two Aib residues.

References

01. Grudzielanek, S., Jansen, R. & Winter, R., J. Mol. Biol., (2005), 351, 879–894.
02. Störkel, S., Schneider, H. M., Müntefering, H.,Kashiwagi, S., Lab. Invest., (1983), 48, 108–111.
03. Swift, B., Hawkins, P. N., Richards, C. & Gregory, R., Diabetic Med., (2002), 19, 881–886.
04. Ivanova, M. I., Sievers, S. A., Sawaya, M. R., Wall, J. S., Eisenberg, D., Proc. Natl. Aca. Sci.USA., (2009), 106, 18990-18995.

DEVELOPMENT OF NOVEL CYCLIC RGD AND NGR PEPTIDE DRUG-CONJUGATES FOR TUMOR TARGETING

Andrea Angelo Pierluigi Tripodi, Ivan Randelovic, Kata Nóra Enyedi, József Tóvári, Gábor Mező
MTA-ELTE Research Group of Peptide Chemistry, Eötvös Loránd University, 1117 Budapest, Hungary
National Institute of Oncology, Department of Experimental Pharmacology, 1122 Budapest, Hungary

PP IX - 308

Introduction

In the last decades, short peptides with sequences like NGR (L-asparaginyl-glycyl-L-arginine) and RGD (L-arginyl-glycyl-L-aspartatic acid) have been proved useful for ligand-directed targeted delivery of many different chemotherapeutic drugs to tumor vasculature. Among the peptide family, RGD peptides are commonly known as antimetastatic agents and are able to decrease the number of spontaneous metastases in in vivo models.[1] The tripeptide sequence can be recognized by integrins which are cell adhesion transmembrane receptors for (ECM) proteins, in particular the $\alpha_v\beta_3$, $\alpha_v\beta_5$ and $\alpha_5\beta_1$ subfamilies which play critical role in tumor-induced angiogenesis and metastasis formation.[2] In addition, it has been demonstrated that NGR-containing peptides are recognized by Aminopeptidase N (APN; CD13) which play a key role in tumor cell migration and metastasis, too.[3][4] In order to this we developed new cyclic NGR and RGD peptide - daunomycin (Dau) conjugates as cytotoxic agents. Furthermore, in order to improve their biochemical properties and facilitate the drug release cathepsin B-cleavable GFLG or LRRY tetrapeptide spacers were also inserted between the homing peptide and drug molecule.

Results and discussion

RGD and NGR cyclic peptides were prepared by SPPS on a Rink-Amide MBHA Resin, using Fmoc/tBu strategy. Daunomycin (Dau) as an anticancer agent was conjugated to an aminooxyacetic acid linker (oxime linkage) connected through an enzyme labile spacer (either GFLG or LRRY) via oxime linkage allowing an efficient drug release in lysosomes of cancer cells. The spacer was elongated with a side chain chloroacetylated lysine that could be conjugated to the thiol functional group of head-to tail cyclic [RGDfC] peptide via chemoselective thioether ligation ([RGDfC{Dau=Aoa-GFLGK(CH_2CO)-NH_2}]; P6 and [RGDfC{Dau=Aoa-LRRYK(CH_2CO)-NH_2}]; P7).
The in vitro cytotoxic effect of Dau-RGD cyclic peptide conjugates was studied by MTT assay on MCF7 human breast cancer and B16-F10 murine melanoma cells, while HT29 human colon cancer cells were used as negative control considering the low expression of $\alpha_v\beta_3$ receptors. Compounds dissolved in serum containing (FBS+) or serum free (FBS-) RPMI 1640 medium were added to the cells and the treatment was carried out for 72 h. Non treated cells in both conditions were used as controls. Results indicated cytotoxic effect of the conjugates on MCF-7 and B16-F10 cells, but not on HT-29 cells suggesting receptor mediated activity. Furthermore, the sequence of the spacer and the applied conditions have influence on the measured cytotoxicity. FACS analysis on MCF7 cells confirmed the effective internalisation.

Table 1. Cytotoxicity of cyclic RGD peptide – Dau conjugates

Code	Spacer seq.	MCF7(IC_{50} µM) 72h		B16 F10(IC_{50} µM) 72h		HT29(IC_{50} µM) 72h	
		FBS+	FBS-	FBS+	FBS-	FBS+	FBS-
P6	GFLG	34.4±3.2	10.8±1.6	9.6±1.1	39.5±2.5	>>50	>>50
P7	LRRY	23.7±2.1	>>50	9.9±1.6	>>50	>>50	>>50

Peptide 6 MCF-7	
Treatment length	**Dau+ living cells (%)**
Control	0
1h	42.7
3h	75.2
6h	84.7
24h	97.2

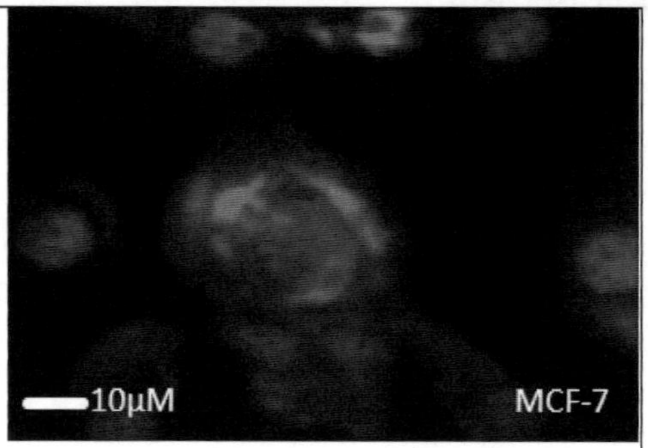

Figure 1. Cellular uptake of ([RGDfC{Dau=Aoa-GFLGK(CH_2CO)-NH_2}] by MCF-7 cells

OP
PP I
PP II
PP VI
PP VII
PP VIII
PP IX
PP X
PP XI
PP XII
PP XIII
PP XIV

In our previous study it was indicated that Dau=Aoa-GFLGK(c[CONH-KNGRE]-GG)-NH_2 conjugate has significant cytotoxic effect on CD13 positive cells (HT-1080 fibrosarcoma), but its synthesis is quite difficult. Therefore, five new conjugates Dau=Aoa-GFLGK(c[CONH-XNGRE]-GG)-NH_2, where X=Ala, Leu, Nle, Pro and Ser) were developed to see whether Lys can be changed in the sequence. The conjugates were characterized by HPLC and MS. Preliminary in vitro studies suggested, that Lys might be replaced by Leu or Pro.

Code	Compounds	RP-HPLC Rt(min)	ESI-MS M(calc)/M(exp)
P16	Dau=Aoa-GFLGK(c[CONH-ANGRE]-GG)-NH_2	22.2	1725.8/1725.8
P17	Dau=Aoa-GFLGK(c[CONH-LNGRE]-GG)-NH_2	22.4	1767.8/1767.4
P19	Dau=Aoa-GFLGK(c[CONH-NleNGRE]-GG)-NH_2	22.6	1767.8/1767.2
P20	Dau=Aoa-GFLGK(c[CONH-PNGRE]-GG)-NH_2	17.3	1751.1/1751.4
P21	Dau=Aoa-GFLGK(c[CONH-SNGRE]-GG)-NH_2	20.1	1741.9/1741.7

Table 2. Characteristics of cyclic NGR peptide – Dau conjugates

Acknowledgement

This project has received funding from the European Union's Horizon 2020 research and innovation program under the Marie Sklodowska-Curie grant agreement No 642004, and from the Hungarian National Science Fund (OTKA 104045).

References

01. Dzimbova, T., et al. J. Pept. Sci., 2012; 18: S122-S122
02. Friedlander, M., et al. Science, 1995; 270: 1500-1502
03. Zou, M., et at. Anticancer Agents Med. Chem., 2012; 12: 239-246
04. Enyedi, K. N., et al. J. Med. Chem., 2015; 58: 1806-1817

THEORETICAL PREDICTION OF THE BINDING ENERGY OF A PROPOSED NON PEPTIDE MIMETIC MOLECULE WITH THE T CELL RECEPTOR (TCR), INVOLVED IN MULTIPLE SCLEROSIS

Mary Patricia Yannakakis[1,3], Haralambos Tzoupis[1], Carmen Simal[1], Efthymia D. Mantzourani[2], James A. Platts[3], Theodore V. Tselios[1]
[1] University of Patras, Department of Chemistry, Rion Patras, 26504, Greece
[2] Cardiff University, Cardiff School of Pharmacy, CF10 3NB, Wales
[3] Cardiff University, Cardiff School of Chemistry, CF10 3AT, Wales

Introduction

Multiple Sclerosis (MS) is an immunologically controlled, inflammatory, demyelinating disease. The disease is believed to be mediated by an autoimmune T cell response directed to the proteins of the myelin sheath, such as Myelin Basic Protein (MBP).[1] The T-cell response is triggered by the formation of the trimolecular complex between the Major Histocompatibility Complex [MHC or HLA (Human Leukocyte Antigen) for humans], the immunodominant myelin protein epitopes and the T Cell Receptor (TCR).[2] In the present work, the design, synthesis and a preliminary in vitro biological evaluation of a non-peptide mimetic analogue (compound 15) was carried out. Compound 15 was rationally designed based on the immunodominant MBP_{83-96} epitope, its interactions with the TCR and HLA receptors as well as its structural orientation in the trimolecular complex.

Results and Discussion

Pharmacophore Modelling: The crystal structure[3] of the trimolecular complex (pdb 1ymm) was used to design the pharmacophore model. The used pharmacophore model (Figure 1), was based on the MBP_{83-96} epitope and the conformational characteristics of the residues involved in the interaction with the HLA and TCR.[4] The pharmacophore search was performed in the ZINC database (Drug Alike, All Clean) using the MOE software. The hits with the preferable orientation and binding with the TCR, were purchased (total 13), from AMBINTER, to perform in vitro experiments. The proposed analogue 15 was designed after optimization of the molecule selected from the pharmacophore search as lead compound and was also tested in vitro.

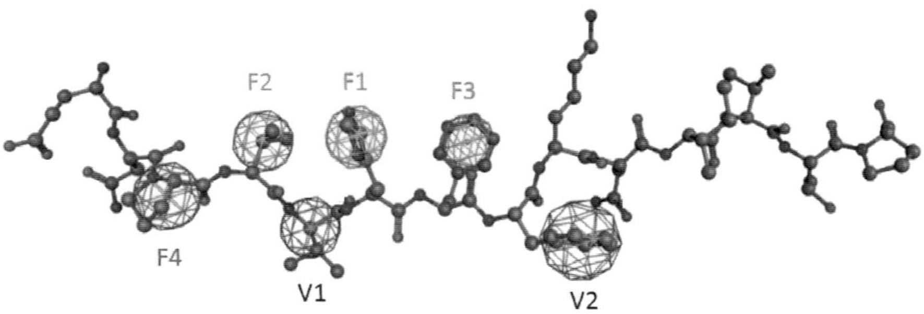

Figure 1: The developed pharmacophore model (spheres) based on the MBP_{83-96} key residues that interact with the TCR (F) and the HLA (V). F4: Pro^{85}, F2: Val^{86}, V1: Val^{87}, F1:His^{88}, F3: Phe^{89}, V2: Phe^{90}.

Molecular Modelling: Molecular Dynamics simulations were carried out for compound 15-TCR complex in explicit solvent using the AMBER12 software. QM/MM analysis followed for compound 15, using DFT and semi-empirical (SE) methodologies.[4] Calculations were performed with the

Gaussian09 and MOPAC2012 software respectively. The SE method PM7 was selected as the most appropriate one to be used for further experiments, to calculate the interaction energy of the Compound 15-TCR complex, as it best reproduced the DFT values obtained for the TCR active site residues.

Chemistry: Compound 15 (Figure 2) was synthesized using the commercially available 3- methyl pyrrolecarboxylate.[5]

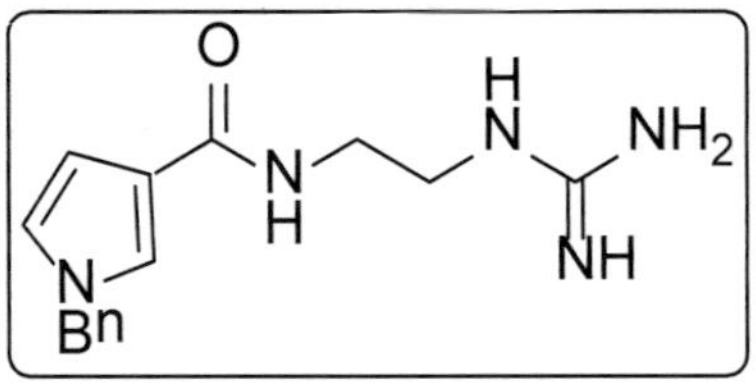

Figure 2: 2D structure of compound 15.

Biological Assay: The in vitro biological evaluation (proliferation assay) was carried out for the 13 compounds selected from the pharmacophore search and synthesized compound 15 with 0.1nM of $MBP_{83\text{-}99}$. The results show that analogue 15 was the most effective TCR antagonist and conferred the highest inhibition of cell proliferation (data not sown).

Conclusion

The scope of this work was the rational design and synthesis of non-peptide mimetics that will bind to the TCR and not to the HLA receptor. For compound 15, MD and QM/MM simulations were performed to explore the interactions with the TCR. The results from the proliferation assay revealed that compound 15 decreases the proliferation of PBMCs in the presence of the immunodominant epitope $MBP_{83\text{-}96}$ and seems to be promising for further investigation as a putative TCR antagonist.

Acknowledgment

This work was financially supported by the "Cooperation" program 09SYN- 609-21, (O. P. Competitiveness & Entrepreneurship (EPAN II), ROP Macedonia - Thrace, ROP Crete and Aegean Islands, ROP Thessaly – Mainland Greece – Epirus, ROP Attica).

References

01. Mantzourani, E.D.; Mavromoustakos, T.M.; Platts, J.A.; Matsoukas, J.M.; Tselios, T.V.; Curr. Med. Chem., 2005, 12, 1521-1535.
02. Yannakakis, M.P.; Tzoupis, H.; Michailidou, E.; Mantzourani, E.; Simal, C.; Tselios, T. J. Mol. Graph. Model., 2016, 68, 78-86.
03. Hahn, M.; Nicholson, M.J.; Pyrdol, J.; Wucherpfennig, K.W. Nat. Immunol., 2005, 6, 490-496.
04. Aldulaijan, S.; Platts, J.A. J. Mol. Graph. Model., 2010, 29, 240-245,
05. Mochona, B.; Le, L.; Gangapuram, M.; Mateeva, N.; Ardley, T.; Redda, K.K. J. heterocycl. chem., 2010, 47(6), 1367-1371,

ANTIMICROBIAL POTENTIAL OF BIOLOGICALLY ACTIVE COMPOUNDS DERIVED FROM BULGARIAN TOAD SKIN SECRETION

Yakimova B.[1], Lazarkevich I.[2], Kussovski V.[2], Engibarov S[2] and Stoineva I.[1]
[1]Laboratory Chemistry and Biophysics of Proteins and Enzymes, Institute of Organic Chemistry with Centre of Phytochemistry, Bugarian Academy of Sciences, 1113 Sofia, Bulgaria,
[2]Institute of Microbiology, Bugarian Academy of Sciences, 1113 Sofia, Bulgaria, Sofia, Bulgaria

Introduction

Peptides with potential antibacterial and antifungal activity play an important role in the system of innate immunity and constitute the first-line defense against invading pathogens. The need of new antimicrobial agents to overcome microbial antibiotic tolerance or resistance stimulates investigations towards novel strategies or novel sources of antimicrobial compounds to fight against microbial infections. Amphibian skin secretions contain a rich chemical arsenal of diverse components, including biogenic amines, alkaloids, peptides and proteins, some of them with antibacterial, antifungal, antiviral, hormone, analgesic, neurotransmitter and antitumor activities.
The goal of this study is to elucidate the antimicrobial potential of skin gland secretion isolated from Bulgarian Bombina variegata toad.

Results and Discussion

Skin secretions from Bulgarian Bombina variegata toad were collected from adult specimens from the region of Ljulin mountain according to the procedure described by Lai et. al [1]. A solution of lyophilized skin secretion was passed through a Centricon centrifugal filter devices of 3 kDa and >10 kDa. The fractions were analyzed by high performance liquid chromatography (HPLC) on C_{18} analytical column (Vydac 238 TP, 25x4.6 mm). The peptides fractions were eluted out of the column by increasing the acetonitrile concentration up to 70% over a period of 90 min with 0.8 ml/min flow rate, and monitored at 214 and 280 nm (Fig.1).

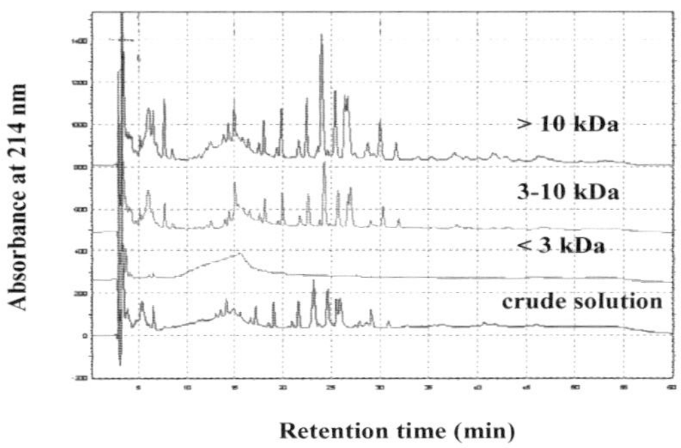

Fig.1. HPLC chromatograms of the skin secretion with different molecular mass of Bombina variegata toad.

Antimicrobial potential of the studied compounds was evaluated using the microdilution broth method described by Andrews J. [2], using 96-well standard microtiter plates.
The results of the present study have clarifyied that the isolated fractions from skin secretion of Bulgarian Bombina variegata toad have the potential to kill a broad range of microorganisms Table 1.

Table 1. Antimicrobial activity of the skin secretion with different molecular mass of Bombina variegata toad against different strains.

Sample	Staphylococcus aureus	Salmonella dublin	Candida albicans
crude solution	+ + +	+	+
< 3 kDa	+	+	+
3-10 kDa	+ + +	+	+ +
> 10 kDa	+ + + +	+ +	+ + +

In summary, evidence is provided that in skin secretion of Bombina variegata toad there are compounds with different molecular mass and expressed antimicrobial activity against G+, G- and fungus. The structural investigations are in progress.

References

01. Lai, R., Liu, H., Lee, W. H., & Zhang, Y. (2002). A novel proline rich bombesin-related peptide (PR-bombesin) from toad Bombina maxima.,Peptides, 23(3), 437-442.
02. Andrews JM. (2001). Determination of minimal inhibitory concentrations. J of Antimicrob Chemothe,48, 5-16.

Yoshio Hamada, Kenji Usui
Faculty of Frontiers of Innovative Research in Science and Technology (FIRST), Konan University, Kobe, Japan
pynden@gmail.com

PP X - 323

Introduction

The selective cleavage of a specific amide bond of peptides at room temperature promises to be a powerful tool for life science research. Previously, we reported a novel N-terminal-degradation reaction of peptides based on our novel prodrug strategy [1]. Our prodrugs have a guanidino-acyl moiety on the amino group of the parent drugs and could release the corresponding parent drugs by forming a heterocyclic compound under physiological conditions (pH 7.4, 37 °C). Our sulfathiazole and phenytoin prodrugs were rapidly converted to the respective parent drugs ($t_{1/2}$ values of 13 min and 40 min, respectively) in pH 7.4 phosphate-buffered saline (PBS) at 37 °C, and seemed suitable as an injectable formulation and an orally administered drug, respectively (Fig. 1A). Since the guanidino-acyl moieties contain an amino acid residue in their structure, we speculated that the N-terminal amino acid residues of peptides could be cleaved at room temperature using our prodrug strategy. In particular, we envisioned that the N-terminal amino acid residues might be cleaved at room temperature or under physiological conditions after the conversion of the residues into guanidino-acyl residues in a peptide (Fig. 1B)[2].

Results and Discussion

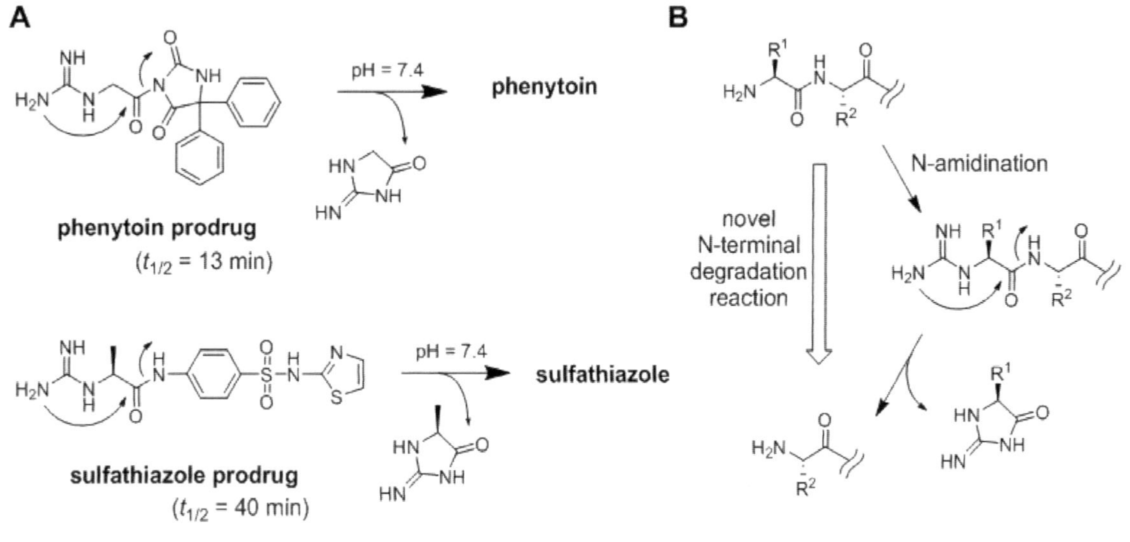

Figure 1. (A) Prodrugs of phenytoin and sulfathiazole. (B) Design of novel N-terminal degradation of peptides via N-amidination.

As model compounds, we designed and synthesized a series of N-amidino-dipeptides 1–8 with a C-terminal anilide using the common solution-phase peptide synthesis method. The peptide bonds were formed using 1-ethyl-3-(3-dimethylaminopropyl)carbodiimide hydrochloride in the presence of 1-hydroxybenzotriazole. The N-amidination reaction was performed using N′,N′-bis(tert-butoxycarbonyl)-1H-pyrazole-1-carboxamidine. The final deprotection steps were performed using trifluoroacetic acid with/without cation scavenger, m-cresol and ethyl-mercaptan.

First, N-amidino-peptide 1 with an alanylalanine sequence was incubated with pH 7.4 PBS at 37 °C and evaluated by HPLC. Although the cleavage rate of the degradation reaction was very slow ($t_{1/2}$ = 35.7 h, Fig 2), we confirmed the first N-terminal degradation of peptides at room temperature. (The N-terminal degradation reaction, Edman degradation, requires acid and heating conditions [3].) Higher pH condition accelerated the N-terminal degradation of peptide 1 ($t_{1/2}$ = 1.5 min in 2% aq NaOH).

We demonstrated that the N-terminal amino acid residues could be successfully cleaved at room temperature in aqueous solutions at neutral-to-alkaline pH. Especially, N-terminal degradation of peptides 1–8 showed rapid cleavage rates with $t_{1/2}$ values from 1–10 min in 2% aq NaOH (Table 1).

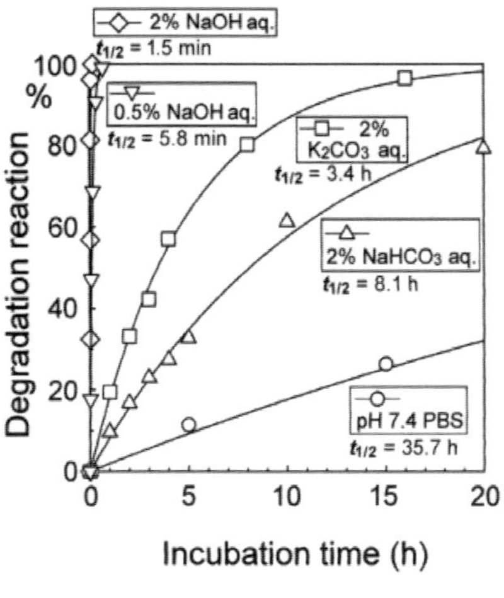

Table 1. N-terminal degradation reaction of *N*-amidino-peptides **1–8**, and their $t_{1/2}$ values.

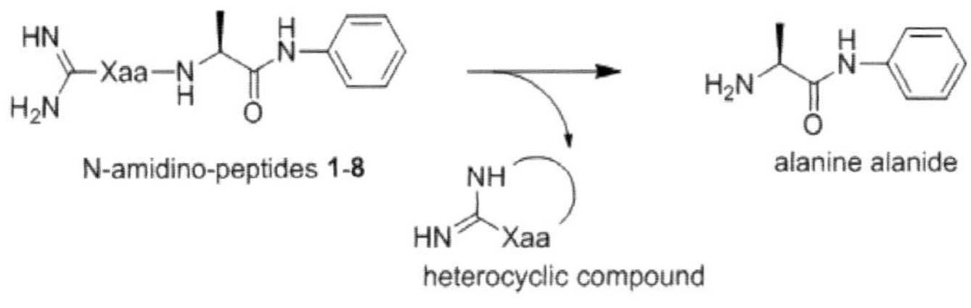

	Xaa	$t_{1/2}$ (min.)		Xaa	$t_{1/2}$ (min.)
1	Ala	1.5	**5**	Cys	3.4
2	Lys	2.8	**6**	Tyr	3.4
3	Glu	2.8	**7**	Val	9.2

References

01. Hamada, Y. Bioorg. Med. Chem. Lett. 2016, 26, 1690.
02. Hamada, Y. Bioorg. Med. Chem. Lett. 2016, 26, 1685.
03. Edman, P. Acta Chem. Scand. 1950, 1, 283.

Koushi Hidaka, Keiko Hojo, Yuko Tsuda
Kobe Gakuin University, Faculty of Pharmaceutical Sciences, Kobe, Japan

PP X - 327

Avidin-biotin affinity is extremely high with a K_d value of 10^{-15}. Utilizing the binding ability, a biotinylated peptide ligand was firstly reported by Hofmann and Kiso to purify its receptor in 1976 [1]. The affinity technology has become popular for many purposes such as purification, staining, and imaging of target biomolecule through the development of biotinylation reagents and avidin analogues such as streptavidin.

Protease inhibitor is a research tool for understanding the role of the enzymatic activity and many of the inhibitors are clinically available for treatment of diseases to control the malignant processing. We previously reported biotin conjugates of an aspartic protease inhibitor, KNI-10006 which has potent inhibitory activity against HIV protease and malarial plasmepsins [2]. The conjugation was accomplished by introducing spacer with several lengths of aminocaproly residues. Results of their HIV protease inhibitory activity in a presence of streptavidin and of the protein recovery yield in the affinity purification suggested that the longer spacer was favorable for simultaneous binding of the two proteins. On the contrary, the derivative with a short aminocaproyl spacer lost the affinity to target protease and the recovery yield as well. Therefore, we thought that if the biotin probe completely lacked the spacer structure might lose the inhibitory activity.

Based on the speculation, we designed a directly biotinylated derivative as a removable inhibitor against HIV protease. bPI-11 was synthesized as one without spacer between biotin and inhibitor (Fig. 1). Biotin was coupled with additional amino group of KNI-10006 derivative by mixed anhydride method. The crude product was purified by preparative HPLC and identified by TOF-MS analysis. bPI-11 inhibited enzymatic activity of HIV-1 protease potently, more than 97% at 5 nM that was attenuated with an addition of streptavidin. Ten equivalent of streptavidin was enough to suppress the inhibitory activity to less than 3% (Fig. 2A). Interpreting the result differently, the enzymatic activity of the inhibited HIV protease was fully recovered by adding streptavidin. We speculated that the binding equilibrium shifted from the protease to streptavidin, moving the inhibitor from the protease to streptavidin, caused this phenomenon and named it ISAAC (inhibitor stripping by avidin affinity competition).

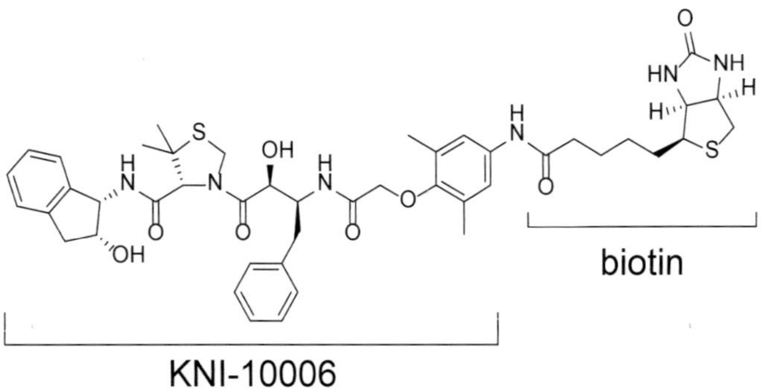

Fig. 1. Structure of bPI-11.

In conventional affinity purification of proteins, acidic buffer is used to denature the bound protein to release from the affinity carrier. In the case of protease, the process requires extra procedures of removing the denaturant and refolding the protein into native conformation to detect the enzymatic activity. Another option is to use ligand for elution in which the active conformation is maintained. However, it is impossible for the eluted protease to perform the enzymatic activity because the ligand is still attached in the active site. On the other hand, ISAAC would make it possible to detect enzymatic activity of affinity purified protease very easily, by adding the streptavidin after the ligand elution using the biotinylated inhibitor.

To realize the concept, we firstly performed affinity binding of recombinant HIV-1 protease from a mixture of cellular culture medium including fetal bovine serum to inhibitor conjugated magnetic beads. Then, we washed the unbound proteins and eluted with a buffer containing bPI-11. The eluent was transferred to 96 well plate, then streptavidin and FRET substrate were added in sequence. We successfully detected an increase of fluorescence resulting from the substrate cleavage. This methodology was also confirmed using human serum (Fig. 2B).

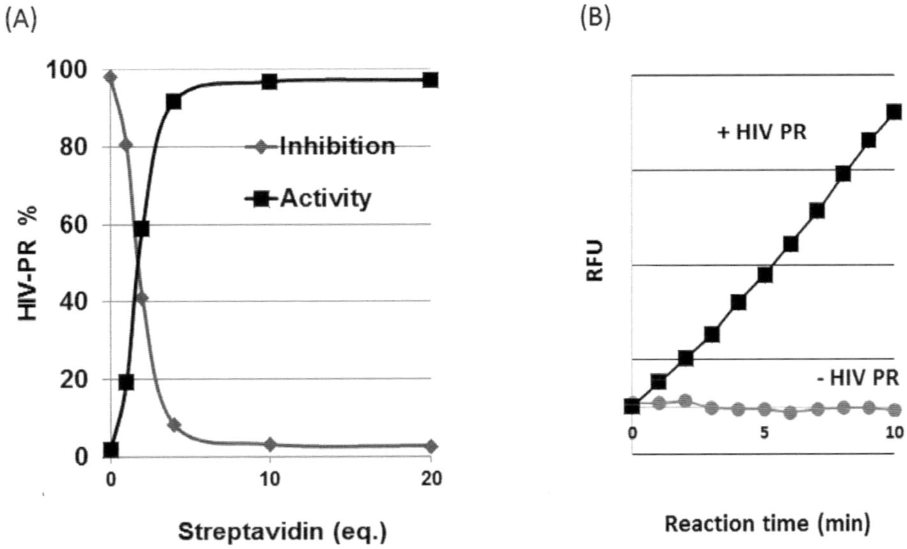

Fig. 2. Recovery of HIV protease activity inhibited (A) and eluted (B) by bPI-11 using ISAAC.

In conclusion, we designed a directly biotinylated protease inhibitor, bPI-11, to be removed after the binding with HIV protease. We succeeded in detecting the enzymatic activity after the affinity purification from human serum. The presented technique allows handling protease as its active form and switching the enzymatic activity OFF to ON, disclosing activity of proteases in natively modified forms purified from biological samples, and opening the door to develop new type of diagnostic agents.

References

01. Hofmann, K. and Kiso, Y. PNAS, 73, 3516-3518 (1976).
02. Nishiuchi, Y. and Teshima, T. (Eds.) Peptide Science 2013 (Proceedings of the 50th Japanese Peptide Symposium), Japanese Peptide Society, Minoh, 2014, p. 73-74.

DESIGN, SYNTHESIS AND APPLICATION OF PREDICTIVE ENANTIOSELECTIVE COUPLING REAGENTS FOR SYNTHESIS OF OPTICALLY PURE PEPTIDES DIRECTLY FROM RACEMIC N-PROTECTED AMINO ACIDS

Zbigniew J. Kaminski, Katarzyna Kasperowicz-Frankowska, Justyna Fraczyk, Beata Kolesinska
Institute of Organic Chemistry, Lodz University of Technology, Zeromskiego 116, 90-924 Lodz, Poland.

Introduction

An easy access to both enantiomeric forms of amino acids could be a crucial factor limiting the progress in the synthesis of peptide analogues. In order to overcome this limitation we developed the predictable enantioselective coupling reagents. The reagents designed according to this concept are based on chiral N-triazinylammonium salt obtained by treatment of achiral 1,3,5-triazine derivatives with optically active tertiary amines. Due to modular structure of predictive coupling reagent and the participation of the chiral component only in stage of carboxylic group activation, following by its departure after fulfilling stereoselective function, it is possible to predict the stereochemical outcome of the process. Coupling experiments using alkaloids (brucine, strychnine, quinine) as chiral components, incorporated single enantiomeric residue into the peptide chains with enantiomeric purity up to 99% directly from racemic N-protected amino acids, with predicted configuration [1]. In all cases, enantioselective syntheses proceed under coupling conditions typical for native achiral triazine coupling reagents. In order to incorporate both enantiomeric forms of building blocks we attempted to transform proline (non toxic and readily accessible in both enantiomeric forms) into suitable chiral components of predictive coupling reagent.

Results and discussion

Treatment of esters of both enantiomers of N-methyl proline with 2-chloro-4,6-dimethoxy-1,3,5-triazine (CDMT) gave appropriate N-triazinylammonium chlorides in both enantiomeric forms.
Figure 1. Predictable coupling reagents from N-methylproline; synthesis and application.

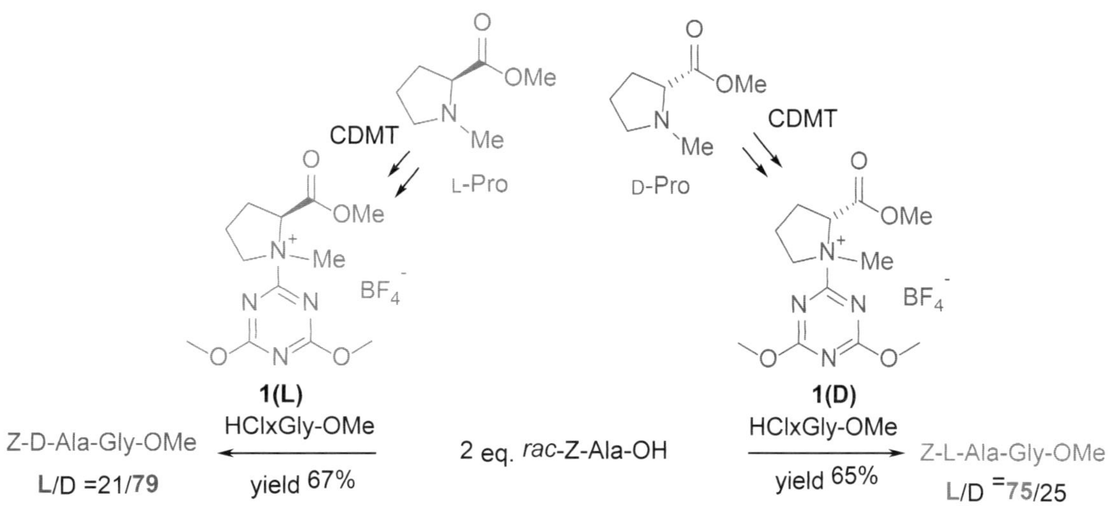

The exchange of chlorine ion into non-nucleophilic BF_4- anion yielded stable coupling reagents 1(L) and 1(D). With 2 equivs. of racemic substrate rac-Z-Ala-OH enantiomerically enriched D peptide (L/D=21/79) was obtained using chiral component derived from L-Pro and respectively L peptide (L/D=75/25) in the coupling involving chiral component prepared from D-Pro (Fig. 1).
In an alternative approach, chiral components of predictive reagent were prepared via transformation of proline into both enantiomers of 2-trichoromethyloxazolidinones 2L and 2D respectively (Fig. 2) [2]. Unfortunately, any attempts to prepare enantioselective reagents by treatment 2L and/or 2D with CDMT in the form of stable tetrafluoroborates failed, therefore in this case carboxylic function was activated in situ by less stable reagents 3L and 3D.
Figure 2. Predictable coupling reagents 2L and 2D from proline; synthesis and application.

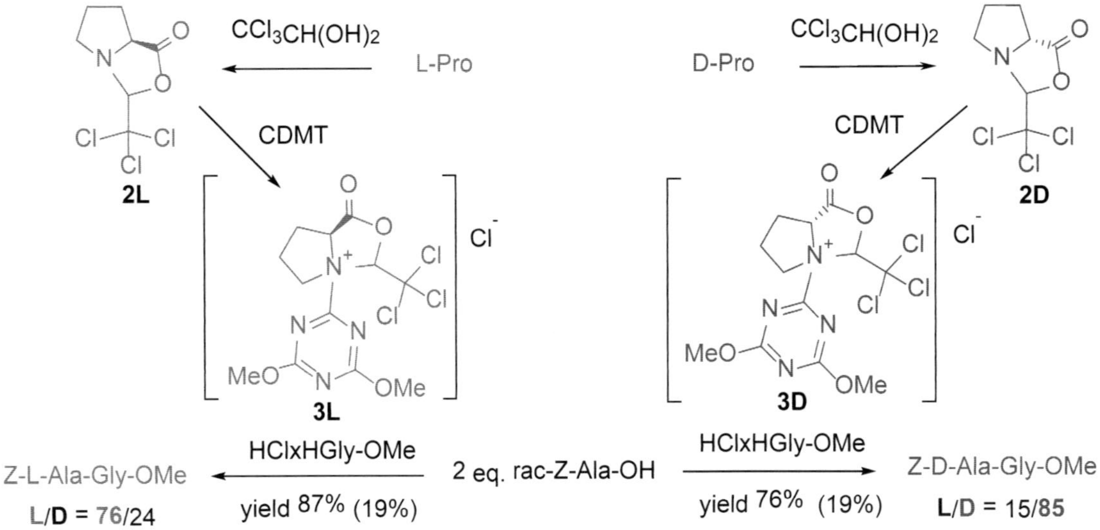

As could be expected, the bicyclic chiral component 2L and 2D with stabilized configuration on bridgehead nitrogen atom were more efficient as enantioselectors, however the presence of the bulky trichloromethyl group hampered the rate of activation. Nevertheless, both experiments confirmed potential of proline derivatives, because even under non-optimized conditions, the results are promising and validate the accuracy of assumptions.

Acknowledgement
Financial support from NSC: project number: 2012/07/N/ST5/01883 (K.K-F) and NCBiR project PBS2/B7/0/2013 (ZK, BK) is gratefully acknowledged.

References

01. a) Kolesinska, B., Kaminski, Z.J., Org. Lett. (2009), 11, 765-768;
b) Kolesinska, B., Kasperowicz, K., Sochacki, M., Mazur, A., Jankowski, S., Kaminski, Z.J., Tetrahedron Lett. (2010), 51, 20-22; c) Kolesinska, B., Kasperowicz-Frankowska, K., Fraczyk, J.,
Kaminski, Z.J., Helv. Chim. Acta, (2012), 95, 2084-2098.
02. Wang, H., Germanas, J.P., Synlett, (1999), 1, 33-35.

Rémi Desmet, Mindaugas Pauzuolis, Emmanuelle Boll, Hervé Drobecq, Laurent Raibaut, Oleg Melnyk
UMR CNRS 8161, Université de Lille, Institut Pasteur de Lille, 59021 Lille, France.

PP X - 330

In serine (Ser) or threonine (Thr) O-acyl isopeptides (R = H or Me respectively in Scheme 1), the peptidyl chain preceding the Ser/Thr residue is connected to this residue through an ester bond involving the Ser/Thr β-hydroxyl group.[1-3] O-acyl isopeptides spontaneously undergo an oxygen to nitrogen acyl shift at pH > 5 which restores a native peptide backbone structure. This modification was used for minimizing peptide aggregation, favor the solubilization of hydrophobic peptides or designing activatable peptidic or protein scaffolds. Given the importance of unprotected O-acyl isopeptides for studying the function of peptides and proteins, we sought to develop a simple method for accessing these compounds by the chemoselective ligation of unprotected O-acyl isopeptide segments in water.

The native chemical ligation (NCL) is a powerful tool for accessing large polypeptides or proteins.[4] Unfortunately, the use of this reaction for accessing directly unprotected O-acyl isopeptides is complicated by the fact that spontaneous O,N-acyl shift occurs in the optimal conditions for the ligation, that is in water at neutral pH. Moreover, the rate of NCL which relies on thiol-thioester exchanges decreases significantly by decreasing the pH. For example, the rate of the reaction of cysteine with Ac-Gly-SPh-pNO_2 is about 100-fold less at pH 5 than at pH 7.[5] For this reason, NCL is usually not used below pH 5. In contrast, SEA ligation is significantly accelerated by decreasing the pH from 7.5 to 5.5.[6,7] We found that SEA ligation can even proceed at pH 3.0, that is at a pH where the O-acyl isopeptide is efficiently protected by protonation.[8] As a consequence, this reaction proved to be useful for coupling chemoselectively unprotected O-acyl isopeptides in water (Scheme 1). The reaction was performed in 0.6 M guanidine hydrochloride (pH 3.0, 37 °C) in the presence of 4-mercaptophenylacetic acid (MPAA[9]) and TCEP (125 mM). The solubility of MPAA in these conditions was estimated to be ~30 mM. We showed that the O-acyl isopeptide unit can be situated on both sides of the ligation junction. The method is also useful for accessing cyclic O-acyl isopeptides by intramolecular SEA ligation.

In conclusion, SEA ligation at pH 3.0 is a useful method for producing O-acyl isopeptides from shorter peptide segments. Keeping the solubilizing effect of the O-acyl isopeptide unit(s) throughout the ligation and the HPLC purification steps and avoiding a post-ligation deprotection procedure constitute significant advantages. This strategy allows the O,N-acyl shift reaction to be performed at a later stage and depending of the final application. The method should be particularly useful for accessing hydrophobic peptides or proteins containing Ser or Thr residues in their sequence.

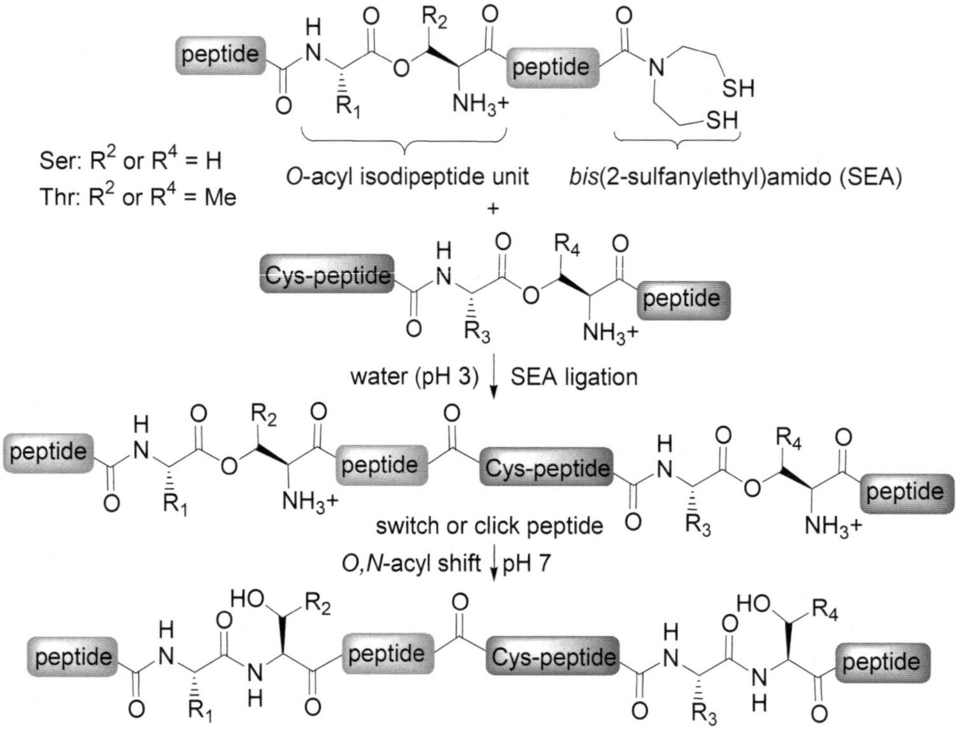

Scheme 1. Synthesis of O-acyl isopeptides using SEA ligation at pH 3.0.

References

01. Sohma, Y.; Sasaki, M.; Hayashi, Y.; Kimura, T.; Kiso, Y. Chem. Commun. 2004, 124-125.
02. Sohma, Y.; Hayashi, Y.; Skwarczynski, M.; Hamada, Y.; Sasaki, M.; Kimura, T.; Kiso, Y. Biopolymers 2004, 76, 344-356.
03. Coin, I.; Dölling, R.; Krause, E.; Bienert, M.; Beyermann, M.; Sferdean, C. D.; Carpino, L. A. J. Org. Chem. 2006, 71, 6171-6177.
04. Dawson, P. E.; Muir, T. W.; Clark-Lewis, I.; Kent, S. B. Science 1994, 266, 776-779.
05. Hondal, R. J.; Nilsson, B. L.; Raines, R. T. J. Am. Chem. Soc. 2001, 123, 5140-5141.
06. Boll, E.; Dheur, J.; Drobecq, H.; Melnyk, O. Org. Lett. 2012, 14, 2222-2225.
07. Raibaut, L.; Cargoet, M.; Ollivier, N.; Chang, Y. M.; Drobecq, H.; Boll, E.; Desmet, R.; Monbaliu, J.-C. M.; Melnyk, O. Chem. Sci. 2016, 7, 2657-2665
08. Desmet, R.; Pauzuolis, M.; Boll, E.; Drobecq, H.; Raibaut, L.; Melnyk, O. Org. Lett. 2015, 17, 3354–3357.
09. Johnson, E. C.; Kent, S. B. J. Am. Chem. Soc. 2006, 128, 6640-6646.

A NOVEL TRACELESS LINKER FOR THE SOLID PHASE SYNTHESIS OF PROTEINS USING BIS(2-SULFANYLETHYL)AMIDO (SEA) CHEMISTRY

Nathalie Ollivier, Raphaël Loval, Annick Blanpain, Rémi Desmet, Oleg Melnyk
UMR CNRS 8161, Université de Lille, Institut Pasteur de Lille, 59021 Lille, France.

PP X - 332

Protein chemical synthesis is made possible by the combination of several essential chemical tools. Solid phase peptide synthesis (SPPS) enables the synthesis of peptide segments by the iterative coupling of protected amino acids to a solid support. Other major tools are chemoselective peptide bond forming reactions which enable the coupling of unprotected peptide segments in water. Among these reactions, Native Chemical Ligation (NCL[1]c3RyeTwv) which is based on the coupling of a C-terminal peptide thioester with an N-terminal cysteinyl peptide is undoubtedly the most popular reaction for protein total synthesis.
The ligation of two peptide segments gives access to polypeptides composed of up to 100 amino acid residues (AA), since the size of the peptide segments produced by SPPS is usually < 50 AA. The synthesis of proteins above this size usually requires the ligation of more than three peptide segments. For example, the production of a polypeptide composed of 300 AA might require the assembly of 6 peptide segments. The major limitations of such complex synthetic schemes are often the limited solubility of the segments and/or intermediates and the significant mass losses that occur during the intermediate purification steps. One potential solution to these problems is to perform the assembly on a solid phase, a strategy that combines the advantages of the NCL reaction and of the SPPS.[2] We have designed such a strategy by exploiting the latent thioester properties of the bis(2-sulfanylethyl)amido group (Fig. 1).[3-7]

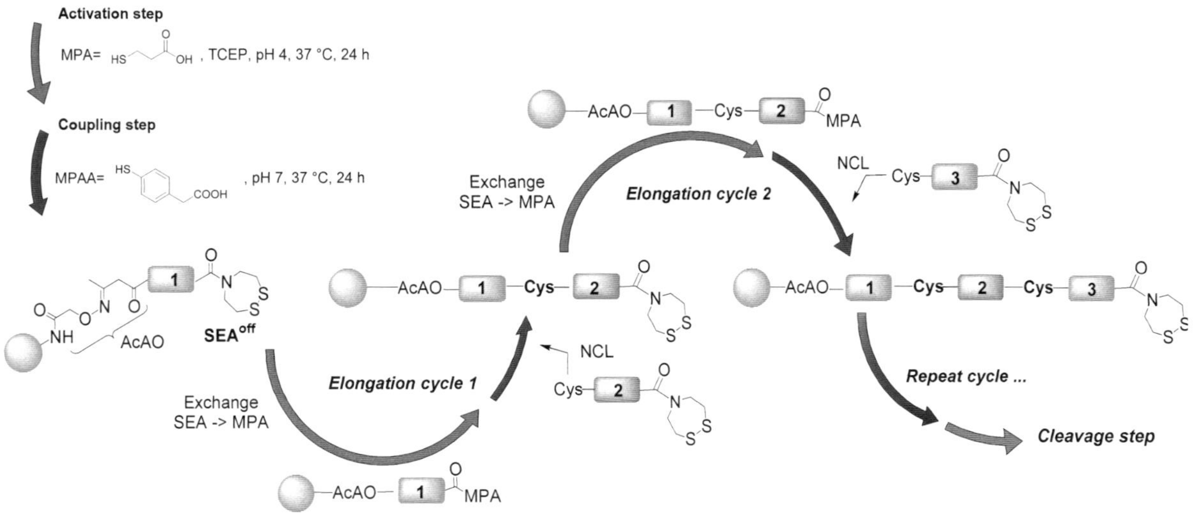

Fig. 1. Principle of the SEA solid phase protein synthesis method using the acetoacetyloxime (AcAO) linker designed in this study.

One major bottleneck that limits the development of solid phase protein synthesis in the N-to-C direction is the design of traceless linkers for attaching the N-terminal segment to the solid phase. The optimal linker must have the following properties: i) the N-terminal functionality should be easily installed during the Fmoc SPPS using affordable and commercially available reagents, ii) the attachment to the solid phase must be chemoselective and compatible with the latent thioester functionality, iii) the linker must be stable during several elongation cycles, iv) the linker should be cleaved using mild conditions in a traceless manner. Previous works used oxime[8] or CuAAC[6] ligations to design linkers relying on 2-sulfonylethyloxycarbonyl (cleavage: pH 11)[6,8] or enamine[9] chemistries (cleavage: 1 M aqueous hydroxylamine, few hours), which required harsh conditions for the final cleavage step.

We found that the acetoacetyloxime linker (Fig. 1) formed by reacting an N-terminal acetoacetyl[10] peptide with a supported aminooxyacetyl handle fulfilled all the above specifications. Importantly, the cleavage occurred in mild conditions in the presence of low concentrations of hydroxylamine (25 mM, pH 3) to provide the target proteins in good yield and purity.

References

01. Dawson, P. E.; Muir, T. W.; Clark-Lewis, I.; Kent, S. B. Science 1994, 266, 776-779.
02. Raibaut, L.; El Mahdi, O.; Melnyk, O. Top. Curr. Chem. 2015, 363, 103-154.
03. Ollivier, N.; Dheur, J.; Mhidia, R.; Blanpain, A.; Melnyk, O. Org. Lett. 2010, 12, 5238-5241.
04. Dheur, J.; Ollivier, N.; Vallin, A.; Melnyk, O. J. Org. Chem. 2011, 76, 3194-3202.
05. Ollivier, N.; Vicogne, J.; Vallin, A.; Drobecq, H.; Desmet, R.; El-Mahdi, O.; Leclercq, B.; Goormachtigh, G.; Fafeur, V.; Melnyk, O. Angew. Chem. Int. Ed. 2012, 51, 209-213.
06. Raibaut, L.; Adihou, H.; Desmet, R.; Delmas, A. F.; Aucagne, V.; Melnyk, O. Chem. Sci. 2013, 4, 4061-4066.
07. Pira, S. L.; El Mahdi, O.; Raibaut, L.; Drobecq, H.; Dheur, J.; Boll, E.; Melnyk, O. Org. Biomol. Chem. 2016, 14, 7211-7216.
08. Canne, L. E.; Botti, P.; Simon, R. J.; Chen, Y.; Dennis, E. A.; Kent, S. B. H. J. Am. Chem. Soc. 1999, 121, 8720-8727.
09. Galibert, M.; Piller, V.; Piller, F.; Aucagne, V.; Delmas, A. F. Chem. Sci. 2015, 6, 3617-3623.
10. Boll, E.; Ebran, J. P.; Drobecq, H.; El-Mahdi, O.; Raibaut, L.; Ollivier, N.; Melnyk, O. Org. Lett. 2015, 17, 130-133.

AN ARRAY OF MOLECULAR RECEPTORS AS A PLATFORM FOR PROBING MOLECULAR FINGERPRINTS OF CYTOSTATIC COMPOUNDS

Malgorzata Walczak, Ksenia Wojtczak, Justyna Fraczyk, Zbigniew J. Kaminski
Institute of Organic Chemistry, Lodz University of Technology, Zeromskiego 116, Lodz, Poland.

Introduction

Peptide microarrays are valuable tools for high-throughput screening, epitope mapping, substrate profiling and probing peptide-ligand interactions. Results of our studies shown that N-lipidated peptides immobilized on cellulose undergo self-organization process leading to formation of binding cavities acting as molecular receptors, selectively interacting with different ligands. Conformational freedom of the peptide fragment and diversity of functional groups of the peptides side chains provide the flexibility and opportunity to adaptation to the shape of bounded ligands [1]. N-Lipidated peptides are immobilized on cellulose via aromatic linker containing fragments of m-phenylenediamine and 1,3,5-triazine derivative. For synthesis of the array of N-lipidated peptides was adopted SPOT methodology with triazine coupling reagents [2].

Results and discussion

In this studies molecular receptors were used to study of interactions with 20 4-[(2-chloroetylo)piperazin-1-ylo]-1,3,5-triazine derivatives (Fig. 1). All tested compounds have anticancer activity [3]. However, the mechanism of anticancer activity is not clear. It has been assumed that compounds 1-20, with 2,4-diamino-6-methoxy-1,3,5-triazine or 2,4,6-triamino-1,3,5-triazine cores can mimic adenine and interact with the ATPase [4]. The hydrophobic pocket of the active center ATPase consists L, I, V, F residues and FT fragment. All these amino acids were applied for synthesis of 112-elements library of tri- and tetrapeptides, acetylated at N-terminus by heptanoyl residue. Prepared arrays of heptanoyled peptides immobilized on the cellulose were used for studies of interactions with derivatives 1-20 according to procedure using reporter dye to determine the strength of binding. The acquired results shown that binding pockets created by N-heptanoylated. peptides are able to selective binding compounds with anticancer activity.

Figure 1. Structures of cytostatic active triazine derivatives 1-20 and chart of their interactions with library of N-heptanoyl peptides immobilized on the cellulose.

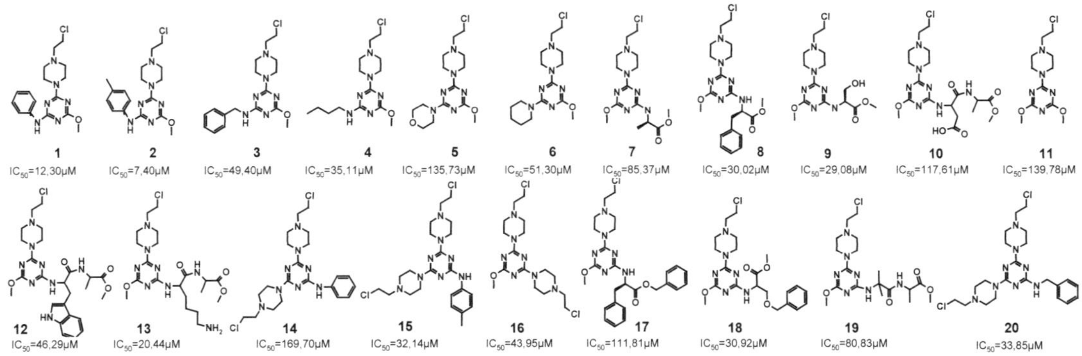

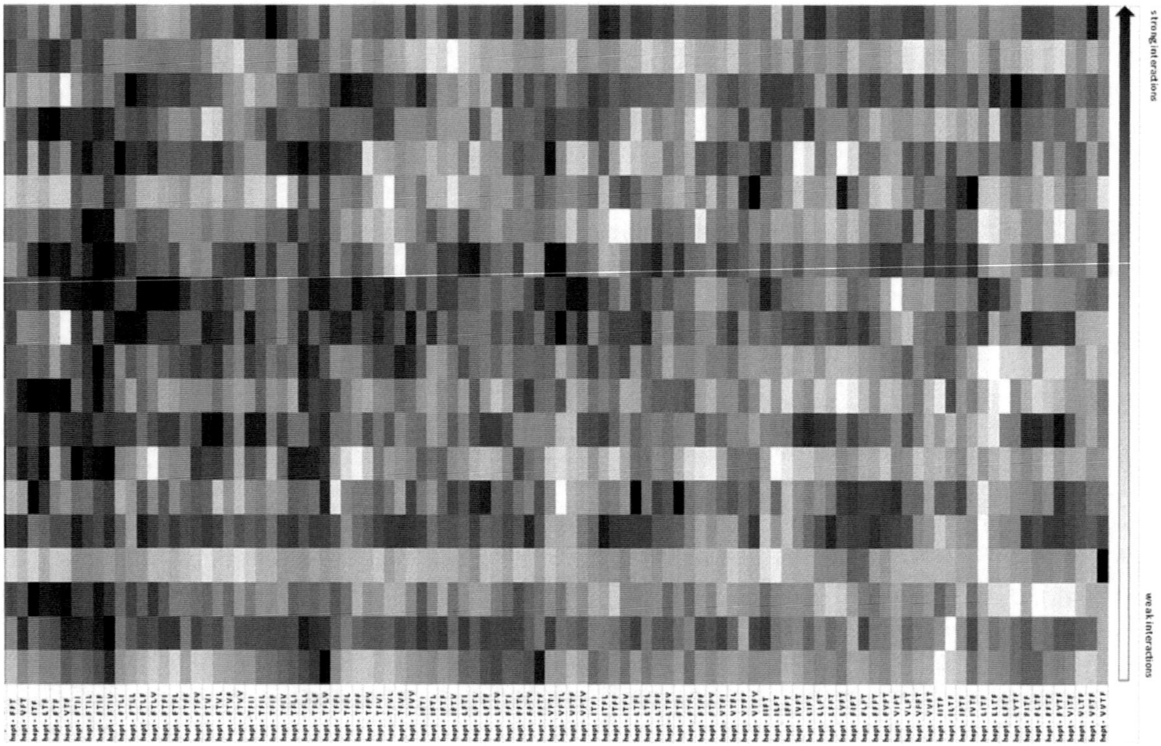

The nonspecific interactions with all tested ligands were observed for molecular receptors: hept.-FT-IL, hept.-FTIF, hept.-FTIV, hept.-TFLF, hept.-TFLV. Screening of the structures of molecular receptors strongly interacted only with the most active ligands 1, 2, 4, 8, 9, 13, 15, 18, 20 allowed the selection of the following molecular receptors hept.-FTI, hept.-LTF, hept.-FTIL, hept.-FTIF, hept.-FTIV, hept.-TFLF and hept.-TFLV. The results shown that arrays of molecular receptors might be used as a platform for probing molecular fingerprints of active compounds with ATPase as a molecular target. It appears that arrays of molecular receptors can be a useful as a novel tool in the search for new pharmaceutically active compounds. Their application can simplify and reduce the costs of the initial phase of the search for new drugs.

Acknowledgements

This work was supported by Ministry of Science and Education N N 405 669540

OP
PP I
PP II
PP VI
PP VII
PP VIII
PP IX
PP X
PP XI
PP XII
PP XIII
PP XIV

References

01. Fraczyk, J., Malawska, B., Kaminski, Z.J., J. Comb. Chem., (2009), 11, 446-451.
02. Kolesinska, B., Rozniakowski, K.K., Fraczyk, J., Relich, I., Papini, A.M., Kaminski, Z.J., Eur. J. Org. Chem., (2015), 401–408.
03. Kolesinska, B., Barszcz, K., Drozdowska, D., Wietrzyk, J., Switalska, M., Kaminski, Z.J.,J. Enzyme Inhib. Med. Chem., (2012), 27, 619-627.
04. Jiang, R., Taly, A., Grutter, T., Trends Biochem. Sci., (2013), 38 20–29.

SYNTHESIS OF MINIGASTRIN-POLYMER CONJUGATES USING MIXED HYDRAZONE/OXIME LIGATION STRATEGY

Maria V. Leko, Pavel S. Chelushkin, Ksenia V. Polyanichko, Marina Yu. Dorosh, Sergey V. Burov
Institute of Macromolecular Compounds RAS, Bolshoi pr. 31, St. Petersburg, 199004, Russia

Introduction

It is known that many types of carcinomas over-express receptors of regulatory peptides. Recently it was shown that minigastrin conjugates with chelating agents are useful for SPECT imaging of CCK-B/gastrin receptor-expressing tumors, including medullary thyroid carcinomas and small cell lung cancers [1]. The diagnostic value of minigastrin analogs can be increased by their conjugation with polymeric carrier resulted in gradual accumulation in tumor tissue (EPR effect). Due to the ability to bind paramagnetic ions of radioactive isotopes these conjugates can be useful both for diagnostics and treatment of oncological diseases (theranostics). Here we describe synthesis of minigastrin analogs containing different N-terminal functional groups and their utility for the conjugation with polymeric chelating agents using hydrazone or oxime ligation strategy.

Results and Discussion

For targeted delivery of radioactive isotopes to cancer cells we have chosen vinylpyrrolidone copolymer with acrolein (VP-Ac) as promising variant of carrier molecule due to its biocompatibility, high solubility in water and simplicity of conjugate synthesis. The literature data show that DOTA is preferable chelating agent owing to the high stability of radiometal-chelate complexes as compared to DTPA and related structures. In order to increase the polymer payload and avoid deprotection problems, we synthesized dendrons with 4 DOTA molecules and reactive hydrazo group. Previously we have shown utility of these building blocks for the modification of VP-Ac copolymers by hydrazone ligation [2].

For the attachment of peptide ligand a set of minigastrin analogs containing N-terminal hydrazide or aminooxy group was synthesized using Rink Amide-MBHA or Rink Amide-ChemMatrix resin (Fig. 1). The presence of N-terminal glutamic acid repeats leads to some problems both at the stage of peptide bond formation and Fmoc group removal. Although coupling efficiency was essentially higher in the case of ChemMatrix resin, subsequent incomplete deprotection in standard conditions resulted in formation of deletion products. Fortunately, these impurities can be completely removed in the course of RP-HPLC purification in 0.1% AcOH buffer.

R-NH-Leu-Glu-Glu-Glu-Glu-Glu-Ala-Tyr-Gly-Trp-Met-Asp-Phe-NH_2

R = -CO-CH_2-O-NH_2; -CO-CH_2-N_2H_3; -CO-$(CH_2)_2$-CO-N_2H_3

Fig. 1. General structure of minigastrin derivatives.

Another significant side process in the course of minigastrin synthesis is methionine oxidation and subsequent incomplete sulfoxide reduction. Both NH_4J and HS-CH_2-CH_2-OH in TFA or NBu_4Br addition to cleavage cocktail seems to be inefficient. However, complete reduction can be achieved by NH_4J addition to minigastrin in NH_4HCO_3 solution.

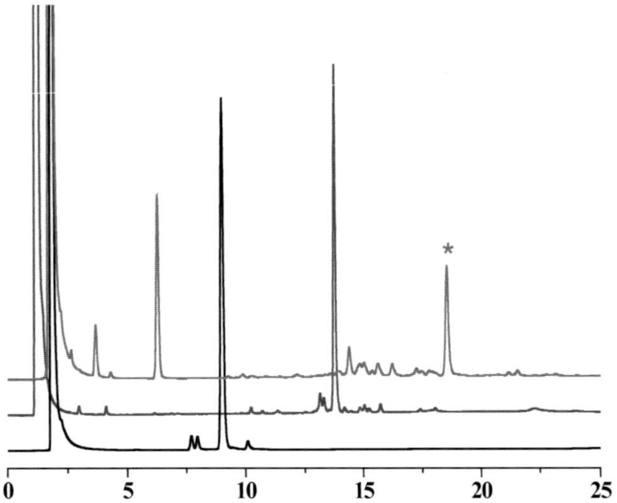

Fig.2. RP HPLC of purified MG (black); crude Aoa-MG (blue) and Boc-Glr-MG (; red) prepared by MG acylation in solution.*

It should be mentioned that attempts of minigastrin modification on solid support by BOC-Aoa-OH, BOC-hydrazide of glutaric acid, Fmoc-Glu(N_2H_2-Fmoc)-OH or BrAc-OH with subsequent treatment with BOC-hydrazide resulted in poor quality of crude peptide. The desired product was prepared by minigastrin cleavage from polymer support, treatment with BOC-Aoa-OSu and deprotection with TFA/DCM (1:1) (Fig. 2).

Peptide-polymer conjugates containing both chelating agent and peptide vector were synthesized by aminooxy-minigastrin attachment to VP-Ac polymer followed by conjugation with dendrons containing 4 DOTA residues (Fig. 3).

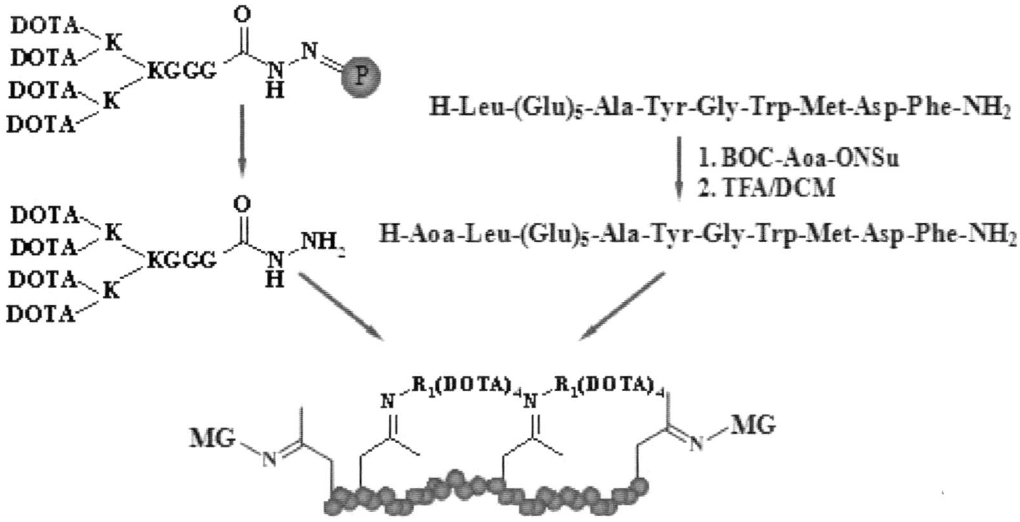

Fig.3. Synthesis of minigastrin conjugates for targeted delivery of radioactive isotopes.

Preliminary experiments have shown the formation of stable complexes between synthesized chelating polymers and radioactive isotopes or paramagnetic ions. The investigation of conjugates utility for SPECT and MRI imaging is now in progress.

References

01. Béhé, M., Becker, W., Gotthardt, M., Angerstein, C., Behr T.M. Eur J Nucl Med Mol Imaging, 30, 1140-1146 (2003).
02. Burov, S., Polyanichko, K., Chelushkin, P., Dorosh, M., Gavrilova, I., Dobrodumov, A., Pokhvoshchev, Yu., Krasikov, V., Panarin E. Dokl. Chem., 466, 18-20, (2016).

NANOBODY CDR3 PEPTIDOMIMETICS AS BREAST CANCER DIAGNOSTICS

Betti Cecilia[1], Krummenacher Sara[1], Martin Charlotte[1], Boeglin Joel[1], Xavier Catarina[2], Devoogdt Nick[2,3], Caveliers Vicky[2], Ballet Steven[1]
[1] Research Group of Organic Chemistry, Vrije Universiteit Brussel, Pleinlaan 2, B-1050 Brussels, Belgium
[2] In Vivo Cellular and Molecular Imaging Lab, Vrije Universiteit Brussel, Laarbeeklaan 103, B-1050 Brussels, Belgium
[3] Cellular and Molecular Immunology Lab, Vrije Universiteit Brussel, Pleinlaan 2, B-1050 Brussels, Belgium

Introduction

Molecular imaging is a non-invasive technique that allows the study of disease-related molecular and cellular events using labeled probes that specifically interact with the biological target of interest. A novel class of promising molecular imaging probes consists of Nanobodies (Nbs).[1] However, the clinical translation of recombinant proteins, such as Nbs, can be problematic due to immune responses or the extremely high costs associated to their development for human use. Hence, our group focuses on the development of CDR3 loop peptidomimetics of promising Nbs. The human epidermal growth factor receptor type 1 (EGFR/HER1) is a receptor that is highly expressed on the cell membrane of many carcinomas. Starting from Nbs that bind to HER1 (Nb7D12/Nb7C12),[2] peptide analogues of their respective CDR3 domains have been synthesized. The crystal structure of Nb7D12-HER1 reveals that the CDR3 domain does not adopt a common secondary structure and that Arg[30] of the CDR1 domain plays a key role in the binding process.[2] To improve the affinity of linear peptide analogues, cyclic CDR3 peptidomimetics bearing a lactam/triazole bridge between residues 105 and 111 were designed. In addition, Gly[101] was replaced by Arg.

Results and Discussion

In order to improve the binding affinity of the previously prepared linear CDR3 peptidomimetics, synthesis of two cyclic peptides were attempted. Based on the crystal structure of Nb7D12-HER1 both the lactam and triazole bridges were inserted between residues 105 and 111 (Fig. A).

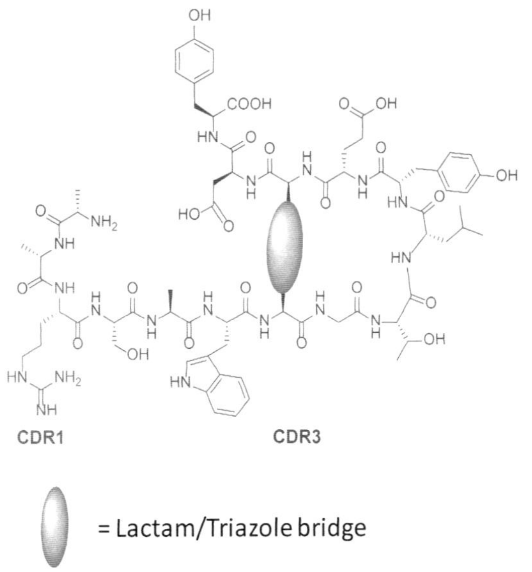

Figure A: Proposed cyclization strategies for discontinuous CDR1/CDR3 loop

The synthesis of the lactam analogue of the CDR3 peptidomimetic was tested on resin and in solution. For both strategies, the linear sequence was synthesized via classic SPPS. Following Allyl/Alloc deprotection, the lactamization was investigated on resin using different coupling reagents but none allowed to obtain the desired cyclic peptide. Next, the lactamization was performed in solution, in presence of a fluorinated solvent, which resulted in a complex mixture containing only a small amount of cyclic peptide. Unfortunately, isolation of the desired peptidomimetic was not successful.
Several synthetic strategies were investigated to afford the triazole analogue of the CDR3 peptidomimetic (Fig. B). The most straightforward one goes via the preparation of the linear sequence by classic SPPS (Fig. B, strategy 1), followed by on resin Huisgen cycloaddition. Despite the different resins (incl. Rink Amide Resin, with high or low loading) and conditions used (Table A, entries 1-4), formation of the triazole was never observed.
Alternatively, the click conditions (Table A, entry 2) were applied to link together two separate segments of the desired peptidomimetic (Fig. B, strategy 2). Therefore, half of the fully protected peptide containing the azido-norleucine residue was added to the resin on which the alkyne-containing segment of the fully protected peptide was attached. Again, none of the investigated conditions gave the desired cycloaddition.

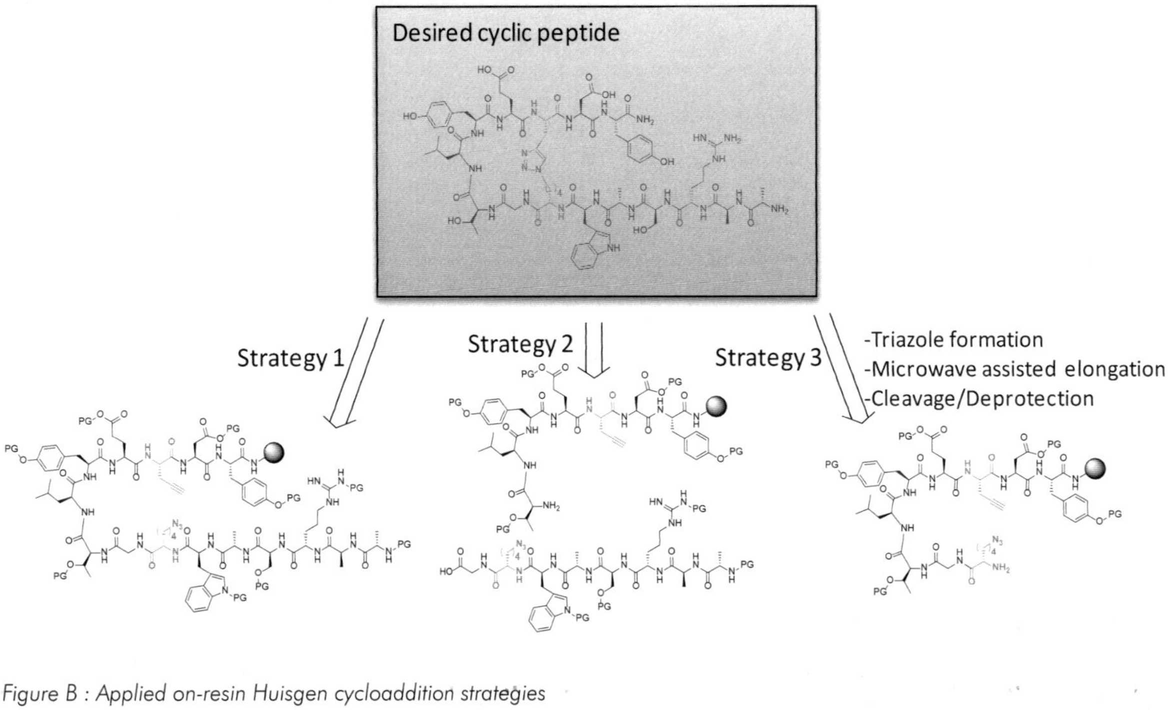

Figure B : Applied on-resin Huisgen cycloaddition strategies

Entry	1	2	3	4
Huisgen cycloaddition conditions[3]	Sodium ascorbate (2eq); CuI (2eq); DIPEA (3eq); DMF (or 2,6 lutidine); r.t.	CuBr (20eq); DIPEA (12eq); DMF:H_2O (9:1) (or THF:H_2O); r.t.	CuBr (12eq); DIPEA (20eq); DMF:THF (1:1); 35°C	CuBr (12eq); DIPEA (6eq); TBTA (0.5eq); DMF:H_2O (9:1) (or tBuOH:H_2O or 2,6 lutidine); r.t.

Table A : Huisgen cycloaddition conditions used

To verify if the Huisgen cycloaddition conditions are efficient only the azido-norleucine residue was added to the resin where the alkyne-containing part of the peptide was attached. CuBr in presence of DIPEA (Table A, entry 2) gave the desired triazole in 4h. In order to determine the length of the peptide which can be linked via a triazole formation to the peptide on resin, fully protected peptides of increasing length (from 1 up to 7 residues) were synthesized and the Huisgen cycloaddition was performed. Interestingly the click reaction worked till 5 residues. Addition of the Arg(Pbf) residue to the sequence completely blocked the reaction.
Gratifyingly, the desired peptidomimetic was obtained via a three step approach (Fig. B, strategy 3). The linear sequence, up to the azido-Nle residue, was synthesized via classic SPPS, followed by on resin Huisgen cycloaddition and final elongation. Note that the click reaction needed 5 days and the final elongation required microwave-assisted peptide synthesis. It is noteworthy to precise that strategies 1 and 2 were also tested in solution without success.

Conclusion

The triazole-bridged cyclic mimic of the discontinuous CDR1/CDR3 loops was successfully synthesized via on resin Huisgen cycloaddition followed by the completion of the sequence using microwave-assisted conditions. In order to evaluate the in vitro binding of the peptidomimetic, the addition of a spacer (such as the 6-aminohexanoic acid) and a chelator (DOTA) at the N-terminus, will be required.[4]

Acknowledgments

We thank the Research Foundation Flanders (FWO Vlaanderen) and the Strategic Research Program – Growth funding of the VUB for the financial support.

References

01. S. Muyldermans et al. Anu. Rev. Biochem., 2013, 83, 1-23
02. K.R. Schmitz et al. Structure, 2013, 21, 1214-1224
03. J.E. Hein et al. Chem. Rev. Soc., 2010, 39, 1302-1315
04. S.R. Kumar et al. Clin. Cancer Res., 2007, 13, 6070-6079

DETECTION OF PROTEASE ACTIVITY USING INTRAMOLECULAR EXCIMER FORMING BISPYRENE PEPTIDE SUBSTRATES

Daisuke Sato, Takuya Kondo, Tamaki Kato
Graduate School of Life Science and Systems Engineering, Kyushu Institute of Technology, 2-4 Hibikino, Wakamatsu-ku, Kitakyushu 808-0196, Japan

PP XI – 347

Pyrene has been exploited for the detection and imaging of protein and nucleic acid targets because pyrene has two unique fluorescence characteristics, which display an ensemble of monomer fluorescence emission peaks (375–405 nm) and an excimer band (460–480 nm) when two pyrene molecules are spatially proximal. Although several protease probes using pyrene have been reported,[1,2] the applications of pyrene monomer/excimer signaling to the detection of protease activity are still rare.[3]

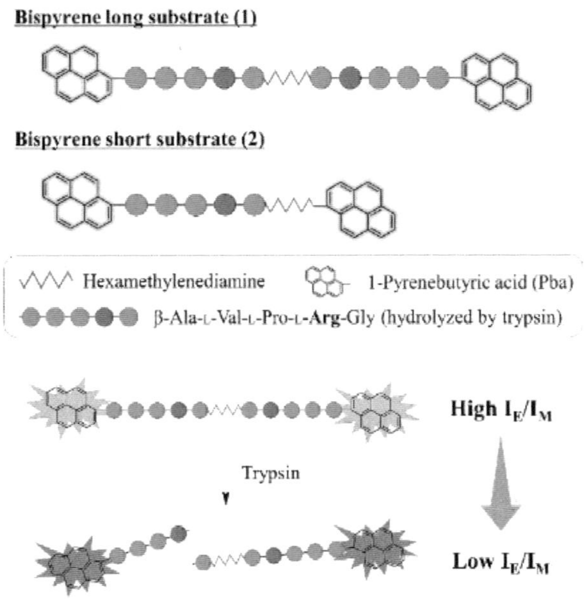

Figure 1. Overview of the bispyrene substrates and their detection mechanism.

In this study, we designed and synthesized two types of pyrene monomer/excimer-based peptide substrates for the detection of trypsin activity: long-type substrate (1) and short-type substrate (2) (Fig. 1). The long-type substrate 1 is comprised of two 1-pyrenebutyric acid (Pba)-linked substrate peptides on both edges of hexamethylenediamine. The short-type substrate 2 is comprised of one Pba-linked substrate peptide and Pba on each edge of hexamethylenediamine respectively. Proximate two pyrene moieties forms excited-state dimers in the substrates, and the substrates emit excimer fluorescence. After tryptic cleavage, these pyrene excimer formations dissociates, and the monomer fluorescence increases as the excimer fluorescence decreases. Hence, the change of monomer/excimer fluorescence ratio (I_E/I_M) allows for the detection of trypsin activity.
For the synthesis of 1, Pba- -Ala-L-Val-L-Pro-L-Arg(Pbf)-Gly-OH synthesized by standard Fmoc-solid-phase peptide synthesis was coupled with both edges of hexamethylenediamine, followed by the

deprotection of Pbf groups. For the synthesis of 2, N-Boc-hexamethylenediamine was coupled with Pba. Pba- -Ala-L-Val-L-Pro-L-Arg(Pbf)-Gly-OH was then linked after the deprotection of Boc group from the previously obtained compound. Finally, Pbf group was deprotected.

Initially, we verified that the intramolecular excimer formation of 1 and 2 was occurred. A total of 2 μM of 1 and 2 were prepared in 50 mM Tris-HCl buffer (pH 8.0) containing 150 mM NaCl, 1 mM $CaCl_2$, and 0.05% Tween 20. Similarly, 4 μM of a Pba solution was also prepared in the buffer as a control sample. All of the solutions contained 4 μM of the pyrene moieties. According to the fluorescence spectra upon excitation at 344 nm prior to the addition of trypsin, the excimer fluorescence (470 nm)/monomer fluorescence (394 nm) ratio (I_E/I_M) was calculated. The substrates displayed higher I_E/I_M with 0.237 for 1 and 0.386 for 2 compared with 0.041 of Pba. Hence, it was demonstrated that 1 and 2 formed intramolecular excimer. Substrate 2 had higher I_E/I_M than 1 because 2 had shorter peptide chain.

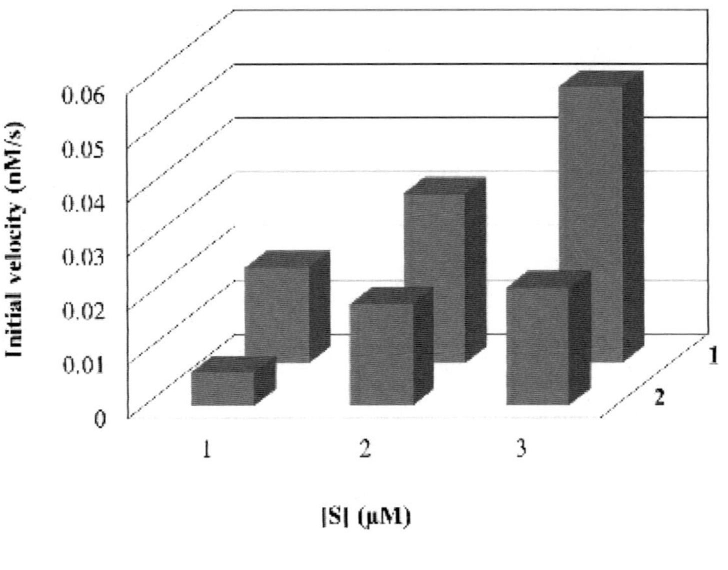

Figure 2. Initial velocities of trypsin hydrolysis.

Next, the increase in the I_E/I_M upon excitation at 344 nm of different concentrations of 1 and 2 during trypsin cleavage was monitored, and the initial velocities were calculated. The final concentration of 1 and 2 was adjusted to 1, 2, and 3 μM, and the final concentration of trypsin was adjusted to 10 nM for 1 or 100 nM for 2 with the buffer. Substrate 1 showed the faster reaction rate than 2 (Fig. 2). One explanation was that 1 had two cleavage sites for trypsin compared with one site in 2, and 1 was the less sterically hindered because of its longer peptide chain. This indicated that the long-type substrate 1 could be preferable to the short-type substrate 2 for the detection of protease activity.

References

01. T. Ahn, J.-S. Kim, H.-I. Choi, C.-H. Yun, Anal. Biochem., 2002, 306, 247–251.
02. N. Xu, Y. Li, H.-W. Li, Y. Wu, Chem. Lett., 2013, 42, 1528–1530.
03. M. Fischbach, U. Resch-Genger, O. Seitz, Angew. Chem. Int. Ed., 2014, 53, 11955–11959.

SYNTHETIC PEPTIDE VACCINE AGAINST HEPATITIS C: THE EFFECT OF ADJUVANTS ON THE PEPTIDE ANTIGEN IMMUNOGENICITY AND ON B-EPITOPE SPECIFICITY OF PRODUCED ANTIBODIES

E.A. Egorova[1], M. V. Melnikova[1,2], A. V. Talanova[1], V. N. Kashirtseva[1], L. V. Kostryukova[1], R. I. Ataullakhanov[3], T.M. Melnikova[2], E. F. Kolesanova[1]
[1] Institute of Biomedical Chemistry , Moscow, Russia;
[2] N.D. Zelinsky Institute of Organic Chemistry Russian Academy of Sciences, Moscow, Russia;
3National Research Center "Institute of Immunology" Federal Medical Biological Agency, Moscow, Russia

PP XII – 352

Summary. Two synthetic peptide antigens composed of two different conserved HCV E2 protein fragments as putative B-epitopes and the same conserved E2 fragment as T-helper epitope were synthesized and checked for their antigenicity with different carrier/adjuvant formulations in mice. The highest immunogenicity with regard to the formation of anti-HCV envelope protein antibodies was achieved for the formulations composed of the peptide antigen-Immunomaks™ conjugates.

Introduction.

Synthetic peptide vaccines have some evident advantages over vaccines based on whole recombinant proteins, killed or alive viruses and microorganisms or their subunit preparations [1]. With regard to the anti-hepatitis C vaccine development, the use of synthetic peptide immunogens allows bypassing the problem of viral main protein antigen variability in different hepatitis C virus (HCV) isolates [2] and of poor immunogenicity of sites that can induce potential virus-neutralizing antibodies [3,4]. However, peptides usually do not demonstrate high immunogenicity by themselves. Specific, highly efficient adjuvants and macromolecular carriers are necessary for enhancing peptide antigen stability and immunogenicity. Our work is devoted to the development of carrier/adjuvant formulations for recently designed peptide constructs composed of conserved putative B- and T-helper epitopes from HCV envelope protein E2 [5].

Experimental.

Peptides YPYRLWHYPGGSTGLIHLHQNIVDVQYLYG-amide (CR4-GG-CR5, or CR4-CR5) and STGLIHLHQNIVDVQYLYGGCPTDCFRKHPEATYS-amide (CR5-GG-CR3, or CR5-CR3) were prepared by SPPS on Rink amide-Wang resin by FastMoc procedure on 433A synthesizer (Applied Biosystems), purified by HPLC and analyzed by MS and MS/MS. Conjugation of peptides with immunomodulator Immunomaks™ (IM) (proteoglycan from potato sprouts [6]), was achieved via peptide N-terminal amino group with the use of squaric acid diethyl ester. SPLT80nanoem (average particle diameter 20 nm) were prepared by microfluidization. Mice (10-12 animals in each group) were immunized three times with 14-day intervals and bled 7 days after the last injection. Titers of anti-peptide and anti-HCV E2 and E1E2 antibodies were determined by ELISA [4].

Results

Earlier designed peptide constructs prepared from HCV E2 conserved fragment showed immunogenicity in mice and rats with Freund's adjuvant [5]. Since this adjuvant is not suitable for human use, we have studied the effect of known and clinically applicable immunomodulators Immunomaks™ and squalene (as SPLT80nanoem) on the immunogenicity of peptide constructs. Results of peptide antigen formulation testing for immunogenicity are shown in Table 1.

Table 1. Parameters of murine immune responses to peptide HCV-derived peptide antigens in different formulations.

Peptide, dosage	Adjuvant/carrier formulation	% mice responded	Antibody titers		
			Anti-peptide	Anti-E2	Anti-E1E2
CR4-CR5, 50 mg	PBS	0	-	-	-
CR4-CR5, 5 mg	SPLT80nanoem in PBS	0	-	-	-
CR4-CR5, 50 mg	IM in PBS	0	-	-	-
CR4-CR5,0.5 mg	IM-CR4-CR5 in PBS	67	1:464	1:22	1:25
CR4-CR5,0.2 mg	IM-CR4-CR5 + SPLT80nanoem in PBS	83	1:211	1:23	1:31
CR5-CR3, 50 mg	PBS	83	1:707	1:28	1:65
CR5-CR3, 5 mg	PBS	0	-	-	-
CR5-CR3, 5 mg	SPLT80nanoem in PBS	0	-	-	-
CR5-CR3, 50 mg	IM in PBS	83	1:346	1:41	1:85
CR5-CR3,0.5 mg	IM-CR5-CR3 in PBS	75	1:293	1:100	1:163
CR5-CR3,0.2 mg	IM-CR5-CR3 + SPLT80nanoem in PBS	83	1:447	1:70	1:96

CR5-CR3 formulations with Immunomaks^{TM} caused the formation of CR3-specific antibodies, while no CR4- or CR5-specific antibodies were detected after CR4-CR5 immunizations. SPLT80nanoem alone as well as mechanical mixing of peptides with Immunomaks^{TM} did not enhance the peptide immunogenicity. The best results with regard to the titers of anti-HCV E2 and E1E2 antibodies and the percentage of responded animals were achieved for peptide conjugates with Immunomaks^{TM}. It was a real immunostimulation rather than an effect of multimerization, since CR5-CR3 conjugated to high molecular weight dextran did not cause immune responses in mice. Hence Immunomaks^{TM} can serve as a clinically applicable both carrier and adjuvant for peptide antigens in anti-HCV vaccine.

Acknowledgments
Research was performed in the frame of the State task for fundamental research development, theme No. 0518-2014-0003 and supported in part by the State Contract No. 14N08.12.0025 from the Russian Ministry of Education and Science. Authors thank Dr. Jean Dubuisson (Institut de biologie de Lille) for the kind donation of HCV E2 and E1E2 proteins.

References

01. Moisa, A.A., Kolesanova, E.F. Biochemistry (Moscow) Suppl. B. Biomed. Chem., 2010, 4, 331.
02. Sobolev, B.N., et al. J. Viral Hepat., 2000, 7, 368.
03. Olenina, L.V. et al. J. Viral Hepat. 2002, 9, 174.
04. Kuzmina, T.I., et al., Biochemistry (Moscow) Suppl. B. Biomed. Chem., 2009, 3, 177.
05. Kolesanova, E.F. et al. Biomed. Khim. 2015, 62,
06. Ataullakhanov, R.I. et al. Immunologiya, 2005, 26, 111.

THE CHALLENGE OF COMPLEXITY: PEPTIDE TOOLS FOR THE DEVELOPMENT OF IMMUNOTHERAPIES

Karsten Schnatbaum, Tobias Knaute, Johannes Zerweck, Maren Eckey, Pavlo Holenya, Florian Kern, Holger Wenschuh, Ulf Reimer*
JPT Peptide Technologies GmbH, 12489 Berlin, Germany

Abstract

To address the challenge of sequence diversity in immunotherapy, a peptide based workflow was established that combines bioinformatic algorithms, high throughput peptide synthesis, innovative peptide presentation approaches and synergistic assay formats

Introduction

Immunotherapy is gaining attention as promising approach to fight cancer as well as infectious diseases. [1]A major challenge in immunotherapy is the selection of optimal antigen sequences to derive efficient therapeutic agents. Among others, this task is hampered by sequence diversity in the target organisms caused by isoforms, splice variants, polymorphisms, mutations, and PTMs.

Methods

We address the challenge of sequence diversity by a peptide library based workflow that combines: Improved bioinformatic algorithms: Algorithms for library design were developed. These are based on the scoring of all possible peptides according to their frequency of occurrence across all sequences to provide the most homogenous overall coverage.[2]The result of the so- called Ultra Concept is illustrated in Figure 1 for the HIV Nef protein. The majority of the 3903 known sequences is covered by only 150 peptides (for antigen specific T-cell stimulation with peptide pools) or 667 peptides (for humoral immune monitoring with peptide microarrays).

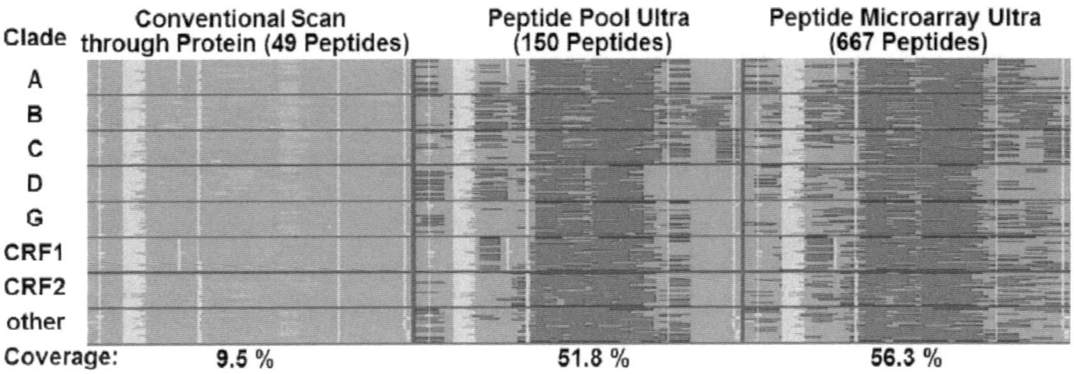

Figure 1: Sequence coverage by different HIV Nef libraries. Red/orange: Sequence parts which are covered by the respective peptide library.

High throughput peptide synthesis, peptide presentation and synergistic assay formats: For B-cell epitope discovery and humoral immune monitoring high density peptide microarrays represent an efficient technology that accommodates vast numbers of sequence
variants and PTMs. Peptides are synthesized by SPOT synthesis[3] and re-immobilized on microarrays[4] in a clean-room environment. Readout after incubation with plasma samples is usually performed by fluorescently labeled secondary antibodies.
For T-cell epitope discovery and cellular immune monitoring peptides selected through the Ultra Concept can be synthesized and presented as individual peptides, matrix pools or antigen spanning pools for application in T-cell assays such as Elispot (Scheme 2).

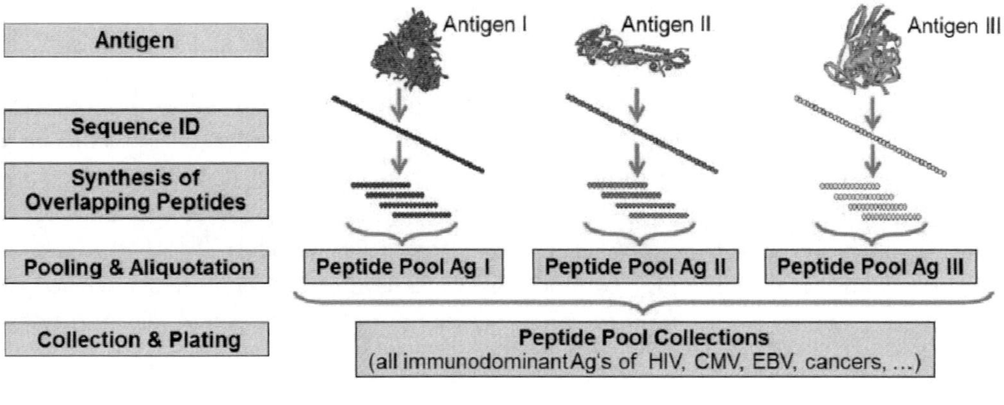

Scheme 2: Preparation of peptide pools to study cellular immunity.

Application Examples

Humoral immune response: To provide guidance for vaccine development, plasma samples from several HIV/SIV vaccination studies were examined with peptide microarrays.[2,5] The analysis of serum samples from the first successful HIV vaccination trial (RV144 trial) with peptide microarrays showed that plasma levels of IgG directed towards the V2 loop of gp120 correlated with a reduced risk of infection.[5a] In a follow-up study, an Ad26 vector-based vaccine stimulated a dose dependent response against the V2 loop.[5b]

Cellular immune response: To increase stimulating efficiency for antigen specific T-cell responses (HLA independence, reduction of assay numbers, sample volume requirements), antigen-spanning overlapping peptide pools were developed. Examples include pools for the ex vivo generation of T cells for HIV[6] and broad-spectrum antiviral (AdV, EBV, CMV, BKV, HHV6)[7] treatment, where a 94% virological and clinical response rate was achieved.

References

01. e.g. (a) Vormehr, M. et al. J. Immunol. Res. 2015, 595363. (b) Kranz, L. M. et al. Nature 2016, 534, 396-401.
02. Stephenson, K. et al. J. Immunol. Meth. 2015, 416, 105-123. (3) Wenschuh, H. et al. Biopolymers 2000, 55, 188-206.
03. (a) Schutkowski, M. et al. Angew. Chem. Int. Ed. Engl. 2004, 43, 2671-2674. (b) Masch, A. et al. Methods Mol. Biol. 2010, 669, 161-172.
04. (a) Gottardo, R. et al. PLoS ONE 2013, 8, e75665. (b) Barouch, D. H. et al. J. Infect. Dis. 2013, 207, 248-256. (c) Barouch, D. H. et al. Science 2015, 349, 320-324. (6) Lam, S. et al. Mol. Ther. 2015, 23, 387-395.
05. Papadopoulou, A. et al. Sci. Transl. Med. 2014, 6, 242ra83.

INNOVATIVE SYNTHESIS AND CD CONFORMATIONAL ANALYSIS OF LIRAGLUTIDE

Ivan Guryanov[1,2], Alex Bondesan[1], Dario Visentini[1], Andrea Orlandin[3], Barbara Biondi[3], Claudio Toniolo[3], Fernando Formaggio[3*], Antonio Ricci[1], Jacopo Zanon[1], Walter Cabri[1]
[1] Fresenius Kabi Anti-Infectives Srl, 45010 Villadose (RO), Italy;
[2] Institute of Chemistry, St. Petersburg State University, 198504 St. Petersburg, Russia;
[3] ICB, Padova Unit, CNR, Department of Chemistry, University of Padova,
35131 Padova, Italy, e-mail: fernando.formaggio@unipd.it

PP XIII – 359

Introduction

Liraglutide is a palmitoylated glucagon-like peptide-1 (GLP-1) analog used for the treatment of type II diabetes. The amino acid sequence of liraglutide (Figure 1) displays 97% identity to that of the native hormone, although the presence of the palmitoyl moiety plays a crucial role in its in vivo behavior [1,2].

H-His-Ala-Glu-Gly-Thr-Phe-Thr-Ser-Asp-Val-Ser-Ser-Tyr-Leu-Glu-Gly-Gln-Ala-Ala-Lys-Glu-Phe-Ile-Ala-Trp-Leu-Val-Arg-Gly-Arg-Gly-OH

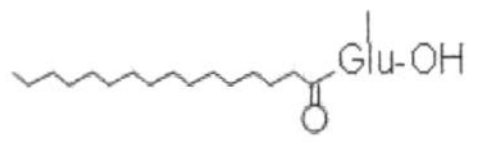

Fig. 1. The amino acid sequence of liraglutide.

The original method of liraglutide synthesis encompasses the preparation of the peptide main-chain by a recombinant approach, followed by its functionalization with N-palmitoyl glutamic acid tert-butyl ester (WO 9808871). Here, we report a new synthetic approach, exclusively based on chemical methods, and the results of a CD conformational analysis in membrane mimetic environments.

Results and Discussion

We propose a new approach for the chemical synthesis of liraglutide. The key step is the synthesis in solution of the Lys/Glu building block, Fmoc-Lys-(Pal-Glu-OtBu)-OH, in which Lys and Glu are linked through their side chains and Glu is N^{α}-palmitoylated. This dipeptide is then inserted into the peptide sequence on solid phase (Figure 2) [3]. As liraglutide is obtained with great purity and high yield, this approach can be particularly attractive for the industrial production.
Lipidation of liraglutide and its aggregation seem to be the main reasons for its enhanced proteolytic resistance [4]. Recently, A CD and light scattering investigation revealed that liraglutide forms reversible aggregates in buffered aqueous solutions, at pH values lower than 6.9 [4]. Therefore, we decided to investigate this behavior in the presence of micelles and phospholipid bilayers, as these environments better mimic that of a living organism. We found that in SDS micelles, and in DOPC (1,2-dioleoyl-sn-glycero-3-phosphocholine) and POPC (1-palmitoyl-2-oleoyl-sn-glycero-3-phosphocholine) vesicles as well, liraglutide is largely α-helical (Figure 3). However, at variance with ref. [4], we did not observe a time-dependent conformational transition, leading to the formation of liraglutide aggregates. The shape of the CD curves did not change even after 4 days. In particular, the 222 nm/208 nm ellipticity ratio, a probe of peptide aggregation, did not increase [3]. We are inclined to ascribe the absence of time-dependent, conformational transitions to the presence of membrane mimetic environments. Thus, in addition to the already reported aggregation [4] and albumin interaction [2], this new finding should be taken into consideration when evaluating the reasons for the prolonged enzymatic resistance in vivo of liraglutide.

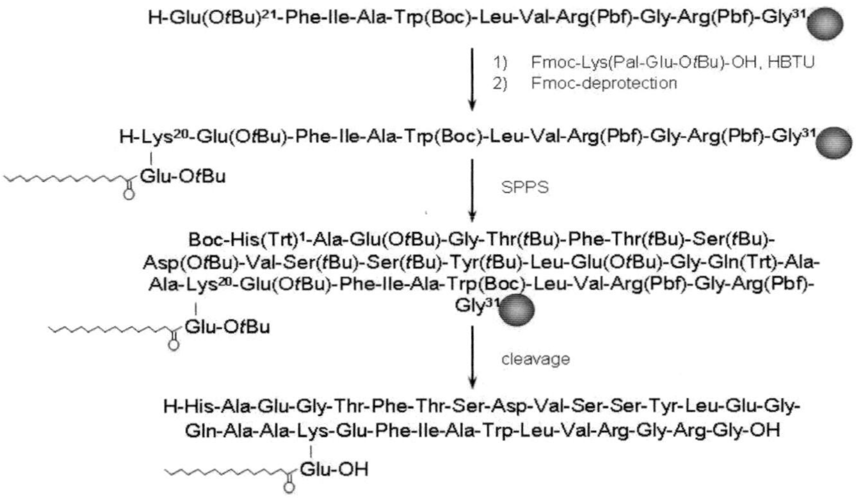

Fig. 2. Scheme of liraglutide synthesis.

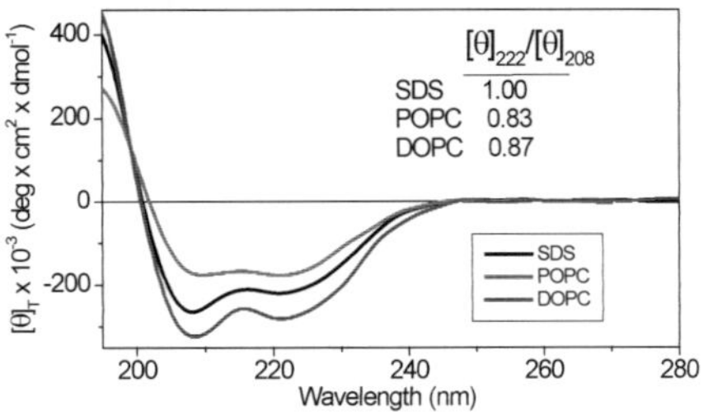

Fig. 3. CD spectra of liraglutide in SDS, POPC, and DOPC solutions (pH 6.7).

References

01. Cada, D. J., Levien T. L., Baker D. E. Hosp. Pharm. 45, 552-563 (2010).
02. Lau, J., Bloch, P., Schäffer, L., Pettersson, I., Spetzler, J., Kofoed J., Madsen, K., Knudsen, L. B., McGuire, J., Steensgaard, D. B., Strauss, M. H., Gram, D. X., Knudsen, S. M., Nielsen, F. S., Thygesen, P., Reedtz-Runge, S., Kruse, T. J. Med. Chem. 58, 7370-7380 (2015).
03. Guryanov, I., Bondesan, A., Visentini, D., Orlandin, A., Biondi, B., Toniolo, C., Formaggio, F., Ricci, A., Zanon, J., Cabri, W. J. Pept. Sci. 22, 471-479 (2016).
04. Wang, Y., Lomakin, A., Kanai, S., Alex, R., Benedek, G. B. Mol. Pharmaceutics 12, 411-419 (2015).

EXPLORING THE MOLECULAR MECHANISM OF HIV-1/GBV-C VIRAL INTERFERENCE USING PEPTIDES DERIVED FROM VIRAL PROTEINS

Rebecca Hoffmann[1], Johanna Schaubächer[2], Barbara Schmidt[2] and Jutta Eichler[1]
[1] Department of Chemistry and Pharmacy, University of Erlangen-Nürnberg, Schuhstraße 19, 91052 Erlangen
[2] Institute of Microbiology, University of Regensburg, Franz-Josef-Strauß-Allee 11, 93053 Regensburg

PP XIV - 369

The human blood borne, non-pathogenic virus GBV-C replicates primarily in lymphocytes [1]. Co-infection of HIV-1 positive individuals with GBV-C has been reported to be beneficial for the patients [2]. This phenomenon, termed viral interference, is thought to be based on the interaction of the GBV-C envelope protein 2 (E2) with the envelope protein (ENV) of HIV-1 (Figure 1, A). Peptides derived from the Nterminus of GBVC E2 were found to inhibit the entry of HIV1 to its host cells

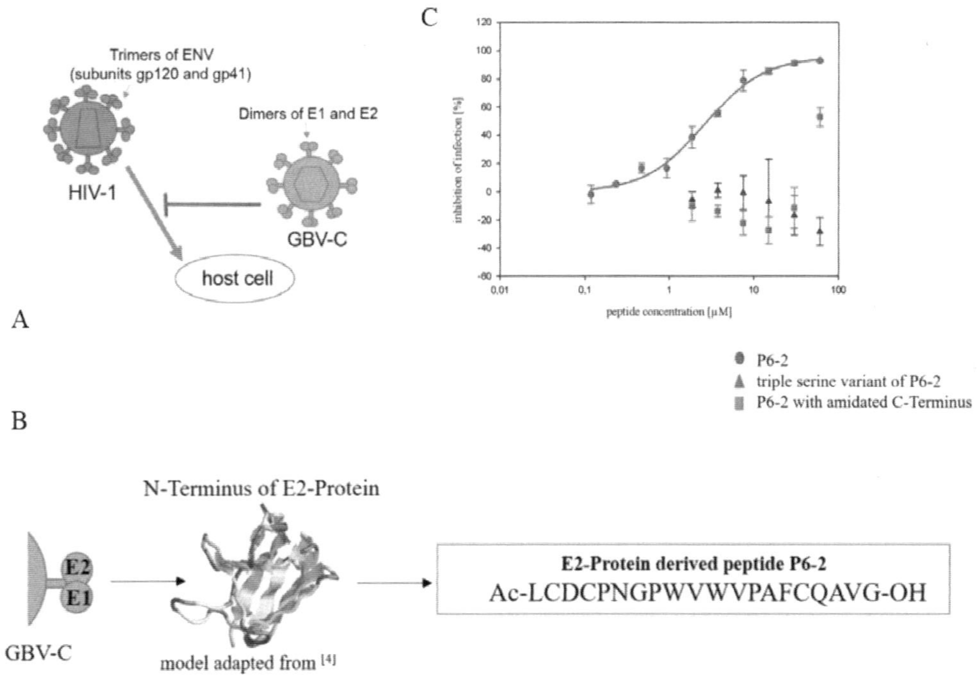

Figure 1: A: Viral interference of GBV-C with HIV-1 B: E2-protein and the derived peptide sequence P6-2. C: Effect of P6-2 and variants on HIV-1 infection.

The peptide P6-2, presenting residues 45-64 of the GBV-C protein E2, inhibits HIV-1 infection at low micromolar concentrations (Figure 1C). Previous studies had shown that the cysteine residues in P6-2 are essential, since their collective replacement with serine completely abrogates its HIV-1 inhibitory activity [4]. Using a range of truncated and substitution variants, we have now further characterized the molecular determinants of the virus neutralizing activity of P6-2. We synthesized and tested several P6-2 variants, in which one, two, or all three cysteine residues were replaced by serine or methionine, resulting in the identification of Cys60 as the most important of the three cysteine residues. Additionally, we found that a negative charge at the peptide's C-terminus is required

for virus neutralization, as P6-2 with an amidated C-terminus is completely inactive (Figure 1C). Activity of the amidated peptide can be restored by replacing the C-terminal glycine residue of P6-2 with aspartic acid, re-introducing the required negative net charge.
Furthermore, we identified the hydrophobic tetrapeptide core (WVWV) of P6-2 as an important element for the HIV-1 inhibitory activity. Individual replacement of these residues with alanine resulted in 10-fold to 100-fold reduction in activity, in which the effects of tryptophan to alanine exchanges were even more dramatic than the valine replacements. Finally, using C-terminally truncated variants of P6-2, we found that omission of the three Cterminal residues is well tolerated, as demonstrated by an unchanged, or even higher, HIV1 inhibitory activity of the truncated peptides.
In summary, our results demonstrate the importance of the cysteine residues of P6-2, in particular Cys60, for its HIV-1 inhibitory activity. Furthermore, the hydrophobic tetrapeptide core, in particular the two tryptophan residues, as well as a C-terminal negative charge, are essential for the activity (Figure 2). Omission of three C-terminal residues of the peptide, on the other hand, is well tolerated, enabling truncation of P6-2 without loss of activity. Based on these results, we aim at improving the HIV-1 inhibitory activity of P6-2.

Figure 2: Schematic presentation of the P6-2 sequence. Residues and parts of the molecule essential for HIV-1 neutralization are highlighted.

References

01. J. Xiang, J. H. McLinden, T. M. Kaufman, E. L. Mohr, N. Bhattarai, Q. Chang, J. T. Stapleton, Virology 2012, 430, 53-62.
02. M. T. Giret, E. G. Kallas, Current HIV/AIDS reports 2012, 9, 26-33. [3]
03. Y. Koedel, K. Eissmann, H. Wend, B. Fleckenstein, H. Reil, Journal of virology 2011, 85, 7037-7047.
04. K. Eissmann, S. Mueller, H. Sticht, S. Jung, P. Zou, S. Jiang, A. Gross, J. Eichler, B. Fleckenstein, H. Reil, PloS one 2013, 8, e54452.

Katarzyna Kalejta[1], Marta Spodzieja[1], Daniel E. Speiser[4], Laurent Derre[2], Justyna Iwaszkiewicz[3], Vincent Zoete[3], Olivier Michielin[3], Sylwia Rodziewicz-Motowidło[1]
[1] University of Gdansk, Department of Chemistry, Wita Stwosza 63, 80-308 Gdansk, Poland,
[2] Urology Research Unit, Urology department, University Hospital of Lausanne (CHUV), Lausanne, Switzerland.
[3] The Swiss Institute of Bioinformatics, Quartier Sorge, Batiment Genopode, CH-1015 Lausanne, Switzerland.
[4] Ludwig Cancer Research, Department of Oncology, Biopole 3, Rte Corniche 9A, CH-1066 Epalinges, Switzerland.

PP XIV - 371

Introduction

Melanoma is the most serious type of skin cancer. The number of cases worldwide has doubled in the past twenty years [1]. Patients with melanoma often have increased numbers of tumor antigen specific T cells that can be beneficial for patients [2]. Indeed, one of the most promising methods to treat melanoma is immunotherapy that supports activation and function of the patient's T cells. The CD160 protein was identified as a co-inhibitory molecule that binds to the herpesvirus entry mediator (HVEM), a TNF receptor superfamily member. CD160 is expressed on the surface of immune cells, including T, B and NK cells [3]. The HVEM-CD160 complex inhibits $CD4^{+}$ T cell activation [4]. Our research is focused on blocking the interaction between CD160 and HVEM proteins to stimulate immune response. Therefore, we characterize the interaction of both proteins by using affinity chromatography, enzyme-linked immunosorbent assay (ELISA) and mass spectrometry. Both protein fragments engaged in the interaction will be characterized, based on which we can evaluate various possible strategies to block the interaction.

Results

In the first stage of the research we decided to determine fragments of protein CD160 which bind with HVEM protein. The CD160 protein was divided into 10 fragments (20-25 amino acids residues in length, overlapping by 8 amino acid residues) which were synthesized and purified. In peptides cysteine residue was replaced by aminobutyric acid.
Those fragments were used to perform affinity chromatography with HVEM protein. In this experiment each peptide was incubated with HVEM immobilized on microcolumn for 2 hours at room temperature. Three fractions were collected:
1 – supernatant fraction – containing excess of peptide,
2 – last wash fraction – to check if all unbound peptide is removed with buffer,
3 – elution fraction – the affinity-bound peptide is eluted from the microcolumn.
All fractions were analyzed by using MALDI TOF/TOF 5800 (ABSciex).
The other method was the hydrogen deuterium exchange mass spectrometry (HDXMS). Experiments were initiated by mixing of protein CD160 with protein HVEM in PBS in H_2O. The mixture was incubated by 2 hours at 25°C. Then, complex CD160-HVEM was portioned. To each portion was added deuterated buffer and incubated at 25°C for varying times (10 s, 1 min, 5 min, 25 min, 2 h and 24 h). The exchanged was quenched by adding the stop buffer in D_2O (150 mM NaCl, 6M guanidine hydrochloride, 1M TCEP, 2M glycine, pH 2.4) and cooled on ice. In order to enhanced digestion efficiency each portion of CD160-HVEM complex was cleaved by adding Aspergillus Saitoi, incubated 1 minute and immediately injected on an immobilized pepsin column. Results were obtained by using UPLC-MS/MS technique (Synapt G2 HDMS, Waters). The results of deuter-

ation level for each fragment of protein CD160 was calculated with DynamX 2.0 software (Waters). The same experiment was performed for protein CD160 as a control.
Results from affinity chromatography and hydrogen deuterium exchange experiments are presented on Figure 1.

Fig. 1. Blue lines - peptides bind to the protein HVEM, obtained from the affinity chromatography; red lines – peptides bind to the HVEM, obtained from H/D experiments.

Discussion

Results from the affinity chromatography show, that 3 fragments of CD160 protein CD160(39-58), CD160(51-70) and CD160(111-130) bind to HVEM protein (Fig. 1 – blue lines). Similar results were obtained from the hydrogen/deuterium study, which suggest that CD160 protein binds to HVEM protein by using three fragments. The first fragment is located in regions 42-47, second in 56-60 and third in 111-118 of CD160 protein (Fig. 1 – red lines).
In conclusion, the results from both experiments suggest, that binding sites are between 42-60 amino acid residue and 111-118 amino acid residue.
We expect that the modified, earlier identified fragments, responsible for protein-protein interactions, may inhibit CD160-HVEM interactions.

Acknowledgements
Project No. PSPB-070/2010 "Design of BTLA inhibitors as new drugs against melanoma" is financed by a grant from Switzerland through the Swiss Contribution to the enlarged European Union.

References

01. V. Gray-Schopfer, C. Wellbrock, R. Marais, Nature, 2007, 445, 851-857
02. D.M. Pardoll, Nat Rev Cancer, 2012, 12(4), 252-64
03. R. Kojima, M. Kajikawa, M. Shiroishi, K. Maenaka, J Mol Biol., 2011, 413(4), 762-72
04. G. Cai, A. Anumanthan, J.A. Brown, E.A. Greenfield, B. Zhu, G.J. Freeman, Nat Immunol, 2008, 9(2), 176-85

M. Rafalik[1], A. Kołodziejczyk[1], P. Czaplewska[2], K. Dabrowska[3], M. Dadlez[3], S. Rodziewicz-Motowidło[1]
[1] Department of Biomedical Chemistry, Faculty of Chemistry, University of Gdansk, Poland
[2] Laboratory of Mass Spectrometry, Intercollegiate Faculty of Biotechnology, University of Gdansk – Medical University of Gdansk, Poland
[3] Mass Spectrometry Laboratory, Institute of Biochemistry and Biophysics, Polish Academy of Sciences, Warsaw, Poland

PP XIV - 375

Introduction

Human cystatin C (hCC) is a small protein found in all human physiological fluids, belonging to the family of papain-like cysteine proteinases. There are increasing number of evidence to suggest that hCC is involved in many processes related to dimerization, oligomerization and amyloid formation. The processes are directly associated with a number of neurodegenerative diseases such as Alzheimer disease or hereditary cystatin C amyloid angiopathy (HCCAA) [1,2]. The neurodegenerative disorders are gaining importance since they affect the independence and quality of life of aging societies. One of ideas of how to prevent dimerization and amyloid formation is immunotherapy. HCC3 is one from the group of antibodies, which binds hCC and reduces the in vitro formation of cystatin C dimers by 60% [3]. Therefore, identification of binding sites in the hCC-HCC3 complex may allow for a search of effective drugs against HCCAA and for understanding the mechanisms of neurodegenerative diseases.

For the epitope identification, many methods, such as affinity chromatography, the epitope excision and extraction based on MS-assisted partial proteolysis of antigen-antibody complexes, the enzyme-linked immunosorbent assay and hydrogen-deuterium exchange combined with mass spectrometry (HDX MS) were used [4]. Only the comparison of all obtained results may allow to identify the epitope, especially when it seems to be the discontinuous one.

Results and Conclusions

The MS-assisted epitope excision and epitope extraction (digestion with trypsin and endoproteinase Asp-N) followed by affinity chromatography were methods we started our research with. Using digestion with Asp-N two fragments of hCC: hCC(40-64) and hCC(87-118), were identified as probably involved in hCC binding with the HCC3 antibody (Fig.1, A). The results of affinity chromatography for the hCC fragments were compatible with the results of previously mentioned enzymatic methods. ELISA tests for the same hCC fragments led to similar results. It seems that the hCC(40-64) peptide interacts with HCC3 antibody the strongest. The hCC(54-70) fragment and hCC(93-120) peptide (the hCC fragments obtained from the enzymatic methods using trypsin) bind with antibodies very well, even at low concentrations so they seem to be also fragments of the epitope sequence. Unfortunately, these fragments were too long to complete the epitope identification as an epitope consist usually of about 8 – 16 amino acid residues. Therefore, the MS-assisted hydrogen-deuterium exchange was the successive method applied. On the basis of the HDX results obtained until now, it seems that the epitope is located around hCC(52-62) and hCC(100-105) (Fig. 1, B). There is only one substantial difference between the results of HDX and previously applied methods. Surprisingly, HDX method shows that part of hCC alpha helix (the sequence from 17 to

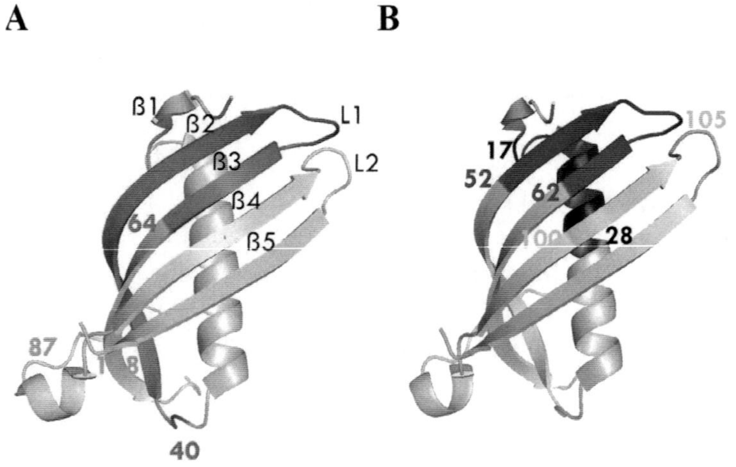

Figure 1. Human cystatin C (PDB 3GAX) with fragments which are possibly involved in interactions with HCC3 based on: A – epitope excision and extraction method (Asp-N), hCC(40-64) – pink, hCC(87-118) – green; B – hydrogen-deuterium exchange with MS, hCC(17-28) – black, hCC(52-62) – blue, hCC(100-105) – orange.

All our results indicate that the epitope for HCC3 – hCC complex is discontinuous. Moreover, location of the epitope around the crucial for dimerization loop L1 (Fig. 1) may explain the inhibitory properties of the antibody HCC3 towards the dimerization of human cystatin C. The mechanism of hCC dimerization (3D domain swapping) consists in the domains (alpha helix, β1 and β2 strands) exchange between two hCC molecules and the loop L1 is an essential part in the mechanism (hinge region) [5, 6].
In our further studies we are going to check using affinity chromatography and ELISA test with synthesized short peptides which correspond to the loops and beta strands near the loops which part or parts of the epitope (Fig. 1, B) is/are involved the most in the interactions with the antibody.

Acknowledgements
Work supported by grant from National Science Center 2011/01/N/ST5/05642, 538-8725-B709-15 and 538-8725-B077-15

References

01. E. Levy, M. Jaskólski, A. Grubb. (2006) Brain Pathol. 16(1): 60-70
02. A. Palsdottir, A. Snorradottir, L. Thorsteinsson. (2006) Brain Pathol., 16, 55-59
03. G. Ostner, V. Lindstrom, AB. Postnikov, TI. Solovyeva, OI. Emilsson, A. Grubb. (2011) Scand. J. Clin. Lab. Invest., 71, 676-682
04. U. Reineke, M. Schutkowski. (2009) Mol. Biol., 524, 87-101
05. M. Jaskólski. (2001) Acta Biochim. Pol. 48, 807-827
06. A. Szymanska, E. Jankowska, M. Orlikowska, I. Behrendt, P. Czaplewska, S. Rodziewicz-Motowidło, (2012) Front. Mol. Neurosci.5, 1-10

European Peptide Society
c/o Professor Paula Gomes
Rua do Campo Alegre, s/n
PT-4169-007 Porto
Portugal

ISBN 978-1-5108-5248-8

9 781510 852488

CURRAN ASSOCIATES INC.
proceedings.com

Society of Economic Geologists Annual Conference (SEG 2023)

Resourcing the Green Transition

London, United Kingdom
26 - 29 August 2023